CHEMICAL EXPOSURE AND DISEASE

Diagnostic and Investigative Techniques

THE PROFESSIONAL AND LAYPERSON'S GUIDE TO UNDERSTANDING, CAUSE AND EFFECT

D1534316

CHEMICAL EXPOSURE AND DISEASE
Diagnostic and Investigative Techniques

THE PROFESSIONAL AND LAYPERSON'S GUIDE TO UNDERSTANDING, CAUSE AND EFFECT

Janette D. Sherman, M.D.
Internal Medicine—Toxicology
Alexandria, Virginia
and
Southfield, Michigan

Princeton Scientific Publishing Company, Inc.
Princeton, New Jersey

Copyright © 1994 by Princeton Scientific Publishing Co., Inc. and
  Janette D. Sherman

ISBN 0-911131-31-0

All rights reserved. No part of this work covered by the copyright hereon may be reproduced or used in any form or by any means—graphic, electronic, or mechanical, including photocopying, recording, taping, or information storage and retrieval systems—without written permission of the publisher

Printed in the United States of America

Princeton Scientific Publishing Co., Inc.
P.O. Box 2155
Princeton, New Jersey 08543

Cover design by Donald O. Nevinger

DEDICATION

This book is dedicated to my husband, Donald Nevinger, whose bright mind, constant good cheer, and support make all things possible.

It is also dedicated to my children, Constance and Charles. May they not suffer the fate of their four grandparents, Wilma, Frank, Ruth and Howard, all of whom died of cancer, although none from smoking.

Additionally, it is dedicated to my friends Orlando, Judy, Corky, John, Patricia, Shore, Phyllis, Wendell, Edward, Charlotte, Dida, Helen, Morrie, Dave, Walter, Monty, Henry, Bob, and Paul, all of whom have suffered the ravages of cancer.

This book is also dedicated to the vast number of people, exposed to hazardous chemicals without their knowledge or consent. May this book help them achieve the knowledge to protect their own health and the health of our collective environment.

PREFACE

We shall require a substantially new manner of thinking, if mankind is to survive.
—Albert Einstein

In the five years that have passed since the first publication of *Chemical Exposure and Disease*, I had hoped that there would be little need for an update or another book on the subject. Sorry to say, there has been no letup in chemical pollution, and no letup in chemically-related illnesses. In fact, taking into consideration the increases in a number of cancers, immunological abnormalities, multiple chemical sensitivities, neurological damage from pesticides and solvents, endocrine disruption and birth defects in both wildlife and humans, and declines in reproductive capacity, there is ever more need to provide information concerning the connections between disease and chemical exposure.

It is a rare week when the popular press does not report a story about a previously unknown or covered-up source of pollution, be it from chemicals or radioactivity, from civilian or military sources. Exposure continues with both old and new synthetic chemical products in the workplace and in the home. The transport of toxic chemicals over roads and rails endangers all in the path. Discharge of toxic chemicals into the air, water, and soil spreads pollution far and wide, endangering many, mostly unaware of the source of contamination.

Mass promotion and use of pesticides, despite three decades of warnings, will put these diseases into the forefront.

The recognition of chemical connections to immunological, endocrine, and reproductive disruption is on the cutting edge of science, and will increase in importance.

One of the few weapons against injury from chemical exposure is knowledge. It is hoped that the information, methods, and examples contained in this book will aid in the recognition of illnesses caused by exposure to chemicals, and, more critically, will provide tools for the prevention of future harm.

Janette D. Sherman, M. D.
Alexandria, Virginia
1994

CONTENTS

CHAPTER

1

INTRODUCTION

My interest in the association between disease and chemicals began in 1970 when the first patients were referred to me for evaluation of Workers' Compensation claims. The task was to determine if the patients were sick or not, and, if sick, whether there was an association between the work they had done and their illnesses.

Initially, I approached the task with a high degree of skepticism; but, after several months of examining such patients, and hearing nearly identical histories from unsophisticated workers, I decided to pursue the field further. Within a short period of time, I had seen 100 patients and was convinced that *prevention* should be possible, and if undertaken would be greatly preferable to continuation of the ex post facto system of workers' compensation. At that time, I had no idea how difficult the idea of prevention would be. In these days of the "bottom line" and cost–benefit analysis, the idea of preservation of health and working ability may seem anachronistic. If a worker is considered merely a part of the "cost of doing business," and the costs of medical care and workers' compensation claims are written off against profits before taxes, then we have failed miserably in our basic concepts of public health and ethics, and our avowed moral and legal position of equal protection for all. Morality and ethics aside, pure economics dictates that trained and experienced workers are critical for the survival of every business and industry, and must be protected.

In 1973, I approached a university department about my findings and was told I did not have a large enough sample size for analysis. I continued to see patients and soon had examined over 400 workers. In frustration, I contacted U.S. Senator Philip Hart of Michigan, who referred my concerns and data to the Health Research Group in Washington, D.C. This group, which is an arm of Ralph Nader's Public Citizen, offered to supply statistical help to analyze my data. The results showed clearly that certain occupations in the auto industry were associated with the development of lung disease, and that the likelihood of developing lung disease increased with the length of time in the industry.

This did not seem surprising, based upon decades of documentation of illness in industrial workers. However, what was surprising in my small study was the finding that the lung disease, in general, was independent of smoking. The latter finding was so contrary to my thinking and training that the data were reevaluated, with the same results.[1]

Since 1973, when the study on workers in the auto industry was released, I have continued to evaluate workers, as well as other persons who have been exposed to chemicals that have caused a variety of illnesses. Over the past 17 years, I have evaluated approximately 8000 individuals and have obtained work/environmental histories from all of them. With the burgeoning increases in the production and use of chemicals, it has become critical that we be able to recognize the effects of these chemicals and be able to *prevent* chemically caused diseases. It is my hope that this book will make the task easier for all.

Total medical care expenditures approached $425 billion in the United States in 1985. This represented 10.7% of the gross national product, amounting to $1,721 for each person.[2] It is estimated that this cost will triple to $1.5 trillion by the year 2000.[3] The U.S. Congressional Office of Technical Assessment estimates that "there are about 6000 deaths annually—about 25 per working day—due to injuries," with "about 10,000 injuries that result in lost work time and about 45,000 that result in restricted activity or require medical attention." The Bureau of Labor Statistics (BLS) estimated 106,000 occupational illnesses in 1983, mostly easily diagnosed conditions such as dermatoses. Serious diseases—respiratory and neurological disorders and cancers—are not generally captured in the BLS records of workplace illnesses. The report continues: "There is so little agreement about the number of workplace-related illnesses that the Office of Technical Assistance (OTA) does not take a position on the controversy about the 'correct' number. Most deaths and injuries occur one at a time or in small numbers in the Nation's more than 4.5 million work places. . . . Arguments about the number of occupationally related diseases may obscure the important fact that occupational illness is preventable."[4] "Occupationally related cancers may comprise as much as 20% or more of total cancer mortality in forthcoming decades. Asbestos alone will probably contribute up to 13–18%. . . . these data do not include effects of radiation, nor the effects of a number of other known chemical carcinogens."[5] The therapy used to treat cancer, much of it preventable, is a growth industry, accounting for costs of $813.2 million in 1987, with projections to $1.4 billion by 1990.[6] It may be argued that many of the companies producing chemicals that cause cancer are the same ones that produce the pharmaceuticals used to treat them.

Although an accurate "body count" is not available,[7] it must be concluded that if we do not take active steps to control environmental and workplace illness and injury, we will soon outstrip our ability to pay for such care.[8]

HISTORICAL PERSPECTIVE

Chemicals cause diseases. This fact has been known since ancient times. Lung disease, in the workers who fashioned fireproof tablecloths from asbestos, was described by Pliny (A.D. 23–79) as "the disease of slaves."

Ramazzini, who lived in Italy from 1633 until 1714, is regarded as the founder of the field of occupational medicine. He investigated the working conditions of

miners of metals, healers by inunction, chemists, potters, tinsmiths, glass workers and mirror makers, painters, sulphur workers, black-smiths, workers with gypsum and lime, apothecaries, cleaners of privies and cesspits, fullers, oil pressers, tanners, cheese makers and other workers in dirty trades, tobacco workers, corpse carriers, midwives, wet-nurses, vintners and brewers, bakers and millers, starch makers, sifters and measurers of grains, stone cutters, laundresses, workers who handle flax, hemp and silk, bathmen, porters, athletes, those who strain their eyes over fine work, voice trainers, singers, farmers, fishermen, soldiers, learned men, nuns, printers, scribes and notaries, confectioners, weavers, coppersmiths, carpenters, grinders of razors and lancets, brick makers, well-diggers, sailors and rowers, hunters and soap makers.

Ramazzini counseled physicians:

When a doctor visits a working class home he should be content to sit on a three-legged stool, if there isn't a gilded chair, and he should take time for his examination: and to the questions recommended by Hippocrates, he should add one more—What is your occupation?[9]

Comparable attention to a person's occupation and its attendant chemical exposure clearly is needed today.

In 1776, Sir Percival Pott described cancer of the scrotum in chimney sweeps who had been exposed to the tars and soot found in their working environment. In 1898, The Chief Inspector of Factories[10] in England forewarned of the asbestos tragedy that was to follow when he reported lung disease in workers handling asbestos. The connection between arsenic, and cancer was described at the turn of this century.[11] The play *Arsenic and Old Lace* reiterated what we already knew about the toxic nature of arsenicals. Lead poisoning, with weakness of the legs and arms, abdominal pains, and brain damage, has become well-known. It was seen originally in miners, painters, and allied workers, and is now seen in the general population as contamination has spread. Currently, public health workers attribute a significant portion of mental retardation in children to the ingestion and inhalation of lead contamination of their environment,[12] although the connection between lead exposure and nervous system damage has been known for decades.[13-18] We need not congratulate ourselves for the speed of our discoveries and remedies, given that Hippocrates recognized the association between colic and lead in about 370 B.C. In nearly every instance, the toxicity of a chemical, be it mercury, arsenic, lead, asbestos, or solvents, was originally described in workers handling the various substances.[19,20]

This process—that is, determining where and in whom illnesses are concentrated, and then proceeding to the source of the cause—was followed by William Farr and John Snow in London in the mid-1800s, and, a century later, by Lois Gibbs and Dr. Beverly Paigan in the Love Canal area.[21,22] In England, when it became apparent that much misery and illness was spread via contaminated water and food, legislation was enacted and vast sums were spent to clean up the filthy sources of infection and to prevent the spread of the agents identified as causative of contagious illness and death. These practices, the control of water supplies and of sewerage, are accepted as minimal precautions throughout most of the industrialized world. The Love Canal residents were not so fortunate, however, as they suffered a myriad of neurological and reproductive problems.[23,24] It took considerable pressure from the residents as well as the pressure of litigation to force closure of the area and the razing of 550 homes. Over a decade later, cleanup efforts to control chemical contamination from the landfill still are not complete.[25]

The beginning of medicine was steeped in the effects of chemical agents. Perusal of the writings of Galen, Paraselsus, Agricola, and Hippocrates attests to this. Hunter's *The Diseases of Occupations* (see reference 9), runs 1259 fascinating pages, and contains a wealth of information about the history of occupational diseases. It illustrates very clearly how much is already known of the connection between occupation and illness, and how much people can do to prevent these illnesses if we only will.

UTILITY OF KNOWLEDGE

Understanding the historical development of information concerning occupationally related diseases is a way to avoid the pitfalls of the past. We need not reinvent the wheel and reject knowledge that has been accumulated over the past decades and centuries. But when has humankind rejected the chance to recreate the wheel?

Hardly a day passes when we are not informed by the media of a toxically associated problem. Offending agents include substances leaching from landfills, from the smokestacks of factories and incinerators, and from those chemicals, such as pesticides, building materials, and consumer products, intentionally used in homes and offices. With the marked increase in the manufacture and use of chemicals, the recognition of these conditions takes on greater importance.

An unfortunate word in medical parlance is "idiopathic," frequently used to denote a disease of unknown cause. It stems from the Greek *idios* (one's own, peculiar) and *pathos* (feeling or suffering). Victims have been left alone to suffer all too often; but the fact that a disease is considered idiopathic is not a reason to allow lack of curiosity to continue. Until investigations led to the identity of and connection between various bacteria, rickettsiae, and viruses, most infectious diseases were classified as idiopathic. Leprosy, yellow fever, and poliomyelitis were considered idiopathic until a causative agent was found for each of them.

The hundreds of cases of hypertension, cancer, and/or lung disease found in a single industry, and often within a single job description, can hardly be characterized as peculiar only to that particular person and thus dismissed as idiopathic. Too often, diseases of undetermined origin are accepted as just that, and little thought or effort is expended in trying to piece a pattern together to get to the basic inciting event or agent.

Thinking in terms of toxicology implies a cause-and-effect relationship. Although we may not always be able to discern the cause of a chemically associated illness, that should not stop us from taking orderly steps to try to determine the cause or causes of a disease or condition.

Utilizing such information sources as the British *Registrar General's Occupational Mortality Supplement*, begun in 1851 and brought up to date every ten years, it has been possible to identify certain trades as having excessive mortalities, and thus to take preventive efforts to maintain a healthy work force. With the vast and impressive computer technologies available in the United States (and similarly industrialized counties), it appears to be a matter of will to make similar compilations of occupationally and environmentally associated illnesses and deaths. These compilations could begin by employing the available data from death certificates, social security records, state workers' compensation claims, and Department of Commerce production and import data. A similar, and relatively simple technique has been used in New Jersey to identify workers exposed to asbestos, with the aim of notifying the workers at risk, and planning for allocation of public health resources.[26] The establishment of a "National Disease Registry," combined with a cross-referenced litigation index, could provide information as to where cases involving workers' compensation and chemical tort claims were filed, by whom, against whom, and their ultimate disposition. This information could then be correlated with data obtained by the Environmental Protection Agency, the Food and Drug Administration, the Occupational Safety and Health Administration, and the Consumer Product Safety Commission concerning air and water quality, waste disposal practices, and chemical, drug, and pesticide production and usage to determine if there is an association between various illness in a population and exposure to chemicals manufactured, used, and/or disposed of in an area.

It is hoped that the resultant information would be utilized to allocate scarce resources in a cost-effective, intelligent, and humane manner.[27] People have asked why this has not been done already. There have been many opinions, ranging from lack of interest and will to an effort to suppress information that could be utilized to restrict industrial and commercial interests. Whatever the reasons, the various costs of the failure to deal with chemicals and disease will accrue, and be borne by future generations, if not our own.

CONCEPT OF CHANGE

Keeping in mind the philosophy that nature provides for a nearly perfectly functioning individual at birth, be that individual a plant, a fish, a bird, or a human, we must deduce that some aspect in the milieu of that person has changed when we find disease in that entity. In general, at birth each human arrives with the basic systems in place for optimal survival. Babies are not born with birth defects without cause, and people do not develop cancer, kidney, or liver failure and lung fibrosis because they think bad thoughts! There is a **cause** in every case, although we may not be able to discover the inciting agent in each victim. To blindly accept the **idiopathic** concept of disease without probing, questioning, and seeking its cause is to delay preventive efforts. This is not to say that all questions may be answered, but to emphasize that we are remiss if we do not at least try to find the answers.

To do investigative toxicology is to question **what has *changed*** in the milieu of the person. This concept is never more evident than when we encounter a child with a malignancy. It is obvious that the time span between conception and the diagnosis of the malignancy in a child is finite. There cannot have been an infinite number of exposures for such a child, and so it is the task of the investigator to identify as many events as possible where the child could have been exposed to a carcinogenic agent, in utero and since birth. This same principle serves when we are investigating liver damage, kidney failure, cancer, nervous dysfunction, and so on, in any person. Investigating an adult's exposure history is more complex than investigating a child's history, but it too is finite. The main difference is that with the longer time span of an adult's life, there is more opportunity for toxicological events to occur.

One must try to obtain as complete an exposure history as possible, especially when there is an extended period of time between the initial exposure to a toxic agent and the diagnosis of a disease. The importance of searching the past for chemical exposures was demonstrated in the people who died of mesothelioma decades after *nonoccupational* exposure to asbestos, which was considered "minor" at the time.[28,29] One family—the husband a shipyard pipe insulator, the wife, and their 34-year-old-daughter—all succumbed to mesothelioma.[30] One might erroneously attribute the clustering of cancer in that family to hereditary causes, but mesothelioma is a "marker" disease, linked to asbestos exposure with few exceptions.[31,32] Had the members of the family developed malignancies of the colon nasopharynx and/or a common form of lung cancer, it is doubtful that a common exposure would have been sought, even though each of those malignancies is also associated with asbestos exposure. We must seek to answer questions whenever we find malignancies in anyone, and especially when they occur in various members of the same family, or in any groups that may share common exposures. For instance, can anyone answer the following? Is the appearance of breast cancer in various members of the same family due to hereditary causes, or is there a common exposure, such as to DES (diethylstilbestrol) residues in meat, other hormonal preparations, or an as yet unidentified chemical? I know of no wide-ranging research that is addressing this question.

THINKING IN TERMS OF CAUSE AND EFFECT

This process of investigation involves thinking in terms of **cause**, **effect**, and **change**. This method of investigation works "both ways"—that is, retrospectively, from the sick person to the cause, and prospectively, from the known toxic effects of various chemicals to the assessment of an exposed population.

The first way, starting with a sick person, is an exercise in retrospective toxicology: determining what exposures the person may have had, and reviewing the known effects of any of the chemical factors in either animals or other humans. The next step is trying to determine if the illness in the individual is compatible with what is known in other organisms, utilizing what is known about effects in animals or other humans.

The second method of investigation, which requires looking for expressions of illness in an exposed population, is prospective and carries the potential for prevention. This process should be instituted when there is a population exposed to a chemical that, because of its structure, use, and/or toxicology, can be predicted to have adverse effects. When animal tests show adverse effects, or there is similarity of structure to a known toxic agent, as well as potential for human exposure, then monitoring of the exposed population should proceed. The "head in the sand" approach, or failure to look for reasonable effects, must be condemned.

Performing a retrospective inquiry—trying to discern whether a person's illness is a result of a toxic exposure or not—frequently falls to the treating physician, who has the least amount of time, perspective, and knowledge (and even interest in some cases) to delve into toxicological matters. I hope that the substance of this book will make the investigation of chemically caused diseases straightforward, and no more difficult than investigating any other condition in a patient.

CURIOSITY AND DUTY TO EXPLORE

If a physician failed to investigate a drug reaction (such as asthma or a rash caused by penicillin) and allowed a patient to be reexposed, he or she could be held liable for failure to diagnose. It is not unreasonable to extend this duty to the investigation and diagnosis of chemically caused diseases, to preventing disease in a person or population where there is a known likelihood of exposure to toxic chemicals with resultant harm. Suppose that a physician has seen several patients who have developed liver failure while working in a dry cleaning plant, and it is known that perchlorethylene is commonly used in the dry cleaning industry, and that this chemical has been reported to cause liver damage in both animals and humans; the physician then has the duty to counsel other workers before they too lose their health. The same principle applies to a physician who sees workers with asbestos-related lung disease, in regard to cautioning them about further exposure.

Although the concept of failure to warn of a hazard has not been tested in nonmedical personnel, does it not follow that a union representative or a corporate representative who fails to warn of a known hazard is also liable if preventable damage occurs?[33-37] Must we wait until this concept of duty to warn is tested in court, or may we proceed on the premise that physicians, union representatives, and corporate personnel have both a moral and an ethical duty to protect against harm when there is a reasonable possibility that it may occur?

If we know that chemicals such as arsenicals, asbestos, chlordane, heptachlor, lindane, PCBs, trichlorethylene, perchlorethylene, lead, and so on, are buried in a landfill, and that these products can vaporize into the air or leach into the water of a population living nearby, do we not have a duty to warn the population? And, finding sick persons in such a population, is it too difficult a task to investigate the cause-and-effect relationships that may exist in such a situation? It is a very short step from the presently accepted duties that are implicit in medical care to theorize that failure to investigate the possibility of a chemically caused disease will be regarded as malpractice. Likewise, once a chemically associated disease is diagnosed, the logical next step—to treat appropriately—may also fall under the scrutiny of medical–legal affairs. If "treatment" is counseling about expected effects or removal from further exposure, then failure to do either may also be considered as failure in one's duty as a physician. As Hippocrates poignantly stated in the fourth century B.C.: If you can do no good, at least do not harm.

The frequent failure to investigate diseases associated with chemical exposure may involve not only a lack of method but, in some cases, an unfortunate lack of will. The "I don't want to be involved" excuse has been offered by some, not only with a sense of intellectual laziness, but sometimes out of concern that one may be called upon to provide facts or an opinion for one side or the other of a legal battle. Lest anyone fear the latter, that is, having to submit medical records, or even give testimony, it involves simply telling "the truth, the whole truth, and nothing but the truth." In that context, it is a simple matter to deal with one's records, and as in all medical care, the records should be clear, comprehensive, and complete. Any lack of expertise or any unwillingness to become "involved" can be solved by practicing the orderly process of information gathering, or by referring the patient to a person or facility experienced in the field. The practitioner certainly should not neglect his or her duty and end up being sued for failure to diagnose and treat the patient! As bad as the effect would be on the practitioner, it would be much worse for the patient— and it is the patient with whom we are concerned.

With the rapid accumulation of data from both human and animal studies, it is difficult, if not impossible, to deny that chemicals are causative factors in human disease and environmental damage. Thus today, although physicians are primarily involved in the treatment of the sick, another aspect of medicine must be learned, just as the advancing fields of virology, immunology, and pharmacology must be kept current. The field of toxicology is merely an extension of pharmacology, as each are disciplines involving the interaction between chemicals (sometimes called drugs) and living systems. The initiatives taken by the legal profession have shown that there is a strong interest in the field of chemically caused diseases. The efforts of the legal profession have resulted not only in recovery for past damages, but also both legislatively imposed and self-imposed restraints, with the prevention of further harm in some instances.

COSTS—ECONOMIC AND HUMAN

The enormous costs of medical care, which are increasing yearly, are expected to grow in the future in nearly every case. Moreover, industrial cost savings now can mean vast outlays for treatment later on. For example, the money "saved" by the asbestos companies in the 1940s and before, by not using adequate respiratory protection, is being spent now in caring for hundreds of respiratory disease and cancer victims. Unfortunately, the costs of disease and death generally accrue to one segment of the population, while the benefits of production accrue to another.

In some notable situations, it was not until the costs of litigation exceeded the benefits of production and use that the risks to workers and consumers were accepted on a serious basis. This litigation served a partially protective function for some victims, although few would argue that cash settlements replace, in any way, lost health, lost lives, or irreparable ecological damage. It was the pressure of litigation that resulted in legislation to remove, restrict, and/or warn of the harmful effects of a number of products and processes in the market-place.[38-40] Actions have been taken against such diverse products as chlordane, chloramphenicol, birth control pills, DES (diethylstilbestrol), heptachlor, kepone, cyclamates, and other chemicals and drugs. This legal pressure was particularly applicable to the asbestos industry, which fought remedial action for years, in the face of overwhelming data, and which now will probably never fully recover from its neglectful history.[41]

The emerging return to cottage industries as big industries out-source manufacturing processes to small shops will undoubtedly result in more chemically caused diseases. This practice, common in the 1800s, had been largely abandoned in the name of better quality and productivity. Many small industries do not have access to industrial hygiene data, and are unable to evaluate what information they are given. Considering that nearly half of the new jobs created between 1979 and 1985 paid less than a poverty-level income of $180/week,[42] it comes as no surprise that many workers are under such severe economic pressure to produce that they ignore cautions.

Already reported in the medical literature is the case of a worker who got sick as a result of chemical exposure as she was performing piecework in her home.[43] The authors emphasized that neither the manufacturer nor the distributor of the materials used provided any warning as to the hazards of the process. They emphasized further the importance of obtaining an occupational history and investigating for chemical exposure. Unless legislation is instituted restricting this spread of industry to the home milieu, there will be adverse effects not only upon the pieceworker, but upon those living in the same dwelling where the manufacturing activity is carried on. We can also expect to see the spread of chemical contamination to the local environment. If this trend of dispersal of manufacturing activities is not contained, we will also witness an increasing need for "do it yourself" toxicology by the owners and workers in such shops.

Months of ineffective and costly treatment were prescribed for the polychlorinated biphenol (PBB)–poisoned victims living on contaminated farms in Michigan,[44] as well as for the families living in the Love Canal area. It was not until the affected persons themselves became aware of the poisonous nature of their environment, and took steps to either remove themselves or remove the contamination, that any relief was obtained. The problems were largely dismissed as mass hysteria in the beginning, and later ignored by persons who either could not or would not learn about the toxicology of the chemicals.

VALUE OF KNOWLEDGE

Compilations of information and access to data bases can simplify decision making by producers of toxic chemicals by imposing prudent and careful decisions on the production of potentially harmful chemical products. When forward-thinking decision makers prevail, we should see a decrease in the litigious need for recovery and compensation. We should also see a cost-effective decrease in the costs of illness and death. Is it not incumbent upon physicians, corporate decision makers, union representatives, legislators, and the public to learn this timely discipline, and to exert their maximum influence to prevent further chemically caused disease?

Few people have either the awareness or the skill to evaluate toxic products that are brought into the home environment, believing that if it is for sale, then it is probably safe to use. Examples abound of problems that have followed the introduction of such chemicals as formaldehyde, cleaning solvents, and pesticides into the house. Given some of the serious consequences of exposure to these products, we can conclude that knowledge of the toxicity of such products would dissuade some consumers from buying them. Retrospectively, knowledge of toxicology could aid discovery of the cause of illness within a family, and make it possible to take definitive action to eliminate a cause rather than to continue with costly and ineffective treatment.

The ability to recognize chemically caused disease and injury has come to the forefront because it is necessary not only to treat these conditions, but, more important, to anticipate the consequences of chemical exposures, and to prevent them. To this end, this book is dedicated to workers in the field of toxic chemicals. This includes not only physicians, nurses, and public health workers, but chemical engineers, manufacturing decision makers, purchasers of potentially toxic products, legislators, and lastly attorneys dealing with sick or deceased clients for whom prevention has come too late.

REFERENCES

1. Sherman, J. D., Wolfe, S., Hricko, A., Mets, M. *A Health Research Group Study on Disease Among Workers in the Auto Industry.* Public Citizen, Inc., Washington D.C., 1973.
2. Kittrell, J. *U.S. Dept. Health and Human Services News*, Library of Congress, Congressional Research Service, July 29, 1986.
3. Specter, M. Health bill seen at 15% of GNP by year 2000, *Washington Post*, p. A-11, June 9, 1987.
4. Office of Technical Assessment. Summary and options, in *Preventing Illness and Injury in the Workplace.* U.S. Congress, Washington, D.C., 1985, pp. 3–5.

5. Bridbord, K., Decoufle, P., Fraumeni, J. F., Hoel, D. G., Hoover, R. N., Rall, D. P., Saffiotti, U., Schneiderman, M. A., Upton, A. C. *Estimates of the Fraction of Cancer in the United States Related to Occupational Factors*. National Cancer Institute, National Institute of Environmental Health Sciences, National Institute for Occupational Safety and Health, Washington, D.C., Sept. 15, 1978.

6. *Medical Tribune* (quoting Frost and Sullivan, Inc.'s, *Cancer Therapy Products Market in the U.S.*), p. 1, Aug. 26, 1987.

7. Whorton, M. D. Accurate occupational illness and injury data in the U.S.: can this enigmatic problem ever be solved? *Amer. J. Pub. Health* 73(9): 1031-32, 1987.

8. Yankauer, A. Public and private prevention. *Amer. J. Pub. Health* 73(9): 1032-33, 1987.

9. Hunter, D. *The Diseases of Occupations*. Little, Brown and Co., Boston, 1969, pp. 34, 35.

10. Chief Inspector of Factories. *Factories and Workshops. Annual Report, Part II*. Her Majesty's Stationery Office, London, 1898, pp. 171-72.

11. Haerting, F. H., Hesse, W. Der lungenkrebs, die bergkrankheit der Schneeberger gruben. *Vjschr. geschicht. Med.* 30: 296-331, 1902.

12. Falk, H. Conclusions of the committee on human health consequences of lead exposure from automobile emissions. *Environ. Health Perspect.* 19: 243-46, 1977.

13. McKahn, C. F. Lead poisoning in children: the cerebral manifestations. *Arch. Neurol. Psychol.* 27: 294, 1932.

14. Albert, R. E., Short, R. E., Sayers, A. J., Strehlow, C. D., Kneip, T. J., Pasternack, B. S., Friedhoff, A. J., Kovan, F., Cimno, J. A. Follow-up of children over-exposed to lead. *Environ. Health Perspect. Experimental Issue* 7: 33, 1974.

15. Moore, M. R., Meredith, P. A., Goldberg, A. A retrospective analysis of blood-lead in mentally retarded children. *Lancet* 717-19, 1977.

16. Needleman, H. L. *Studies in Children Exposed to Low Lead Levels*. U.S. Environmental Protection Agency, Health Effects Research Laboratory, EPA-600/S1-81-066, 1981.

17. Fulton, M., Thomson, G., Hunter, R., Raab, G., Laxen, D., Hepburn, W. Influence of blood lead on the ability and attainment of children in Edinburgh. *Lancet* 1: 1221-25, 1987.

18. Hamilton, A. *Industrial Poisons in the United States*. Macmillan Co., New York, 1925, pp. 19-205.

19. van Ziemssen, H. *Cyclopedia of the Practice of Medicine*, Vol. 19. William Wood & Co., New York, 1897, p. 233.

20. Hamilton, A. *Exploring the Dangerous Trades*. Little, Brown and Co., Boston, 1943.

21. Zinsser, H. *Rats, Lice and History*. Little, Brown and Co., Boston, 1934, 1935, 1963.

22. Hunter, Donald. *The Diseases of Occupations*. Little, Brown and Co., Boston, 1969, pp. 94, 97.

23. Vianna, N. J., Polan, A. K. Incidence of low birth weight among Love Canal residents. *Science* 226: 1217-19, 1984.

24. Personal communication with Lois Gibbs and Dr. Beverly Paigen, 1980.

25. *Detroit Free Press*, p. 9-A, June 25, 1987.

26. Stanbury, M., Rosenman, K. D. A methodology for identifying workers exposed to asbestos since 1940. *Amer. J. Pub. Health* 77(7): 854-55, 1987.

27. Glotta, R., Sherman, J. D. Learning the lessons of the asbestos tragedy: a reform proposal. *Trial* 19(11): 68-72, 134, 1983.

28. Greenberg, M., Davies, T. A. L. Mesothelioma register 1967-68. *Brit. J. Ind. Med.* 31:91-104, 1974.

29. Anderson, H. A., Lillis, R., Daum, S. M., et al. Household-contact asbestos neoplastic risk. *Ann. N.Y. Acad. Sci.* 271:311-23, 1976.

30. Li, F. P., Lokich, J., Lapey, J., Neptune, W. B., Wilkins, E. W., Jr. Familial mesothelioma after intense asbestos exposure at home. *J.A.M.A.* 240(5): 467, 1978.

31. Wagner, J. C. Experimental production of mesothelial tumors of the pleura by implantation of dusts in laboratory animals. *Nature* 196(4850): 180–81, 1962.

32. Hueper, W. C. Cancer induction by polyurethane and polysilicone plastics. *J. Nat. Cancer Inst.* 33: 1005–27, 1964.

33. Brown, M. H. Portrait of a polluter: Hooker Chemical Company. *Amicus.* Natural Resources Defense Council, New York. 1(3): 20–29, 1980.

34. Bad news: cover-ups uncovered. *Calypso Log*, The Cousteau Society, 1(6): 3, 1979.

35. *Wall Street Journal.* Health-risk suit cites Occidental Petroleum, units, p. 13, Dec. 19, 1979.

36. MacArthur, M. Regulating hazardous waste sites. *Paper, Film and Foil Converter*, 108–10, Feb. 1980.

37. *Wall Street Journal.* Occidental Petroleum settles SEC charge on inadequate environmental disclosure, July 3, 1980.

38. Purdy, M. Pollution police seek out, arrest offenders of environmental laws. *Detroit Free Press*, p. 6A, June 8, 1987.

39. Ankeny, R. Homicide charges jolt boardrooms. *Detroit News*, p. 8A, Apr. 18, 1985.

40. Lawton, M., Mehri, C. The victimization of America. *Congress Watcher*, pp. 3–6, Feb./Mar. 1986.

41. Brodeur, P. *Outrageous Misconduct.* Pantheon Division of Random House, New York, 1985.

42. Hightower, J. *New York Times*, p. D-1, June 22, 1987.

43. Liss, G. M., Halperin, W. E., Landrigan, P. J. Occupational asthma in a home pieceworker. *Arch. Environ. Health* 41(6): 359–62, 1987.

44. Egginton, Joyce. *The Poisoning of Michigan.* W. W. Norton and Co., New York, 1980.

CHAPTER

2

CLUES TO CHEMICALLY CAUSED DISEASES

Many medical conditions that are caused by chemicals have aspects in common with "ordinary diseases of life." The problem lies in recognizing the contribution of each. Although there are overlapping situations, many chemically associated diseases form a pattern, or produce clues as to causation. These patterns include:

1. Sudden appearance of disease in a previously healthy and functioning individual that is not explained by traditional diagnostic criteria.
2. Increase in specific diseases in a population of workers with common exposures.
3. Clustering of diseases in a confined population group.
4. Unusual or uncommon diseases in a specific group.
5. Diseases in humans that are similar to those produced in animals with similar exposures.
6. The appearance (or disappearance) of a disease temporally associated with a change of occupation and/or environment.

Where one or more of the above patterns occur, the possibility of finding a chemically associated disease increases. A common chemical etiological agent is not so different in concept from a common infective agent in producing recognizable patterns of disease. As with communicable agents, such as bacteria, viruses, and rickettsiae, the common chemical agent should be determined, with the object not only of treatment, but prevention of additional harm.

For instance, all of the above categories can be represented by a group of agricultural workers who develop nausea, vomiting, diarrhea, and lacrimation after being exposed to organophosphate pesticides.

A painter who develops weakness, headache, and abdominal pains should be considered as suffering from lead poisoning until proven otherwise. When an adult plastics worker, with no previous history of respiratory disease, develops sudden difficult breathing, an inquiry into chemical exposures must be made. If urethane plastics are being handled, the most likely cause of the sudden appearance of breathing problems is one of the isocyanates, TDI, or MDI (toluene diisocyanate or methylene diisocyanate.)

If a group of garage mechanics develops headaches and flusing of the face, carbon monoxide poisoning must be considered as the most likely explanation for their common symptoms.

Liver or kidney disease in a group of dry cleaner or degreaser workers should point to solvent toxicity.

The clustering of diseases can be recognized in both occupational and residential settings. Notable examples include the following:

- Leukemia deaths in the children in Woburn, Massachusetts[1]
- Reproductive failures and neurologic disorders in the Love Canal residents[2]
- Infertility and low sperm counts in the DBCP production[3-5] and lead workers[6]
- Liver cancers in the vinyl chloride–exposed workers[7,8]
- Kidney cancers in the benzidine dye workers[9,10]
- Silicosis and death of the Gauley tunnel workers[11]

The above examples also include a number of unusual or uncommon diseases in these exposed groups.

It was angiosarcomas of the liver that led to the control of vinyl chloride, but it was also found that there was a higher incidence of birth defects in children born in areas where vinyl chloride is produced.[12] These studies point up the importance of looking for other, allied conditions when one problem is identified. A chemical such as vinyl chloride, that causes cancer is more than likely to have other biological effects; and, in this case, it interfered with fetal growth. Vinyl chloride is a small-sized reactive chemical that is readily taken into the body where it reacts with cells. Other chemicals, determined to have a biological effect such as mutagenicity, teratogenicity, carcinogenicity, or neurotoxicity, must be studied to ascertain associated biological actions.

The appearance of aplastic anemia, an uncommon disease, sometimes transforming into leukemia, presented a problem as to cause until it was determined that some of the patients had taken the antibiotic drug chloramphenicol, or the anti-inflammatory drugs phenylbutazone or its chemical cousin, oxyphenbutazone. This knowledge led to marked restriction of use of the first two drugs,[13] as well as the outright removal of oxyphenbutazone (Tandearil®) from the market.[14]

Drugs as well as occupational and/or environmental factors must be considered in causation of disease. One could have been misled by reasoning that these cases of aplastic anemia were caused by infections or injuries, when in reality, it was the drugs used to treat the infections and inflammatory conditions that were the inciting agents for the disease.

Drugs, pesticides, and many industrial and consumer products require testing under various U.S. federal laws before they can be sold in commerce. These testing efforts are intended to predict human health effects before they are used.

However, these data may also be used after the fact, to confirm adverse effects in humans, particularly when the effects are similar in several species of animals.

If all or most of the members of a family, or of an office force, develop irritation of the eyes, upper respiratory symptoms, skin disorders, and/or headache, the possibility of off-gassing or formaldehyde, plasticizers, or other chemicals from building, furnishings, and/or insulation materials must be sought. This is, of course, in addition to ruling out a common infective agent. This is especially true when people live or work in closed or poorly ventilated buildings. These upper respiratory symptoms are commonly attributed to "colds" or the "flu," but with careful questioning and examination, there are differences between infectious and chemically caused problems. Commonly, people report less illness when they are away from the offending work site or dwelling, have fewer problems when windows and doors are open, and, conversely, have increased symptomatology when the buildings are closed up and poorly ventilated.

When a previously healthy, well-functioning person develops a symptom unusual to him or her, a chemically caused reason should be sought, as one would investigate any symptom and its cause.

One patient, living in a home insulated with urea formaldehyde foam insulation (UFFI), developed severe, unremitting headaches. Before the cause of his headaches was learned, he was subjected to three arteriographic procedures that were not without significant hazard. Needless to say, he was not free of his headaches until he moved from the contaminated home. Another patient, previously healthy, who worked with formaldehyde–resin–impregnated paper in the manufacture of coffee filters, was subjected to two arteriograms, to investigate the cause of his headaches, before any occupational and/or environmental history was obtained.

The recognition of chemically caused diseases is not only a necessity in terms of public health and preventive medicine, but it must be done to conserve our limited medical resources and prevent further harm. Multiple drugs, laboratory tests, and invasive procedures, which are not only costly but can be hazardous in and of themselves, have been tried on patients without producing relief because the identity of the offending chemical agent was not recognized. Such failure to identify the true cause of many persons' illnesses has contributed to their burdens, both medically and financially, without providing for the elimination of the causative chemical or chemicals.

This field of medicine, which deals with chemicals as causative agents of disease, is little different from other scientific areas, in that it does not exist in a vacuum, and it builds upon that which is known. The aspect of medicine called diagnosis is but an exercise in collecting information, and ultimately matching the signs and symptoms with the most likely cause. In other words, much is done on the basis of "most likely" association, and, quite crudely, playing the numbers. When one hears hoofbeats outside a window in the United States, the sound probably comes from horses, not zebras. When a woman younger than 50 years of age stops menstruating, the first factor to consider is pregnancy, not an endocrinological abnormality. When a man develops blood in his urine, one should rule out a urinary tract infection or tumor, before attributing the hematuria to "strain." As one learns, in medical school, common diseases are common.

Similarly, chemically exposed persons frequently have symptoms in common; and the finding of similar symptoms in a group of people may point to a common etiological source.

Some conditions are so typically associated with chemical agents that the conditions, all by themselves, point the way for inquiry. In particular, a person presenting with a diagnosis of mesothelioma or bladder cancer virtually demands inquiry as to the source and identity of a carcinogenic agent. In areas where there is a concentration of a particular industry, such as a petrochemical cracking plant, a dye works, a mine, or smelter, a paint facility, or a pesticide manufacturing plant, a constellation of signs and symptoms will accumulate in the population sooner or later. If we have knowledge of the processes and the materials used in each industry, and a degree of alertness, the search for etiology in the affected population will be made simpler.

CANCER AS A MARKER DISEASE

There are few positive things that one can say about cancer. It is better not to develop it, and it is better to be cured of it! However, cancer does exist, and it may be used as a **marker disease,** a "red flag," to provide clues as to causes.

Cancer is a valuable indicator of adverse effects because there is a definable end point. A person may have a "little" emphysema from exposure to silica or toluene di-isocyanate (TDI), or "mild" kidney damage from a heavy metal or solvent exposure, but a person never has a "small amount" of cancer. Just as few would claim that a person could have a "little" venereal disease, or a woman could be a "little pregnant", there is no such thing as a "little cancer." The cancer may be small, and it may be curable, but it is a cancer. For this reason, and throughout this book, I will emphasize the use of cancer and carcinogens as a method of recognizing and investigating the association between chemicals and disease. In the same manner, diseases of the kidney, liver, lung, heart, or nervous system, induced by chemicals, can be investigated.

Quite often, after speaking to groups, and, curiously, even after addressing physicians, I am greeted with the remark: "Everything causes cancer if you take enough of it," or "I can't be bothered, there is too much to know." This is a very unfortunate, and counterproductive attitude, especially among trusted health professionals who should know better. It is of even greater harm to the public, who, seeing disaster after disaster reported in the press and on television, succumb to the feeling that there is too much information to handle, and that nothing can be done. Accompanying these attitudes is the false hope of an all-encompassing "cancer cure," with little understanding of the multiple causes and expressions of malignancy. As long as these erroneous ideas persist, little will be done to control carcinogenic hazards, and the wastage of human health and economic resources will continue. It is incumbent upon all citizens to become educated about the effects of the chemicals that have become so important a part of our lives, and to make informed, rational choices as to their uses and place in our environment.

Consider the progression from normal cell growth to cancer:

I. Most workers in the field subscribe to the theory that a chemical induces a malignant **change** in a cell by interacting with or damaging the basic genetic, cellular structure involving DNA (desoxyribonucleic acid).

The cell can then respond in one of two ways:

1. Death, in which case cancer cannot develop because growth ceases.
2. A change in growth and form that becomes self-perpetuating and self-replicating.

The latter, if continued, will result in a neoplasm (meaning new growth), either benign or malignant, and, if malignant, independent of the control of the body. A way of tracing the interaction of chemicals with DNA via the formation of chromosomal abnormalities and adducts has been developed. These markers allow the identification and monitoring of people who have developed changes in their cells because of exposure to chemicals. It is thought that those with chromosomal abnormalities are at increased risk for malignant disease, although the degree of risk is not known.[15-18]

II. A second concept of carcinogenesis is that it takes but **one** hit on a cell to transform that cell into a malignant one. If that altered cell survives, and becomes self-perpetuating, then a neoplastic growth results.

Obviously, when there is a large or continued exposure to a carcinogen or exposure to multiple carcinogens, more than one cell can be hit and thus altered, resulting in an increased probability that one or more of the altered cells will survive and become expressed as a malignancy.

III. Implicit in this concept of one hit is that of **no-threshold.** That is, if one cell can be changed to a malignant one with the potential for autonomous growth, then there is no level of exposure without potential for harm.

IV. **Latency** is the span of time that occurs between the initial exposure to a carcinogen and the diagnosis of a malignancy. It may be shortened when there are multiple exposures to carcinogens, or when an exposure is extensive, and thus there is the opportunity for multiple cells to be affected. Latency may be hastened in rapidly growing cells, such as those of the bone marrow (expressed as leukemia), and in the young, whose cells are dividing as a consequence of growth and maturation.

Cancer is the second leading cause of death in children (accidents being first), and a malignant stimulus can have occurred only within the finite span of a child's life, as measured from conception to diagnosis. The older the person, the more opportunities that person has had to be exposed to carcinogenic agents, and for those agents to interact with the DNA of his or her cells. Young workers, whom I define as being younger than 50 years of age, and especially those who have had a limited number of jobs with definite exposures, represent a group whose exposures can be investigated with relative confidence.

V. A fifth concept of carcinogenesis is that not all carcinogens are created equal. In general, it may take decades for a person to develop a malignancy following exposure to certain substances, such as asbestos. The more powerful carcinogens, such as formaldehyde, a number of chlorinated pesticides, vinyl chloride, bis-chloro methyl ether, and some plastic components, can cause cancer in an animal or a human after relatively brief periods, ranging from months to years. Many of these powerful carcinogens have in common the presence of, or the conversion to, free-radicals that can gain access to DNA.[19-21]

VI. Although certain cancers are frequently associated with specific carcinogens, it must be understood that chemical carcinogens are not site-specific. Many people operate under the misconception that a chemical carcinogen causes only one type of cancer in a single organ.

For example, asbestos is readily associated with lung cancer, but it also causes cancer of the pleural and parietal mesothelial cells, and has been associated with cancers of the intestines, nasopharnyx, and more recently lymphoma and leukemia. Thus, asbestos is an example of a carcinogen that causes malignancies in all three cell types (squamous, adenomatous, and mesothelial) and in multiple organ systems. Similarly, arsenic has been shown to cause cancers of the skin, lung and intestinal tract; and vinyl chloride to cause cancers of the liver, brain, and lung.

VII. Carcinogens may cause cancer in one organ system in animals and in another organ system in humans. Thus, although there are differences in the way that various species react to a carcinogenic stimulus, carcinogens are not-organ-specific in cross species, just as they are not site-specific in the same species.

The appearance of cancer in wild or domestic animals may herald the occurrence of carcinogenic agents in the environment.[22]

VIII. There is a correlation between chemicals that are mutagenic in test systems and those that are carcinogenic in animals and/or humans. This follows logically, in that a change in the DNA is necessary for either of these processes to occur. A mutagenic effect occurs in the basic genetic code of the germ cell and is transferable through subsequent generations. Not all mutagens are carcinogens, and vice versa, but there is such a strong correlation by way of both mechanism and existing test data that mutagenic testing is required under most regulatory statutes. When positive, it must trigger additional mandatory testing.

Unfortunately, a negative mutagenicity test does not assure safety as far as carcinogenicity is concerned, and animal testing still needs to be done.

IX. There is a correlation in action between chemicals that are mutagenic, carcinogenic, and teratogenic, in that the genotoxic effects of a chemical may be expressed in a number of ways. The last, teratogenicity, requires that the fetus be exposed to a chemical at a critical time of development, resulting in alteration of either form or function. The earlier the exposure to a chemical teratogen, the more likely it is that the fetus will not survive, or will show marked organ abnormalities. The later during the period before birth that the exposure occurs, the more likely it is that function, rather than body form, will be adversely affected.

Carcinogenic chemicals cause forms of damage to an organism other than malignancy. Notable examples are fibrotic lung disease from asbestos exposure and both liver and kidney damage from exposure to various solvents.

THINK LIKE A CHEMIST

Predictability

Chemistry is an orderly science, with knowledge built upon prior knowledge. The structure of a chemical determines not only its intrinsic reactivity, but its interaction with cells, and what effects it may have upon an organism. Most actions can be predicted from knowledge of structure of the chemical, combined with information about similar chemicals. For instance, it was predicted that the transuranic element astatine would behave similarly to the other members of the halide family to which it belongs. Like iodine, it was predicted to accumulate in the thyroid gland, and testing showed that it did.

Familiarity with the chemical elements allows one to predict that chemicals from a single family of elements will distribute themselves within a body in a similar manner. Strontium (whether radioactive or not) will behave like calcium and seek bony tissue. If the element is radioactive, as is strontium 90, malignancy may result due to radiation of the tissues absorbing that element. Brominated organic chemicals will behave in a similar manner to chlorinated ones. By knowing the chemical actions of the polychlorinated biphenols (PCBs), it was predicted that the polybrominated biphenols (PBBs) would prove to be carcinogenic as well. Early work begun in the 1970s provided confirmation that came in 1980.[23-25] Understanding the similarity of the biological effects of the PCBs and PBBs, and considering the large data base on the effects of PCBs, one must conclude that humans exposed to the latter are at increased risk for cancer.[26]

Designer Chemicals

Many pharmaceuticals and trade chemicals are designed from others of the same chemical family by varying one part of the chemical structure. This "me-too" process may retain function while not infringing upon prior patents. Producers of the antihistamines, anesthetic agents, and steroidal products have followed this path, making a multiplicity of products, but with few differences in function or toxicity. The same method has been followed by those developing agricultural pesticides, with respect to action as well as structure. Even one who is not a chemist can look at the structure of a chemical and knowing the structure and function of another, similar compound—predict with considerable accuracy both the action and the probable toxicology of the chemical under question.

Confirmation of action and toxicity is obtained with definitive testing. Although it is the responsibility of producers of new chemicals to determine toxicity before releasing a product into commerce, not all are adequately tested. Where chemicals have been released without adequate testing, their effects may be reasonably predicted on the basis of the known effects of similar chemicals. Several chemicals that demonstrate this concept are shown in the accompanying figures.

PCBs and PBBs (Figure 2-1)

Since 1929, PCBs (polychlorinated biphenols) have been marketed in the United States for a variety of purposes, dependent in part upon their lubricant and fire-resistant qualities, molecular weight and size, resistance to decomposition, and

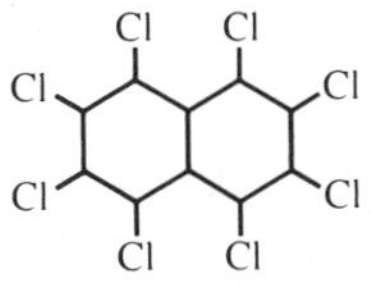

Polychlorinated Biphenyl

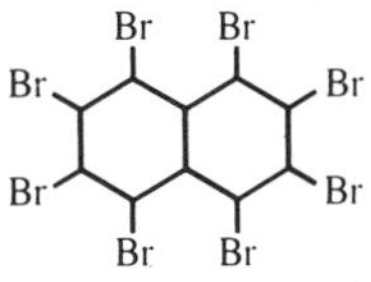

Polybrominated Biphenyl

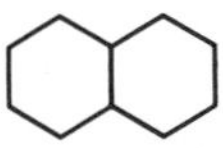

Biphenyl Nucleus

(1 to 4 chlorine or bromine atoms may be substituted on the carbons of each ring.)

Figure 2-1. Biphenyl compounds.

so on. The PCBs have displayed a number of toxic effects upon the liver, skin, blood, and nervous, reproductive, immunological, and other systems, including both enzyme induction and carcinogenicity. The use and sale of the PCBs were eventually restricted.[27] It should have come as no surprise that the brominated congeners of PCBs, produced from 1970 to 1974 in the United States, and causing catastrophic damage to the farm animals and citizens of Michigan, should prove of equal or greater toxicity. Indeed, one major chemical corporation decided against the commercial production of PBBs, based upon testing and upon structure–activity relationships that were similar to the PCBs.[28, 29] In 1972 and 1973, before PBB production was stopped, the only U.S. manufacturer produced 6.2 million pounds of the persistent and toxic product, which was incorporated into thermoplastics for such uses as radio, TV, hand tool, shaver, and business machine housings. Only 1/6000 of this amount (500 to 1000 pounds), incorporated by mistake into animal feed, accounted for the known severe environmental damage in Michigan.[30] Production of PBBs was halted in 1973, with domestic supplies exhausted in 1975,[31] and importation restricted.[32] The fate of the 6.2 million pounds of PBBs that were utilized is largely unknown. Unknown, too, is the medical status of workers who handled the PBB-containing plastics.

Chlorinated Phenols (Figure 2-2)

The chlorinated phenols were proposed as fungicides in the mid-1930s,[33, 34] and soon thereafter workers handling the products began to show a number of skin, kidney, and liver problems.[35-37] Their herbicidal properties became known during World War II, and production increased, followed by more disease reports.[38-40] Although some of the effects were found to be secondary to the contamination of the chlorinated phenols with dibenzo-dioxins and dibenzo-furans,[41] the various chemicals share a number of toxic effects, all related to structure.[42] Lest anyone think this is just an exercise in structure–activity relationships, it must be noted that the first two chemicals, 2,4-D and 2,4,5-T, are what make up Agent Orange, the herbicide used to defoliate large areas of Vietnam. The two herbicides are also widely used in the United States, 2,4-D being available for home yard use and 2,4,5-T largely restricted to agricultural use. Pentachlorophenol (''penta'') is a widely used wood-treatment product, and trichlorophenol is an intermediate in the production of 2,4,5-T and of the antibacterial agent hexachlorophene.

Designer Pharamaceuticals

A simple exercise in structural–activity relationships is provided by some of the antihistamines (see Figure 2-3). Substitution of a chlorine for a bromine atom,

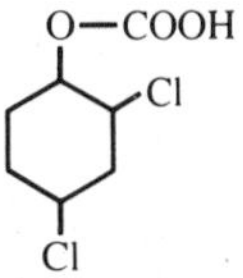

O—COOH

2,4-Dichlorophenoxy Acetic Acid
2,4-D

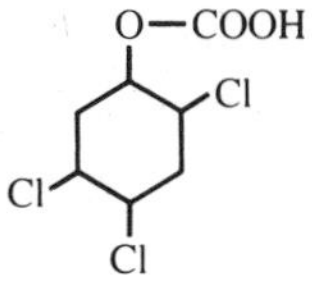

O—COOH

2,4,5-Trichlorophenoxy Acetic Acid
2,4,5-T

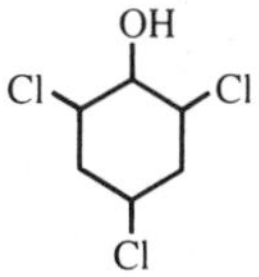

OH

Trichlorophenol

OH

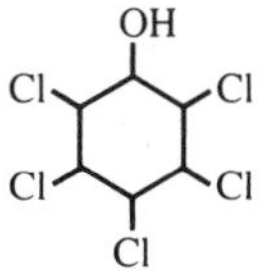

Pentachlorophenol

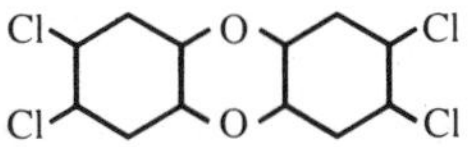

2,3,7,8-Tetrachloro-Dibenzo-p-Dioxin
(TCDD)

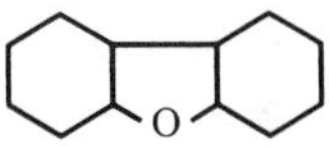

Dibenzo Furan Nucleus

Figure 2-2. Chlorinated phenol, dioxin and furan structures.

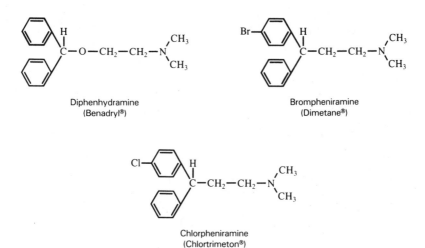

Diphenhydramine
(Benadryl®)

Brompheniramine
(Dimetane®)

Chlorpheniramine
(Chlortrimeton®)

Figure 2-3. Structurally related antihistamines.

as with chlorpheniramine and brompheniramine, will change the action little. If a person experiences drowsiness, dryness, or some other reaction to one of these, it is likely that he or she will experience the same reaction to all of the chemicals in the group. Likewise, elimination of the oxygen atom in the side chain of the diphenhydramine compound does not abolish the antihistamine effect, but does result in a somewhat less sedative product.

Reactions to organophosphate compounds will be similar irrespective of the differences in the individual products, as shown in Figure 2-4. This class of chemical was originally developed to be nerve gases, for use in warfare, because of their disruption of nerve transmission. Two of the examples cited here are used as insecticides; the third has been used clinically to treat glaucoma. Their effects are similar. What is critical is the basic chemical structure, as well as the side chains attached to the phosphate atom, which determines the action and toxicity characteristic of organophosphates.

The first, parathion, is credited with having caused more deaths than any other organophosphate. Malathion, one of the most commonly used pesticides, was broadcast over hundreds of acres of California and Florida to control several varieties of fruit flies. The last, isofluorate, has been utilized selectively to control glaucoma. Additional discussion of the various pesticides will be found in Chapters 7–9 and 12. The important thing to understand at this juncture is that the structure of a chemical determines not only function but toxicity as well.

In Figure 2-5 are diagrams of some simple halogen compounds that demonstrate the relationship between structure, function, and predictable toxicity. These chemicals are fat-soluble, and many function as commercial solvents. Thus, knowing the structure of a chemical allows one to predict not only its usefulness in commerce, but also its probable toxicity. Carbon tetrachloride,

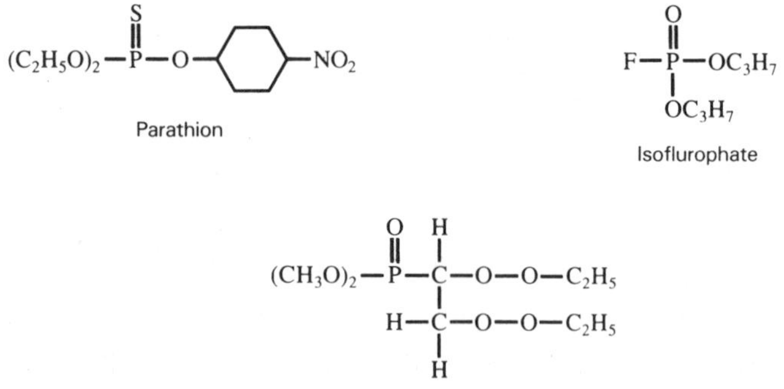

Figure 2-4. Organophosphates.

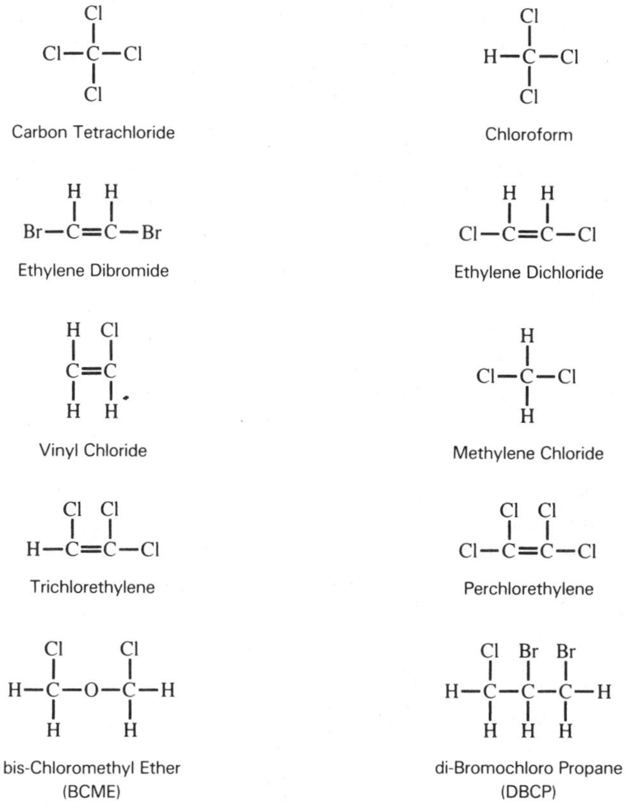

Figure 2-5. Some simple halogen compounds.

for example, is recognized as a potent liver toxin, as are other members of this group.

Looking at the structures of the chemicals in Figure 2.5, and remembering that chloroform has been used as an anesthetic for years, one should not be surprised to learn that exposure to vinyl chloride, carbon tetrachloride, trichlorethylene and perchlorethylene can also produce confusion, sleepiness, anesthesia, and death. This property is recognizable and predictable, based upon chemical structure. This action has been demonstrated in animals under controlled conditions, but, unfortunately, also in workers cleaning storage tanks, operating degreasing machines, and painting, and in persons who became addicted to sniffing glues and paints. The carcinogenicity of these associated chemicals has also been shown.

These chemicals are commercially useful, high-production-volume products that have also been distributed into the air and water. Trichloroethylene and perchlorethylene have been identified in a number of water supplies, including those of the residents of the Love Canal area in upper New York state, in Woburn, Massachusetts, and in the Mississippi Delta area. These geographic areas share a number of medical problems, including pregnancy complications, cancer, and liver and kidney disease. Although it has been extremely difficult to trace the source of the contamination in these various areas, leakage from underground storage tanks and discharge from industrial sources are considered the most common means of contamination.

Generally the impetus to identify these chemicals in the environment came not from prospective public health investigations, but, unfortunately, from residents who became aware of medical problems within their own communities. Love Canal residents, Lois Gibbs and Louella Kenny, with the professional help of Dr. Beverly Paigan provided the investigation and documentation needed to evacuate the Love Canal area. In Woburn, Massachusetts,[43] it was Anne Anderson's awareness of eight children with leukemia within a half-mile radius the started the investigation into possible cause. Her child was 3 years old when he was diagnosed as having leukemia in 1972. He died in 1981. The investigation and documentation of possible causes were initiated by Ms. Anderson, with the help of Rev. Bruce Young from the Episcopal Church, and reporter Charles Ryan from the *Daily Times*. Her son's physician, Dr. John Truman, from the Massachusetts General Hospital also became interested in pursuing the cause of the multiple malignancies that had occurred in the neighborhood.

Unfortunately, these events demonstrate that few *prospective* studies are done for either workers or communities where toxic chemicals are known to exist. The value of *prospective* and *preventive* toxicology cannot be overemphasized. It can be a simple and inexpensive process, using, in part, knowledge of the basic chemical structure of various products.

FROM MOUSE TO MAN

The process of translating the adverse effects of chemicals, from studies performed in animals to humans, requires some explanation. It hardly seems necessary to point out that it is unethical, not to mention illegal, to experiment upon humans without their consent. The Nuremberg Convention, reflecting upon the relatively recent Nazi "medical" experiments upon humans, condemned these practices absolutely.

Some "preclinical" studies are performed upon groups of volunteer subjects. Currently, in the United States, these tests must be done with the written *informed consent* of the person being tested. This means that the test subject must be informed, prior to the test, of the nature of the test and the anticipated effects of the drug or chemical. The record-keeping requirements are strict, and subject to review by human experiment committees. In the past, a number of drug companies contracted with investigators who used prison inmates to test various medications for side effects. Some chemical companies that manufacture toxic chemicals use "volunteers" within their own ranks to test for effects, though it may be argued that this practice may not yield objective results for a variety of reasons, including the concern that "control" groups within a single company may not consist of unexposed persons. Furthermore, the additional exposure of employees who work with or around similar or additional chemicals makes the interpretation of results very difficult.

Pharmaceutical agents and industrial chemicals have caused harm and death that in some cases could have been avoided had the products been tested adequately prior to their release into commerce. As a result, scientists have generally agreed that it is preferable, and less costly in terms of the ethical burden, medical care, and litigation, to test the effects of chemicals upon groups of animals first, before they are introduced into the workplace, the home, the pharmacopoeia, and the general environment. Regulations have been devised in the United States to cover the testing of a number of product categories and include the (Toxic Substance Control Act;[44] regulations of the Food and Drug Administration;[45,46] and the Federal Insecticide, Fungicide and Rodenticide Act[47,48]). These laws were passed in response to documented problems, and while not corrective of past practices, are intended to provide for prevention of harm, utilizing scientific procedures and principles.

Various national and international regulatory bodies have agreed that the results of testing done on groups of animals will be applicable to humans. These data gain validity when similar results are produced in multiple species of animals, and are further validated when the results produced in a species of animal, similarly exposed, are comparable to those found in humans.

Persons ignorant of the basis of toxicological testing complain that the dosages given to test animals are unrealistic. A population of animals being tested for a response to a given chemical is small in relation to the human population at risk; and because of the general principle that response increases within a given population with increasing dose, test animals must be given amounts of chemicals considered in excess of that to which a human would be exposed. The parameters of the doses are not picked at random; they are chosen to provide realistic responses. Additionally, for the information to be usable, statistical validation also requires that large doses be used where the animal population is small in relation to the human population.

When considered on the basis of body surface, the dose to produce an effect in an animal or a human is approximately the same. On the basis of body weight, a human is probably ten times more likely to suffer an adverse effect. Because small animals have shorter life spans than humans, toxic effects can be evaluated within comparable time frames.[49]

All chemicals known to be carcinogenic in humans have produced cancer in at least one species of animal. (It has been said that the arsenicals are the only carcinogenic class shown to cause cancer in humans but not animals.) This is not the case; cancers have been reported in mammals and fish exposed to arsenicals.[50-54] When a chemical has been shown to produce malignant changes in two different species of animals, it must be regarded as a potential human carcinogen, and human exposure should be minimized to the greatest extent possible.

Investigators should carry out testing in a sequential fashion, looking for various acute, subchronic, and chronic effects.

SUFFICIENT NUMBERS

In general, the basic toxicology of the chemical being tested is taken into account, so that there will be a sufficient number of animals at the start of an experiment to survive to the end of the test period. If we start with 50 animals in the course of testing for carcinogenic or other toxic potential, and 25 die of acute and/or chronic effects, intercurrent disease, or accident before the logical end of the experiment, or, in the case of cancer testing, before the end of the expected life span of the animals, then there will be insufficient numbers at the end to determine if the chemical is capable of causing the looked-for effect or not.

As in many other endeavors, common sense is an invaluable guide. Under several of the statutes cited above, ''Good Laboratory Practice'' is spelled out, regarding proper procedures for testing of chemicals under laboratory conditions. Used with common sense, these measures will provide meaningful and statistically useful information at minimal cost. Animal testing is not inexpensive, and it has many critics; but if we are to utilize it to prevent further damage to humankind and the biosphere, the tests must be carried out with care and foresight. An acute testing procedure on 200 rodents has been estimated to cost about $250,000. This generally is the cost of caring for fewer than ten cancer victims for a year, and certainly is less than the lifetime expense of care for a child born with birth defects.

DETERMINING DOSE

To estimate a chemical's relative toxicity, an attempt is made to determine the amount that will kill 50% of one or more species of animals. This result is called the LD-50 (lethal dose for 50% of the animals). This determination is important in order to compare one chemical with another in terms of relative toxic effect. This procedure is indispensable in determining the measure of safety for a pharmaceutical product when comparing the toxic dose to the therapeutic dose.

When the amount of drug needed for a therapeutic effect approaches the toxic dose, there is very little leeway for error, and the patient must be monitored carefully; or one may decide to choose a drug with less risk. Where there is a 100% difference, as opposed to a 10% difference, between the therapeutic and toxic doses, there is less likelihood of a dose-related adverse effect, and the drug may be considered "safer."

Likewise, a chemical that showed an LD-50 with a relatively small dose would be considered more hazardous in an industrial setting than one that had a larger dose. Thus these measurements are of more than academic interest. Those chemicals deemed to be relatively more toxic must be handled in a more prudent manner, or substitute products and/or processes must be sought.

TESTING

Acute Testing

In acute testing, a single dose of a chemical is given, if possible by a route of administration similar to that which humans may experience. Thus a test chemical is administered orally, intravenously, by inhalation, or applied to the skin. The test animals are then observed for various periods of time (from immediate observation to up to approximately two weeks following dosing), to determine any signs of toxicity. Observations include looking for any weight loss, nausea, vomiting, diarrhea, loss of fur, change in behavior, convulsions, trouble in breathing, and so on. Animals that die, as well as those that survive to the end of the experimental period and are sacrificed, are autopsied, their organs examined, and their tissues examined by histological techniques. Certainly all tissues expected to be affected, based upon the chemical characteristics of the chemical or of similar chemicals, must be examined. As in all animal testing, the results are compared to a control group of unexposed animals. This type of test is usually done on small animals such as mice, rats, rabbits, and dogs.

Subchronic Testing

Subchronic testing is carried out for at least 90 days, on at least two species of animals, with at least one route of exposure comparable to that expected in humans. The dosages employed are based upon the information determined in the LD-50 experiments. Three doses of a chemical are administered. One expected to produce toxic effects (but not immediate death), one producing no noticeable effect, and one intermediate dose between these two parameters. The animals are observed for signs of illness, as in the acute testing experiments. Data are collected from blood, urine, enzyme, and other tests, and at the end of the experiment, the animals as well as their organs are examined for weight, gross appearance, and any histological changes. Once again, the animals receiving the test chemical are compared to the control group of animals.

Chronic Testing

One then proceeds to chronic testing, which is done using dosage levels determined in the acute and subchronic studies; that is, enough of a dose to give an effect, but not so much that the animals die before the end of the test period. Depending upon the life span of the test animal, the experimental period may extend from 12 to 18 months or more. In some cases, particularly when one is testing for possible carcinogenic effects, it is appropriate to extend the test period for the natural lifetime of the animals. Many tests done in the past were flawed by doses that were too high and so killed the animals before either chronic or carcinogenic effects could develop, or, conversely, by experiments that were terminated too soon, before long-term effects could develop.

Teratogenic Testing

Studies of chemicals that may produce birth defects are done on at least two species, usually rats and rabbits, with the dosing done during the first trimester of pregnancy, which is the time when major organogenesis takes place. At least two dose levels are given, each being below that which causes overt toxicity in the mother. The fetuses are removed by cesarean section and examined for gross and microscopic abnormalities of development and structure. Some chemicals kill the developing fetus; so as in fertility testing, the uterus of the mother animal is examined to determine the number of implantation sites, and to be certain that the animal was pregnant, and thus to ascertain the original number of fetuses. It follows, then, that the number of implantation sites minus the number of animals born equals the number of fetuses reabsorbed or aborted.

Anatomical deformities are produced in animals by exposing the foetuses to teratogens early in the period of development, usually the first trimester. The earlier the exposure to a toxic chemical, during the period of tissue differentiation and organ development, the greater the damage, and the more likely that the offspring will not survive.

Knowledge of the predictable fetal events allows for retrospective teratogenic evaluation. When an animal or a human is born with a birth defect, one can frequently "calculate backward" to determine the timing of the insult by knowing the sequence and concurrence of organ system development.

One of the most notable teratogens was thalidomide, a product that was marketed in Europe as a sedative, and, when taken by pregnant women, resulted in the birth of children suffering various limb defects, including absent limbs (phocomelia).[55,56] After the reports of these terrible deformities in humans were made, testing was undertaken in laboratory animals, and proved positive in rabbits.[57]

A number of pharmaceutical agents and consumer products have been implicated in birth defects.[58] Unfortunately, in many cases human reports have predated animal testing, underscoring the importance of adequate testing of products before they are put on the market.

Neural tube defects (that is, the failure of the spinal cord to close during fetal life), resulting in various defects such as spina bifida, as well as cleft palate and other defects, have been suspected to have been caused by such chemicals as the herbicide 2,4,5-T (2,4,5-trichlorophenoxy acetic acid) as well as contaminating dioxins, including TCDD (tertachlorodibenzo-p-dioxin) although de facto proof is extraordinarily difficult.[59] That these chemicals are able to produce defects in test animals is cause for concern in humans.[60] (See previous discussion of these chemicals in this chapter.) Until "proof positive" is available, it is prudent to avoid exposure to these chemicals.

Eugene Smith's photography of a mercury-damaged young woman being cared for by her elderly mother stays in any viewer's memory. Neurological affliction can occur in those exposed either before or after birth,[61-63] pointing up the importance of testing for known effects in the unborn.

As of 1981, it was estimated that there were 287,000 babies born with defects each year in the United States, that is, about 9% of all births. Additionally, there were another 560,000 infant deaths, spontaneous abortions, stillbirths, and miscarriages due to defects in fetal development.[64] There is little in the way of information gathering and reseach to determine the cause of these losses. If 10% of the defects and losses could be prevented by proper testing and control of chemicals, there are few who would argue against it.

Reproduction is a complex matter, with factors occurring throughout the lifetime of each parent that may affect the offspring. Assuming that the parental stock (animal or human) reached adult reproductive state in a normal anatomic and physiological state, some of the adverse events are as shown in Table 2-1.

Behavioral Testing

Exposure to a toxic agent that occurs late in pregnancy is less likely to produce an anatomic abnormality than exposure in early pregnancy, but such exposures can result in other toxic effects to the newborn, including neurological abnormalities, behavioral changes, and decreased mentation.[65,66] Frequently, these changes cannot be measured until the offspring has grown and reached maturity.

Table 2-1.   Events Influencing Reproductive Outcome.

Event	Parent Male	Parent Female	Offspring
Failure of ovulation due to endocrine dysfunction or interference		X	
Defective ova genetic		X	
Defective sperm genetic, form or number	X		
Failure of implantation		X	
Loss of fetus (abortion)		X	
Exposure during gestation		X	X
Exposure via lactation			X
Exposure postpartum			X

To determine the effect on growth, behavior, and survival, laboratory studies are conducted on animals exposed in the second and third trimesters of pregnancy and comapred to unexposed animals. The testing is carried out in a manner similar to teratology testing, and the examiner looks for such characteristics as learning ability, sociability, grooming behavior, reproductive behavior, lethargy, aggressiveness, and so on, as compared to the usual behavior seen in unexposed animals.

The obvious problems stemming from the use of street drugs such as heroin and morphine, as well as the overjudicious use of drugs during delivery, can result in babies born with central nervous system depression, resulting in depressed respiration and secondary brain damage. Less well tested are chemical exposures comparable to those occurring in the home and the workplace. Despite the fact that 40% of workers are women, most occupational testing has been concerned only with the male. The reproductive hazards to both sexes need to be addressed.[67,68]

Fertility Testing

To determine whether fertility is influenced or not, a chemical is administered to the test female animal for 14 days prior to mating, and to the test male animal for 60 to 80 days prior to mating. Each of the parent animals is then mated with the opposite-sex untreated animal. The entire period of pregnancy is then examined and compared to untreated normals. That is, the number of animals that become pregnant, the number of implantation sites, the number and condition of stillborn and live animals, and their postnatal behavior and development are recorded. These data are compared to fertility records of untreated animals.

Failure of shrimp to mate may be due to contamination of their water with a chemical such as DDT that interferes with reproductive steroid formation and results in failure of the females to ovulate, and thus attract the males. While this may be of interest from a purely behavioral point of view, it is of intense interest to the shrimp farmer who depends upon this reproductive activity to produce food.

Canadian mink farmers, feeding their animals waste offal from chicken farms, found that their animals failed to reproduce. When it was determined that the estrogenic chemical DES (diethylstilbestrol), which had been used as a growth promoter for the poultry, was responsible for this phenomenon, the practice was discontinued. Hearings before the U.S. House of Representatives in 1951 documented multiple reproductive problems with the use of DES in the mink, and underscored the need for testing of drug preparations prior to marketing, whether intended for animal or human use.[69] The existence of data prior to the 1950s connecting DES with interference in reproduction should have excluded this chemical from human food and medical use until it had been properly and extensively tested under laboratory conditions. Unfortunately, between 500,000 and 2,000,000 women were given DES during pregnancy.[70] Resultant sexual dysfunction and cancer in the children of these women who were given DES could have been prevented by attention to the information obtained from the experiences of the mink farmers, as well as other data that were available in the 1950s. Data exist that arouse grave concern, not only for genital malformations and malignancies in female children but in the male offspring of DES mothers as well.[71-78] It should be underscored that all of the DES-associated abnormalities produced in laboratory animals have been detected in humans clinically.[79]

Mutagenic Testing

Mutagenicity is the ability of a chemical to cause a change in the basic genetic material, with the potential for its transmission to the offspring. If the mutagenic event occurs in either the egg or the sperm, the result can be complete failure of fertilization, anatomic abnormality, and/or death of the progeny. Some mutational events are subtle and not readily apparent at birth, but they interfere with development, and/or cause learning and behavioral problems in animals and humans, and altered form and function in submammalian animals and plants.

For ease of handling, cost, and efficiency of obtaining results, mutagenic testing is commonly done on submammalian species, such as fruit flies, bacteria, neurospora, and even plants. The changes produced by sublethal doses of a number of herbicides are indicative of damage to the basic DNA of the plants, and herald concern for human populations that are exposed to the same agents, either by using them or by living downwind from sources of contamination. The dominant lethal assay for mutagenic effects is a technique that looks for the induction of structural and numerical chromosomal abnormalities. This procedure has relevance to humans, in that many human autosomal defects— such as those found in persons born with trisomy or Down's syndrome—as well as conditions associated with leukemia and lymphoma are associated with chromosomal mutations.[80,81] Because any form of treatment for these largely irreversible conditions is highly unlikely, identification and avoidance of mutagenic chemicals is the only practical course.[82]

There is an association between mutagenesis, carcinogenesis, and teratogenesis. Each involves an interference with the basic building block of life, that is, the DNA (desoxyribonucleic acid), which results in changes in form and/or function. Thus, chemical testing on submammalian species has relevance in assessing the potential for harm to humans. A correlation between changes in the DNA and other adverse effects, including heart disease and aging in general, has been found.[83,84] These concepts underlie the efforts of scientists and regulatory agencies to translate the results of chemical tests on animals to humans, for purposes of decision making and, ultimately, protection of the human population and the biosphere.[85-94]

REFERENCES

1. Anderson v. W. R. Grace Co., 82-1672-S (D. Mass. Sept. 22, 1986).
2. Gibbs, L. M. *Love Canal: My Story*. State University of New York Press, Albany, 1982.
3. Whorton, D., Krauss, R. M., Marshall, S., Milby, T. H. Infertility in male pesticide workers. *Lancet* ii: 1259-61, 1977.
4. Whorton, D., Milby, T. H., Krauss, R. M., Stubbs, H. A. Testicular function in DBCP exposed pesticide workers. *J. Occup. Med.* 21: 161-66, 1979.
5. U.S. Occupational Safety and Health Administration. Occupational exposure to 1,2-dibromo-3-chloropropane (DBCP). Emergency temporary standard, hearing. *Fed. Regist.* 42: 45536-49, 1977.
6. Assennato, G., Paci, C., Baser, M. E., Molinini, R., Candela, R. G., Altamura, B. M., Giorgino, R. Sperm count suppression without endocrine dysfunction in lead-exposed men. *Arch. Environ. Health* 42(2): 124-27, 1987.
7. Creech, J. L., Johnson, M. N. Angiosarcoma of the liver in the manufacture of polyvinyl chloride. *J. Occup. Med.* 16: 150-51, 1974.
8. Heath, C. W., Jr., Falk, H., Creech, J. L. Characteristics of cases of angiosarcoma of the liver among vinyl chloride workers in the United States. *Ann. Acad. Sci.* 246: 231-36, 1975.
9. Hamilton, A. *Industrial Poisons in the United States*. Macmillan, New York, 1925, p. 493.
10. Goldblatt, M. W. *Brit. J. Indust. Med.* 6: 65, 1949.
11. Page, J. A., O'Brien, M. W. *Bitter Wages*. Grossman Publishers, New York, 1973, pp. 59-63.
12. Infante, P. F. Oncogenic and mutagenic risks in communities with polyvinyl chloride production facilities. *Ann. N. Y. Acad. Sci.* 271: 49-57, 1976.
13. *Physician's Desk Reference (PDR)*. Medical Economics Co., Oradell, N.J. 1987, pp. 953-54, 1469-71.
14. "Dear Doctor" letter from Geigy Pharmaceuticals, Summit, N.J., dated Apr. 4, 1985.
15. Maugh, T.H. Tracking exposure to toxic substances. *Science* 226: 1184-85, 1984.
16. Picciano, D. Cytogenetic investigation of occupational exposure to epichlorohydrin. *Mut. Res.* 66: 169-73, 1979.
17. Kapp, R. W., Picciano, D. J., Jacobson, C. B. Y-chromosomal nondisjunction in dibromochloropropane exposed workmen. *Mut. Res.* 64: 47-51, 1979.

18. Picciano, D. Cytogenetic study of workers exposed to benzene. *Environ. Res.* 19: 33–38, 1979.

19. Troll, W., Wiesner, R. The role of oxygen radicals as a possible mechanism of tumor promotion. *Ann. Rev. Pharmacol. Toxicol.* 25: 509–28, 1985.

20. Upton, A. C., Clayson, D. B., Jansen, J. D., Rosenkranz, H. S., Williams, G. M. Report of the ICPEMC Task Group 5 on the differentiation between genotoxic and non-genotoxic carcinogens. *Mut. Res.* 133: 1–49, 1984.

21. Cerutti, P. A. Preoxidant states and tumor promotion. *Science* 227: 375–81, 1985.

22. Foran, J. Focus; On *International Joint Commission Activities* 12(1): 11–12, Mar./Apr. 1987.

23. Kimbrough, R. D., Burse, V. W., Liddle, J. A. Persistent liver lesions in rats after a single oral dose of polybrominated biphenols (Firemaster FF-1) and concomitant PBB tissue levels. *Environ. Health Perspect.* 23: 265–73, 1978.

24. Kimbrough, R. D., Groce, D. F., Korver, M. P., Burse, V. W. Induction of liver tumors in female Sherman strain rats by polybrominated biphenols (Firemaster FF-1, lot No. 7042). *J. Nat. Cancer Inst.* 55: 1453–58, 1975.

25. Moore, R. W., Dannan, G. A., Aust, S. D. Structure–function relationships for the pharmacological and toxicological effects and metabolism of polybrominated biphenyl congeners, Chap. 10 in *Molecular Basis of Environmental Toxicity*, Bhatnagar, R. S., Ed. Ann Arbor Science Publishers, Ann Arbor, Mich. 1980, pp. 173–212.

26. Szeluga, C. v. Velsicol Chemical Co. C-762-00-423-NO (1980).

27. P. L. 94–469, Sect. 6(e). 90 Stat. 2003 (1976).

28. Moore, R. W., Dannan, G. A., Aust, S. D. Structure–function relationships (see ref. 25).

29. Wartiz, R. S., Aftosmis, J. G., Culik, R., Dashiell, O. L., Faunce, M. M. Griffith, F. D., Hornberger, C. S., Lee, K. P., Sherman, H., Tayfun, F. O. *Toxicological Evaluations of Octabromobiphenyl and Hexabromobiphenyl.* Haskel Laboratory, E. I. DuPont de Nemours & Co., Newark, Del., 1972.

30. Getty, S. M., Rickert, D. E., Trapp, A. L. Polybrominated biphenyl (PBB) toxicosis: an environmental accident. *CRC Critical Review in Environmental Controls*, 309–323, 1977.

31. IARC Monographs on the Evaluation of the Carcinogenic Risk of Chemicals to Humans International Agency for Research in Cancer, Lyon, France. 18: 197–224, 1978.

32. *Fed. Regist.* 45(208): 70728–33, Oct. 24, 1980.

33. U.S. Patent 1,991, 329 (1935). Mills, L. E. to Dow Chemical Co.

34. U.S. Patent 2,039,434 (1936). Mills, L. E. to Dow Chemical Co.

35. Anon. Queries and minor notes. *J.A.M.A.* 106: 2092, 1936.

36. Stingily, K. O., *Southern Med. J.* 33: 1268, 1940.

37. Butler, M. G. *Arch. Derm. Syph.* 35: 251, 1937.

38. Baader, E. W., Bauer, H. J. *Ind. Med. Surg.* 20: 286, 1951.

39. President's Science Advisory Committee. *Report on 2,4,5-T.* Office of Science and Technology, Washington, D.C. 1971.

40. Hofmann, H. T. *Arch. Exp. Pathol. Pharamacol.* 232: 228, 1957.

41. Kimmig, J., Schulz, K. H. *Dermatologica* 115: 540, 1957.

42. *IARC Monographs on the Evaluation of the Carcinogenic Risk of Chemicals to Man*, 15: 41–102, 111–38, 273–300, 1977.

43. DiPerna, P. *Cluster Mystery: Epidemic and the Children of Woburn, Mass.* C. V. Mosby Co., New York, 1985, 400 pp.

44. 15 USC 2603 (4).

45. Pub. L. 59–384, 34 Stat. 768.

46. Pub. L. 75–717, 52 Stat. 1040.

47. Pub. L. 80–104, 61 Stat. 163.

48. Pub. L. 92–516, 86 Stat. 973.
49. Doull, J., Klaassen, C. D., Amdur, M. O., Eds. *Casarett and Doull's Toxicology: The Basic Science of Poisons*, 2nd ed. Macmillan Publishing Co., New York, 1980, pp. 17–27.
50. Prell, H. *Arch. Gewerbepathol. Gewerbehyg.*, 1937.
51. Halver, J. E. *Progress in Studies on Contaminated Trout Rations and Trout Hepatoma*. National Institute of Health Report, 1962.
52. Kraybill, H. F., Shimkin, M. G. Carcinogenesis related to food contaminated by processing and fungal metabolites. *Adv. Cancer Res.* 8: 191–248. Academic Press, New York, 1964.
53. Osswald, H., Goerttler, K. Arsenic induced leukemia in mice after diaplacental and postnatal application. *Deutsch. Gesell, Pathol.* 55: 289–93, 1971.
54. Nieberle, K. Endemic cancer in the bone of sheep. *Z. Krebsforsch.* 49: 137, 1973.
55. Lenz, W. Diskussionsbemerkung auf der Hersttagung der Reinisch—Westf. Kinder arzteververeinigung in Dusseldorf, in Faigle, J. W., Keberle, H., Reiss, W., Schmid, K. The metabolic fate of thalidomide. *Experentia* 18: 389–97, 1962.
56. Knightley, P., Evans, H., Potter, E., Wallace, M. *Suffer the Children: The Story of Thalidomide*. Viking Press, New York, 1979.
57. Staples, R. E., Holtkamp, D. E. Effects of parental thalidomide treatment on gestation and fetal development. *Exp. Molec. Pathol.* 2: 81–106, 1963.
58. O'Brien, T. E., McManus, C. E. Drugs and the fetus: a consumers' guide by generic and brand name. *Birth & Family J.* 5(2): 58–86, 1978.
59. Hanify, J. A., Metcalf, P., Nobbs, C. L. Worsley, K. J. Aerial spraying of 2,4,5,T and human birth malformations: an epidemiological investigation. *Science* 212: 349–50, 1981.
60. *IARC Monographs on the Evaluation of the Carcinogenic Risk of Chemicals to Man* 15: 73–76, 284–87, 1977.
61. Kurland, L. T., Faro, S. N., Siedler, H. Minamata disease: the outbreak of a neurological disorder in Minamata, Japan, and its relationship to the ingestion of seafood contaminated by mercuric compounds. *World Neurol.* 1: 370–95, 1960.
62. Tatesu, S., Harada, M. Mental deficiency resulting from intoxication in the prenatal period. *Adv. Neurol. Sci. (Tokyo)* 12: 181–90, 1968.
63. Suzuki, T., Matsumoto, N., Miyama, T. Placental transfer of mercury chloride, phenyl mercury acetate, and methyl mercury acetate in mice. *Indian Health* 5: 149–55, 1967.
64. *March of Dimes Birth Defects Foundation: Facts/1981*. March of Dimes Birth Defects Foundation, New York, 1981.
65. Anon. Unclassified mental retardation. *Lancet*, 250–51. 1979.
66. Kolata, G. B. Behavioral teratology: birth defects of the mind. *Science* 202: 732–34, 1978.
67. Legator, M. S. Hollaender, A., Eds. Occupational monitoring for genetic hazards. *N.Y. Acad. Sci.* 269: 1–45, 1975.
68. Bingham, E., Ed. *Proceedings: Conference on Women in the Workplace*. Society for Occupational and Environmental Health, Washington, D.C. 1976, pp. 1–364.
69. H.R. 846, H.R. 1568, H.R. 2591, H.R. 2592, H.R. 2776, H.R. 2777.
70. Noller, K. L., Fish, C. R. *Med. Clin. N. Amer.* 58: 793–810, 1974.
71. Shimkin, M. B., Grady, H. G., Andervont, H. B. Induction of testicular tumors and other effects of stilbesterol–cholesterol pellets in strain-C mice. *J. Nat. Cancer Inst.* 2:65–80, 1942.
72. Andervont, H. B., Shimkin, M. B., Canter, H. Y. Susceptibility of seven inbred strains and the F_1 hybrids to estrogen-induced testicular tumors and occurrence of spontaneous testicular tumors in strain BALB/c mice. *J. Nat. Cancer Inst.* 25: 1069–96, 1960.
73. Miller, R. W. Editorial: transplacental chemical carcinogenesis in man. *J. Nat. Cancer Inst.* 47: 1169–71, 1971.
74. McLachlan, J. A., Newbold, R. R., Bullock, B. Reproductive tract lesions in male mice exposed prenatally to diethylstilbestrol. *Science* 190: 991–92, 1975.

75. Gill, W. B., Schumacher, G. F. B., Bibbo, M. Structural and functional abnormalities in the sex organs of male offspring of mothers treated with diethylstilbestrol (DES) *J. Reprod. Med.* 16(4): 147–53, 1976.

76. Gill, W. B., Schumacher, G. F. B., Bibbo, M. Pathological semen and anatomical abnormalities of the genital tract in human male subjects exposed to diethylstilbestrol in utero. *J. Urol.* 117: 477–80, 1977.

77. Whitehead, E. D., Leiter, E. Genital abnormalities and abnormal semen analysis in male patients exposed to diethylstilbestrol in utero. *J. Urol.* 125: 47–50, 1981.

78. Conley, G. R., Sant, G. R., Ucci, A. A., Mitcheson, H. D. Seminoma and epididymal cysts in a young man with known diethylstilbestrol exposure in utero. *J.A.M.A.* 249(10): 1325–26, 1983.

79. Arai, Y., Mori, T., Suzuki, Y., Bern, H. Long-term effects of perinatal exposure to sex steroids and diethylstilbestrol on the reproductive system of male mammals. *Internat. Rev. Cytol.* 84: 235–67, 1983.

80. Epstein, S. S., Arnold, E., Andrea, J., Bass, W. Bishop, Y. Detection of chemical mutagens by the dominant lethal assay in the mouse. *Toxicol. Appl. Pharmacol.* 23: 288–325, 1972.

81. Yunis, J. J. The chromosomal basis of human neoplasia. *Science* 221: 227–35, 1983.

82. Drake, J. W., Chrmn. Environmental mutagenic hazards. *Amer. Assoc. Adv. Sci.* 187: 503–514, 1975.

83. Benditt, E. P. The origin of atherosclerosis. *Scientif. Amer.* 236(2): 74–85, 1977.

84. Burnet, F. M. *Intrinsic Mutagenesis: A Genetic Approach to Aging.* Medical and Technical Pub., Lancaster, England, 1974.

85. Miller, E., Miller, J. *Chemical Mutagens,* Hollaender, A., Ed. Plenum, New York, 1970.

86. Ames, B. N., Durston, W. E., Yamasaki, E., Lee. F. D. Carcinogens are mutagens: a simple test system combining liver homogenates for activation and bacteria for detection. *Proc. Nat. Acad. Sci.* 70: 2281–85, 1973.

87. McCann, J., Choi, E., Yamasaki, E., Ames, B. N. Detection of carcinogens as mutagens in the *Salmonella*/microsome test: assay of 300 chemicals. *Proc. Nat. Acad. Sci.* 72(12): 5135–39. 1979.

88. Miller, R. W. Relation between cancer and congenital defects: an epidemiologic evaluation. *J. Nat. Cancer Inst.* 40: 1079–85. 1968.

89. Enslein, K., Lander, T. R., Strange, J. R. Teratogenesis: a statistical structure–activity model. *Terat. Carcin. & Mutagen.* 3: 289–309, 1983.

90. Enslein, K., Craig, P. N. Carcinogenesis: a predictive structure–activity model. *J. Toxicol. Environ. Health* 10: 511–20, 1982.

91. Ames, B. *Environmental Chemicals Causing Cancer and Genetic Birth Defects: Developing a Strategy to Minimize Human Exposure.* Institute of Governmental Studies, Univ. Of Calif., Berkely, 1978, pp. 1–28.

92. Doll, R. Strategy for detection of cancer hazards to man. *Nature* 265(5595): 589–96, 1977.

93. Hiatt, H. H., Watson, J. D., Winsten, J. A., Eds. *Origins of Human Cancer,* Vol. 4, books A, B, C, Cold Spring Conferences on cell proliferation. Cold Spring Harbor, New York, 1977.

94. U.S. Environ. Protection Agency. *Project Summary: Mutagenesis Screening of Pesticides Using Drosophila,* EPA-600/S1-81-017, 1981.

CHAPTER

3

REVIEW-OF-SYSTEMS APPROACH

The **Review of Systems** approach to obtaining information from a patient is inherent in all good medical practice. It is a method designed to ensure that no information concerning any body system is omitted when one obtains a medical history. It is learned by all physicians and nurses in training, and is available in a number of standard texts on diagnosis. It may be learned by nonmedical personnel, including attorneys and patients themselves. There is nothing mysterious or esoteric about obtaining a comprehensive medical history. It merely requires patience and attention to detail while one follows a sequential system of inquiry.

Based upon my 22 years of practice, and evaluation of over 8000 patients, I believe that at least 60% of a correct diagnosis rests upon an accurate and comprehensive history, with the rest of the diagnosis depending upon the physical examination and tests. In this technological age, the importance of obtaining basic information often is ignored. Also, in this age of high technology and expensive tests, the value of a face-to-face interview is downgraded, not only in terms of time spent for services, but also in reimbursement for the effort required to get a comprehensive history. This Review-of-Systems approach, which has served medicine well since its beginning, is very useful in the investigation of diseases associated with chemicals.

For simplicity's sake, and to ensure a reproducible method of obtaining information, the history taker starts at the top of the head and goes to the feet in an orderly fashion. An experienced history taker will make sure that this process is done consistently and carefully, to lessen the likelihood of overlooking vital information.

When recounting an event or describing a medical problem, it is easy for a sick person to get sidetracked and omit information that may seem unimportant at the time. There are no unimportant data—both negative and positive information must be recorded. It is as important to know what symptoms and diseases patients have not had, as it is to know those they have experienced. These negatives increase in importance with the passage of time, providing a base line of health status, particularly when a person develops an abnormality.

To avoid the pitfalls of omission of information, one should start with a checklist and proceed in a consistent manner. The patient must be allowed to recount all information that seems relevant to him or her; but it is the interviewer who is in control of obtaining the history, and it is incumbent upon the interviewer to cover all the informational categories.

Questions should proceed sequentially, but should begin with an open question to determine just what the patient/client considers to be his or her prime concern. It is well to ask as open-ended questions as possible to avoid suggesting symptoms to a patient. While in training, I used to ask, "What brought you to the hospital?" After one patient answered, "A truck," I started asking, "What's troubling you most?"

The initial question serves two purposes:

1. It allows the person to tell his or her concerns and worries, and it signals to the patient that the interviewer is willing to listen.
2. It allows the interviewer to assess the patient's demeanor and his or her ability to relate information.

Difficulties of communicating with a person who is deaf, or who speaks a different language, are obvious. Attempting to speak more slowly and loudly to the first is helpful, although it is too often tried with the latter as well!

Always try to obtain the history from the person him- or herself directly. Many persons with a language or hearing impairment can be communicated with by using a combination of "pigeon" English and sign language, but an interpreter, family member, or coworker can be useful also.

In recording the history, indicate who is supplying the information—the patient/client or an accompanying person. This is particularly important in legal matters.

UNCONSCIOUS, SERIOUSLY ILL, AND DECEASED PERSONS

Unconscious and seriously ill patients are a special concern. As much information as possible should be obtained directly from the patient, but obviously not at the expense of initiating care. While a deceased patient's situation may seem less urgent, it generally is wise to gather information as soon as possible.

Information in some of these situations must be gathered from second-hand sources. These may include:

Family members
Coworkers
Medical records
Family physician

Industrial or other clinic

Hospitals

SAMPLING

While these occasions are not strictly a part of the review-of-systems approach to gathering medical information, incidents may occur in which critical information is available that will be unobtainable later. This happens when there is an acute event such as a pesticide or other toxic chemical exposure. At such times, it is essential to preserve samples of both clothing and body fluids for later examination.

Because plastics and some preservatives may interfere with testing, all samples should be collected in glass containers, capped with an aluminum foil–lined screw-top cap. It is acceptable to use a rubber-stoppered glass collection system, but the sample should not be stored in contact with plastic or rubber. Samples should be labeled, dated, and *frozen*.

This procedure is adequate for:

Blood

Urine

Stool

Gastric contents

Surgical or biopsy specimens

When a specific poisoning is suspected, appropriate samples should be collected, and testing should proceed as with other laboratory tests.

Clothing samples should be saved in glass with metal screw-top containers, not in plastic bags. The cheapest and most readily available of these containers are one-gallon mayonnaise jars, available from most restaurants and hospital kitchens. A supply of these clean jars is adequate for preservation of clothing samples. Clothing environmental samples usually need not be frozen, but are only kept tightly closed until they can be assayed.

Later, in a toxicological investigation, it may be necessary to collect additional body fluid or tissue samples, especially when one is looking for chemicals retained in the bone or fat stores. These samples may be obtained at the time of autopsy or even from an exhumed body.

In general, the sample need not be large. A half-ounce to an ounce of tissue may suffice for gas chromatographic or mass spectrometric test purposes.

SAMPLE HISTORY QUESTIONS

Sample questions are offered below. The wording may be changed to suit the need of the particular situation, but the sequential and orderly process should be adhered to. In all cases, follow a sequence, and be orderly and comprehensive.

All data should be quantified as to time, duration, and severity, when known. Secure obvious identifying information:

Name

Address

Telephone number

Birth date

Social security number

Date of the interview

General

- Starting when you were well, what did you notice first in the way of illness?
- When did this occur?
- How often did this occur?
- How long did it last?
- Did the condition worsen?
- What happened next?
- Did you seek medical care? Describe the care.

This method of inquiry is based upon the assumption that most persons start life in a well and functioning condition, and attempts to determine **what has changed.**

Central Nervous System

- Do you have headaches? Describe them: Where are they located? When did they begin? How often do they occur? How long do they persist? Does anything precipitate them? Does anything relieve them?
- Do you have dizziness or loss of balance? Have you fallen with these conditions?
- Have you ever had a seizure, or fit?
- Have you ever had a head injury? Have you been knocked unconscious? Describe the situation, including where, when, and any treatment rendered.

- Have you been aware of any memory loss?
- Have you received any medical care or treatment for any condition relating to your head?
- Do you have any sleep problems?

A family member may relate that a person gets lost or cannot remember recent events, while the patient may deny such problems. By the time a patient becomes aware of memory loss, it frequently is quite severe.

The way in which a patient relates complaints is also pertinent: He or she may overemphasize, or deny, a medical problem. The person may also be very "concrete" in his or her replies, giving a clue to possible organic brain damage.

Eyes

- Have you noticed any change or loss of your vision? When did you first become aware of it? Has this change progressed?
- Do you wear glasses?
- Have you had any injuries to your eyes? Relate the situation, including where, when, and any treatment.
- Have you received any first aid or medical treatment for your eyes?

For example, it was customary in the past to give anesthetic eye drops to welders who had sustained "flash burns," and then allow them to return to welding. This resulted in persons whose normal pain sensation was depressed, and who sustained eye damage as a result of additional burns or foreign material that they could not detect.

Ears, Nose, and Throat

- Have you had any loss of hearing? When did you first become aware of this? Has it progressed?
- Do you have ringing or roaring in your ears?
- Do you use a hearing aid? If so, for how long?

Hearing loss can occur not only from exposure to loud and continual noise, but from a number of chemicals and medications. Tuberculosis patients treated with dihydrostreptomycin are but one of the better known groups that sustained chemically caused hearing loss.

Occasionally, the patient/client may not be aware of a hearing loss, but a spouse may relate the complaint!

- Do you have crusting, running, or stuffiness of your nose?

- Describe what starts this, how long it occurs, and the duration of the problem.
- Did you have any of these problems as a child?

Allergic problems frequently begin early in life, and are often seasonal. Irritative ENT problems are usually associated with a specific exposure. The sudden onset of a runny nose and eyes in an otherwise healthy farm worker, or a person living adjacent to agricultural land, may be the first clue to organophosphate pesticide exposure.

- Do you have nosebleeds? Describe the occurrence and amount.

This symptom may be indicative not only of local disease of the nose, but of a clotting disorder as well. Usually, an intermittent discharge of a small amount of blood indicates irritation, while a continual discharge of blood indicates an anatomical abnormality, such as a tumor. A sudden, large amount of blood coupled with a history of high blood pressure may indicate a hypertensive crisis. A severe nosebleed may indicate a blood dyscrasia, with failure of the clotting mechanisms.

Exposure to chromium-containing products can result in perforation of the nasal septum, with crusting and bleeding of the nose, and a change in voice to a "nasal" quality.

- Do you notice any change in your sense of smell?
- Do you have sore throats, hoarseness, or phlegm in your throat? Quantify the timing, duration, and frequency.
- Have you noticed any change in your voice?
- Do you have any difficulty swallowing?
- Have you noticed any lumps in your throat or sores in your mouth?

Chest

- Have you had any chest diseases in the past? And if so, when, and what sort of treatment did you receive (home care, hospital, sanitorium, and any medications)? How long were you sick?
- Do you have, or have you had any of the following? And if so, when did it begin? How often does it occur? How long does it last? What treatment have you used?

  Cough: When did it start? How long has it lasted? Do you cough up phlegm? Describe any phlegm (color, amount, consistency, and the amount coughed up in 24 hours—1 cupful, one teaspoonful, etc.).

  Chest pain: Where is it located? When did it begin? Is it sharp or dull? Is it worse with deep breathing?

  Shortness of breath: When did you first notice it? Was the onset sudden or gradual? Has it gotten any different with the passage of time? Does it occur with exertion? Do you get short of breath at night? Do you have to sit up to catch your breath? Do you sleep flat in bed at night, or do you have to prop yourself up in bed in order to breathe? How long have you done this? How often do you prop up, and how many pillows do you use? When did these symptoms begin, and how often do they occur?

- Have you had any of the following, and if so when and how often?
  Pneumonia?
  Tuberculosis?
  Bronchitis?
  Pleurisy?
  Lung cancer?
  Any surgery on your chest?
  Any other lung disease?

Cardiovascular

- Have you ever had hypertension (high blood pressure)? If so, when did you first know you had it, and what medications have you taken for it?
- Have you had any heart disease such as:
  Rheumatic fever?
  Hardening of the arteries (arteriosclerosis)?
  Heart attack?
  Surgery on your heart?
- Do you have any chest pain? Describe its onset, frequency, and character. Is it worse with exertion?

Gastrointestinal

- How is your appetite?
- Have you lost or gained any weight in the last year? If so, how many pounds?
- Are you troubled with any of the following, and if so, when, how often, and the duration?
  Nausea?
  Vomiting?
  Diarrhea?
  Constipation?
  Jaundice or hepatitis?
  Diabetes mellitus (sugar diabetes)?
  Blood in your stool?

Nausea, vomiting, and diarrhea are frequently signs of organophosphate pesticide poisoning, whereas chronic constipation in a painter may be a sign of lead poisoning. Jaundice may be the result of exposure to a solvent that is toxic to the liver.

Genital-Urinary

- Do you have any of the following, and if so, when did it begin, how often does it occur, and how long has it been present?
     Pain or blood with urination?
     Increased frequency or volume of urination?
     Difficulty starting your stream?
     Problems with erection or inability to conceive? (See also gynecological discussion below.)
- Have you had any medical care or surgery for disease of the:
     Kidneys?
     Bladder?
     Prostate?

Gynecological

- How old were you when your menses began?

The onset of menstruation in a 9-year-old girl may be the first clue of exposure to a chemical with estrogenic properties, such as used in animal feed.

- When was your last menses? How often do menses occur, how long do they last, and how do you characterize the flow: heavy, medium, or light?
- Have you had genital sores or rashes or symptoms of dysuria (painful urination)?

The use of "feminine hygiene" sprays, bath preparations, and the like can cause sensitivity reactions.

- If your menses have stopped, when was that, and was it spontaneous or after surgery? If surgery, what was the reason for the surgery?
- Have you had a discharge from your breasts, any masses, or a marked change in size or consistency?

The combination of cessation of menstruation and discharge from the breast in a nonpregnant woman points to a pituitary tumor. One cause of this syndrome is the use of birth control pills. A change in texture and size of the breasts, in either a female or a male, may be due to exposure to estrogen-containing products, such as animal feeds.

• Do you have any anatomical abnormalities of your genital tract?

Children born of mothers who were given DES (diethylstilbestrol) during pregnancy showed varying abnormalities of the uterus, cervix, and fallopian tubes in females, and of the penis, testicles, and sperm in males. When these abnormalities, which understandably may interfere with fertility, are encountered, a search for cause must be undertaken.

• Have you been pregnant? How many times, when, and what were the outcomes [live births, birth defects, stillbirths or miscarriages, and status of the infant(s)]?

Reproductive

If there has been a problem with reproductive events in either male or female subjects, more detailed information as to each pregnancy and/or birth is needed. This should include dates of all births (especially if there has been a change in fertility), any complications during the pregnancy, and outcome, including subsequent development. Also obtain information as to any defect at birth or in an aborted fetus.

Back

• Have you had any pains, stiffness, or injuries to your back?

Constant backache, coupled with urinary difficulties, may be the first sign of metastatic prostate cancer.

Extremities

• Do you have any problems or injuries of your hands, feet, arms, or legs?
• Do you have any of the following, and if so, when did it begin, and how long has it persisted? Has there been any change in any of the symptoms?
  Pain?
  Numbness or tingling?
  Swelling?
  Redness or lack of color?
  Coldness?
  Weakness?
  Inability to perform a function that you could do previously?

Skin

- Have you had any of the following, and, if so, when?
  Rashes?
  Redness?
  Blisters?
  Skin cancer?
  Changes in color or texture?
  Any other eruptions or changes in your skin?
- When was the onset?
- Does it come and go?
- Has anything made it worse or better?
- What do *you* suspect is the cause?

Skin abnormalities may herald internal diseases as well, particularly malignancies. It is not uncommon for a patient to have a skin malignancy diagnosed shortly before diagnosis of a visceral one.[1] Additionally, although sunlight is heavily touted as the main cause of skin cancer, other causes, such as exposure to industrial chemicals or pesticides, are rarely inquired after. Similar to the practice for cancer at any site, a thorough exposure history must be obtained.

Surgeries and Past Hospitalizations

In addition to recording symptoms, it is necessary to obtain information as to past medical care.

- List all hospitalizations and surgeries you have had, including dates, place, and reason (if you know).

Medications

- List all medications that you are currently taking, as well as how often, the duration, and the reason, if you know.
- List the same for medications you have taken within the last several years, if available.

If a bleeding disorder or blood dyscrasia is being investigated, a full medication history should be obtained, going back as far in time as the patient can remember or document. Drugs such as chloramphenicol and phenylbutazone, and industrial chemicals such as benzene, some pesticides, and other organic solvents, have been associated with bone marrow damage.

Table 3-1. Example of Medication History.

Medication name	Dosage	Frequency	From	Until	Reason
aspirin	2 tabs	4 ×/day	1985	current	back pain
Norinyl 1/80	1 tab	daily	1982	1984	birth control
"water pill"	1 tab	daily	1979	current	blood pressure
tetracycline	1 tab	4 × /day	4 days		sore throat
"muscle relaxant"	1 cap	occas.	2 years		sleep
INH	1 tab	daily	1969	1971	TB
Hydrodiuril	1 tab	daily	1985	current	diuretic
etc.					

A medication history may resemble that shown in Table 3-1.

The information gained on a medication history is invaluable in providing the following:

1. Data as to current treatment(s), or no treatment.
2. Data as to previous conditions and treatments.
3. Combinations of medications that may or may not be compatible with one other.
4. Reactions to drugs that may be causally related to the person's current or past illness.
5. Information as to whether the person is taking the drug correctly and for the indicated uses. (For example, the use of a muscle relaxant on a sporadic basis is unlikely to be effective for sleep induction and, in addition, exposes the person to a needless drug.)

The Social and Family History items give information about the place of individuals in their milieu and their family, and how they function therein.

Social History

- Where were you born?
- What was your total number of years of schooling, and where was it completed?
- Have you had any military service? If yes, when were you in the service, which branch, and where were you stationed?

The dates of military service provide a time reference indicating when the person was well enough to have passed a physical examination for duty.

- Give the marital status, ages, and number of persons in your household.

Table 3-2. Health Status of Family Members.

Person	Age if alive	Age at death	Cause of death/Major problems
Mother			
Father			
Brother(s)			
Sisters(s)			
Spouse			
Children			

Family History

- For each of your blood relations, give their current age and state of health. If any are deceased, give the age at death, and cause of death.

For reproductive and community toxicology problems, the health status of spouse and children must be included. (Table 3-2 shows how such information may be compiled.)

Habits History

- Do you smoke cigarettes? If so, when did you start (age or year)? How many do you smoke a day, and for how long?
- If you smoke cigars or a pipe, give similar information (ounces of tobacco/day or week for pipe smoking).
- Do you drink alcoholic beverages? Beer, wine, or liquor? If so, how much per day or week?
- Do you use any "street" drugs such as marijuana, "uppers," "downers," heroin, cocaine, or others? If so, how much, how often, and for what duration?

Work History

The Review of Systems approach, taught for years in medical schools and hospitals, is especially important in regard to what a person may have done for upwards of one-third of his or her lifetime. It is distressing to see the work history seldom obtained in actual practice, with the exception of teaching hospitals. Even there, it is unusual to find an occupational and/or environmental history, with the exception of a line or two, which may state only "retired." The importance of obtaining a work history cannot be overemphasized.[3]

In terms of exposure to toxicological agents, and especially carcinogens, it is important to ascertain what individuals have done for a living since they ended their formal education (and even prior to that for persons who worked "part-time" jobs while completing their education). For many people, the work history begins when they leave school; however, many rural people begin working on the family farm as children. To inquire only about the last job is not sufficient.

For each job you must ask several questions:

- What was your job classification? What did you actually do? Describe the tasks that you did and the materials that you used. How long did you do that (from 19__ until 19__)? (In reviewing the work history, record the information in chronological order. This method helps prevent the omission of any job periods, and it accounts for time away from work for layoff, injury, or illness.)
- What was it like where you worked? Clean or dirty? If dirty, describe the conditions (i.e., smoke, fumes, dust, dirt, damp, chemicals, hot, cold, etc.).
- Describe the ventilation in general, and did it work? Describe any industrial hygiene measures that were used, such as suction devices, fans, masks and/or respirators, protective clothing, gloves, and so on. (Be sure to differentiate between masks and respirators. Masks are usually made of paper or lightweight plastic and may filter varying sizes of particles. Respirators are usually made of rubber with varying kinds of filters. These devices require a tight fit on the face and may filter out gases and fumes, depending upon the characteristics of the filters. The contaminants on some jobs are so toxic and/or difficult to remove that an independent air supply may be required.) Ask whether the person could cough or spit up contaminant even after using such devices.
- Describe any chemicals or processes that used chemicals, such as milling with cutting fluids, painting, degreasing, foundry operations, textile dyes or paper finishing chemicals, welding rods, and so forth.
- If the person related that he or she was injured or became sick on the job, particular attention must be paid to this. Ask: "What were you doing at the time of the incident? Describe the actual events as they occurred (date, time, place, sequence of events)." Ask: "What happened next? How did you feel?"
- Ask whether other workers have been affected similarly.

Home Employment, Family Exposure, and Hobbies

Although chemical exposures in the home rarely approach that of the workplace, a history of any significant chemical use, in the course of either hobbies or home industry, must be obtained. With the out-sourcing of many industries, these exposures can be expected to increase. Home exposure to such products as paints, dyes, photographic chemicals, or ceramics on a regular basis and/or at significant levels should be discussed.

Occasionally, the patient may receive exposure secondarily to another's occupation, particularly if carried on in the home milieu. Asbestos carried into the home on work garments proved to be the source of exposure in a number of people who developed mesothelioma. Some hobbies employ toxic chemicals and may be a significant source of exposure, because of either their intensity or the duration of exposure. These hobbies may include painting, jewelry making, photographic developing, stained glass soldering, fabric dyeing, and so on.[2]

Summation of the Review of Systems

At the end of the history, it is a good idea to say to the person, "Do you want to tell me anything about your health/condition that I may have forgotten to ask you?" Frequently the patient will either add new information or emphasize a particular problem that is of prime concern. Always give the patient an opportunity to tell you his or her concerns. Finally, the review-of-systems information may be supplemented by findings on physical examination, such as scars and deformities, when additional questions may be asked.

UTILITY OF THE ROS APPROACH

A novice or a person not trained in medicine may question why information is needed about other organ systems than the one with the main complaint. In most cases, toxicological illnesses, like other medical conditions, have symptoms referable to more than one organ system, and it is these combinations of signs and symptoms that give clues as to etiology.

Consider a person with a chief complaint of vomiting and diarrhea with a possible toxicological exposure. Knowing only the gastrointestinal symptoms points to a locally acting and/or infectious problem. Accompanying symptoms of wheezing, watery eyes, and a runny nose point to a systemic condition such as organophosphate poisoning.

It is necessary to access the physical condition of the entire person. Frequently, persons with chemically caused diseases, and especially those exposed to neurotoxins, present with vague complaints of fatigue, sleep disturbances, loss of appetite, and weakness. It is all too easy to dismiss such complaints as neurotic and miss finding the cause. Exposure to neurotoxic chemicals should be investigated in these situations.

Much of the above is **subjective**—that is, known only by the patient—but much can be verified from medical and work records, and by observers and other affected individuals.

Frequently, a **pattern** emerges when one obtains histories from a number of individuals with common chemical exposures. It is such patterns, and combinations of problems, that lead an investigator to determine what chemical(s) may be responsible for illness in an isolated person, a work force, or a community.

These interview questions have been couched in the form and terminology taught to physicians and nurses. The same form can be used by a layperson, whether an attorney, a patient, a union representative, or an affected community member. Where a number of persons are affected, the data can be tabulated. Frequently, patterns of disease and incidence are thus revealed. In fact, this method was employed by the affected community members living in the Love Canal and Woburn areas, in New York and Massachusetts, when they became concerned about medical problems in their families and neighbors.

It often takes effort to get full information from people, for a number of reasons. Workers who have been performing their jobs for a long time may just assume that "everyone" knows about those jobs. Conversely, many workers do not have knowledge about what materials they have been handling, even when they have done the same job for years. Householders do not often think about or know the various chemicals used in and around their homes, and frequently assume that if a product has been offered for sale, it is "safe."

DUTY TO INQUIRE

This history may seem like an excessive amount of information to gather on individual patients, but with experience it can be obtained in about an hour, with very sick and/or complicated patients requiring more time and effort. Some information that the patient may not know at the time of the interview can be supplied later. Many physicians may balk at the idea of taking from 1 to $1\frac{1}{2}$ hours to obtain a history from a patient. The same physician, if a surgeon, would not question the need to follow orderly and sequential steps to perform a gall bladder operation, repair a fractured hip, or explore the damage caused by a gunshot wound to the abdomen. The same methodical approach, with attention to detail, assures the successful investigation of toxicological disease just as it does a successful operation or repair of an injury. The injuries caused by toxic agents are just as serious as the ones caused by crushing objects or pathogenic organisms, and the treatment should be no less meticulous and careful.

Until recently, many physicians have avoided investigating chemical causes of illness. With the information now available in both the scientific and the lay literature, and in the press, it can be argued that a physician who neglects to take a thorough environmental/occupational history from a patient with complaints compatible with chemical exposure will be criticized for failure to investigate the condition. If an individual's removal from chemical exposure is necessary to get him or her well, and such removal is not advised, the physician may also be critized for failure to adequately treat the patient. If a number of persons are exposed to a chemical hazard, it is necessary to make a correct diagnosis in the sentinel case or cases in order to prevent damage to others. Chemical agents as causes of disease simply cannot be ignored, and persons in authority can no longer fail to become educated as to their effects and ways to investigate them.

Chemical agents have joined parasites, bacteria, and viruses as documented and common causes of illness and death.

REFERENCES

1, Anon. Certain skin eruptions warrant internal cancer check in aged. *Skin and Allergy News* 18(1): 2, 24–25, 1987.

2. McCann, M., Barazani, G. *Health Hazards in the Arts and Crafts.* Society for Occupational and Environmental Health, Washington, D.C., 1980.

3. Guidotti, T. L., Chrmn. Taking the occupational history. *Ann. Int. Med.* 99: 641–651, 1983.

CHAPTER

4

BASIS OF TOXIC EFFECTS

The action, and thus the toxicity, of a chemical depends not only on its basic chemical structure but also on its physical and chemical characteristics. Whether a chemical belongs to the halide group, the rare earths, the polycyclics, or the heavy metals will determine its reactivity (or nonreactivity). The structure of organic chemicals is particularly important because structure determines a chemical's function, use in commerce, and resultant toxicology.

The following outline is a guide to assessing a chemical for possible toxic effects. This outline can serve as a guide for assessing effects in specific clinical cases as well as for assessing the prospective effects of a chemical. Prospective assessment applies not only to new chemicals, but to new uses, and new exposures of chemicals already in use, whether the exposures occur to humans or to the environment. Prospective assessment presents the opportunity to prevent harm.

Unfortunately, when one is dealing with after-the-fact exposures, it is rare to have complete data available, particularly, on dosage and duration of exposure. In fact, it is extraordinarily rare for an industry or a workplace to conduct either long-term or ongoing monitoring of chemical levels. Still, knowing as much as one can about a chemical, one can formulate some ideas about its effects on a target population.

FACTORS IN ASSESSING TOXICITY

- Physical properties
    Molecular weight
    Solid, liquid, or gas
    Volatility
    Solubility
    Reactivity
    Oil/water partitioning
    Heavier/lighter than air

    Size
    Shape
    Charge
    pH
- Mode of entry into the body
    Dermal (with effect in the skin)
    Transdermal (with effect in the skin and/or beyond)
    Ingestion
    Inhalation
    Injection
    One or more of the above
- Dosage (if determinable)
    Single, intermittent, or continuous
    Acute, subacute, or chronic
    Time of exposure(s)
    Frequency of dosage
    Amount
- Duration of exposure period(s)
    Extrinsic
      Continuous, single
      Discontinuous, multiple
    Intrinsic
- Deposition, storage, and partitioning in body
    Which tissue(s)
    Fat-soluble
    Water-soluble
    Protein binding
    Bone binding
- Transit or route through the body (from point of entry to excretion)
    Lungs—to bloodstream—to kidneys
    Skin—to blood—to liver—to gut
    Gut—to liver—to blood—to kidneys
    Gut—to fat stores—to milk
    Skin—to blood—to kidneys
- Excretion
    Dermal
    Pulmonary
    Liver
    Intestinal
    Renal
    Milk
    Saliva

- Metabolism and secondary products
    Enhanced toxicity
    Decreased toxicity
    Synergism(s)
    Additive effects
    Detoxification mechanisms
- Biologic activity
    Enzymatic
    Hormonal
    Cellular
    Immunological
    Blockage/occlusion
    Genetic
    Irritative
    Cell death
- Presence of other chemicals and/or contaminants
- Status of the host
    Age
    Sex
    Nutrition
    Immunological
    Medical conditions
    Circadian rhythms
    Ambient temperature surrounding host
    Physical activity
    Timing and sequence of exposure

The factors listed above are not intended to be all-inclusive, but are a guide used to assess possible damage from a chemical either retrospectively or prospectively.

DISCUSSION OF OUTLINE GUIDE

Physical Properties

A molecule with a small molecular weight will, in general, gain entry to the body more easily than one with a larger molecular weight.

In general, the physical state of a chemical will determine how it gains entry into the body. Particulates and gases are inhaled more readily than liquids, unless the liquid has been heated to the gaseous state, or the vapor pressure of the liquid allows for evaporation and/or volatilization at ambient temperature.

The use of solvents, whether in open containers or spread on surfaces (as in painting or degreasing), will allow increased exposure because of increased surface area available for volatilization.

Some chemicals, such as mercury, volatilize directly to the gaseous state from the solid state. Spread out in tiny droplets, as when spilled on a floor or workbench, mercury will release more vapor than will an equal quantity held in a container, because of the greater surface area available for evaporation. This problem becomes especially critical when a spill occurs on a rough surface, such as a concrete floor, because the droplets are difficult to collect. Thus, significant concentrations of neurotoxic mercury can volatilize into the air.

Persons working below areas where a chemical that is heavier than air is in use may be exposed to the chemical as it gradually settles. This exposure has occurred in chrome-plating plants where the tanks were elevated above the workers; in battery-recharging facilities where arsenicals are emitted and settle near the floor; and in factories where heavier toxic materials, particularly dusts, may settle through several levels of a building.

Conversely, lighter-than-air chemicals will rise and circulate in a room, and pass by the breathing zone of a worker as the chemical escapes. Lacking suitable ventilation, the concentrations of these chemicals increase near the ceiling.

Workers cleaning tanks and pits containing solvents such as trichloroethylene and perchloroethylene have been overcome and killed because these chemicals are heavier than air, and thus concentrate in the bottom of the tanks. This property allows toxic concentrations of the chemical to buildup, as well as permitting displacement of oxygen-containing air, which is needed to sustain life.

For a particle to be inhaled, it must be small enough to pass into the respiratory tract. Originally, it was thought that long, slender asbestos particles could not be inhaled; but, unfortunately, in studies in rats and rabbits it was found that particles as long as 15 μm not only were inhaled, but caused greater fibrosis than shorter fibers.[1,2]

Sharp and jagged particles adhere more readily and are more easily trapped than smooth ones, as has been shown with silica, fiberglass, and asbestos.

The presence of reactive groups or unstable intramolecular bonds in a chemical permits reaction with other chemicals and also with living matter.

The solubility characteristics of a chemical will determine whether it can be carried in a water-based or an oil-based solution. Products such as cutting fluids or pesticide formulations, to which emulsifying agents have been added, allow both water-soluble and oil-soluble components to be carried onto and into a person's body.

The electrical charge of a chemical will determine adherence and/or binding, as well as reactivity.

The pH of a chemical influences tissue damage and solubility. The caustic effects of lye are quite familiar, whereas the action of the lime in cement dust is less well-known but occurs in part by the same mechanism as that of lye, its alkalinity. Muriatic (hydrochloric) acid, used to clean concrete surfaces and, in the steel industry, to descale metal, owes its action to its low pH. Less obvious, but also due to its acidic nature, is the action of sulfur dioxide, which is given off in the burning of sulfur-containing fuels. Along with the various oxides of nitrogen, sulfur dioxide has been associated with the problem of acid rain.

Temperature increases the rate of evaporation of a volatile chemical, while increased ambient temperature is associated with both increased rate and increased depth of respiration, thus increasing the uptake of a chemical. Therefore, a worker handling solvents in a hot workplace will have a greater likelihood of absorbing a toxic dose than the same person working in a cooler work site.

Mode of Entry into the Body

It used to be thought that the skin was an impenetrable barrier to the entry of chemicals into the body. This is not so, however; both oil-soluble chemicals, such as steroidal hormones, and water-soluble ones, such as arsenic, can be taken into the body, producing effects both locally and at distant sites.

Industrial dermatoses have been recognized for centuries, and frequently herald or accompany internal illnesses. On the skin, these chemicals include both irritants and toxins, and have caused redness, itching, scaling, and sores to countless workers.

The degreasing action of solvents allows for secondary infection with fungi and bacteria due to the breakdown of naturally protective oil-containing barriers. The chemicals that break down the natural oiliness of the skin can also increase penetration, not only of the parent solvent but of chemicals dissolved in the solvent.

The use of emulsifiers or wetting agents, commonly found in such products as soluble oil cutting fluids or in pesticidal preparations, increases the uptake of the components of the parent compound through intact skin. Thus, it is important to consider not only the parent mixture, but also foreign contaminants that may have been added as the parent mixture was in use, plus the breakdown products of the original fluid. Not all "inert ingredients" are biologically inactive.

The transdermal (penetration through the intact skin) mode of entry of toxic chemicals is little recognized, although this method is utilized to intentionally deliver a variety of drugs to patients. Preparations such as nitroglycerine, the anti-motion-sickness scopolamine, hormones such as estrogens,[3] and nicotine-containing gum (Nicobid[R]) are intentionally delivered via the transdermal route (intact skin and/or mucous membrane) to provide an intended result. Of concern, however, are those and similar chemicals that are absorbed with an unintended result. The headaches suffered by workers who were dermally exposed to dynamite are well-known.[4,5] The skin remains a significant route of exposure for many industrial chemicals, such as arsenic, chromium, vinyl chloride, and acrylonitrile.[6]

Solvents such as 1,1,1-trichloroethane, after being applied to the skin, can be recovered in the blood, exhaled air, and the urine.[7] Classical hyperadrenalism (Cushing's syndrome) can be produced by cortisone-containing creams rubbed onto the body. In fact, all steroids and most fat-soluble chemicals can be absorbed through the intact skin.

When a fat-soluble chemical is dissolved in a solvent, its penetration is increased. Because of the large surface area of the skin and the even larger surface area of the lungs, chemicals that gain entry to them are rapidly and easily absorbed and distributed to other parts of the body via the bloodstream and lymph drainage.

The oral route of entry of a chemical is obvious, as when a drug is purposely taken. The entry of chemical contaminants, as in food and water, is also straightforward. However, a route often ignored is the oral route of entry of chemicals that were originally inhaled; for example, asbestos, coke dust, and plastic particles may be inhaled, then coughed up, and swallowed with saliva. This is a significant route of exposure for the gut.

Gases, solids in the form of dusts, and liquid vapors can all be inhaled, and do damage in any part of the respiratory tract from the nose to the deepest part of the lungs, including the alveoli and the pleura. From the lungs, chemicals can be transported to various parts of the body (see below), either by direct penetration out of the lungs or via the blood and lymphatic systems.

Most injections of chemicals are obvious, as when a person in given a medication by hypodermic. When animals are tested to obtain toxicological data, this route is frequently used because it is a relatively certain way to determine the dose of a substance.

Although the injection method of exposure is usually reserved for intentional testing in animals, this mode of exposure is critical for the sick. Within the last two decades, parenteral solutions and blood have ceased to be packaged in glass, but are available in plastic containers. These containers have been shown to leach plastic components into the blood and/or intravenous fluids stored therein.[8] Moreover, many of the plastics-forming chemicals are biologically active.[9] Conceivably, biological reactions could occur between xenobiotic chemicals such as the components of plastic containers and the blood cells or viruses contained therein, resulting in alteration of these live elements.[10] When plastic components are injected into the body of a sick patient, along with the intended blood or intravenous fluids, reactions from the plastics can be expected to occur within the body of the patient as well.[11] If the plastic component has the ability to effect a change in the DNA of a virus or a bacterium, or to interact with the cells of the host, it is not inconceivable that this reaction could weaken or interfere with the recovery of an already sick person.

Considering the varying immunological patterns of the virus associated with the development of acquired immunodeficiency syndrome (AIDS), and its prevalence in persons receiving blood products, it is theoretically possible that not only viruses but xenobiotic chemicals play a part in the development of this condition, the latter by either direct effect upon the genetics of the virus or by alteration of the host's immunological system.

At the very least, it appears unwise to administer intravenous products in containers that can leach biologically active products to immunologically compromised patients. Furthermore, knowing the biological activity of these products, it would be wise for researchers and physicians to monitor environmental contaminants in persons with various immunological deficiencies to determine whether xenobiotic chemicals may be initiating and/or promoting these conditions. It may prove wise to re-introduce glass containers for intravenous fluid use.

In the workplace, exposure frequently occurs by multiple routes. The solvent handler is exposed by inhalation and through the skin. The gas station mechanic receives exposures by all routes: skin, inhalation, and ingestion of contaminated food that he or she may eat on the job.

Dosage

Dosage, in the human milieu, is the most difficult of determinations to make because it is usually not until damage has occurred and been recognized that an interest is taken in it. Monitoring is rarely done on a continuous basis; so most measurements are representative only of that point in time when the sample was collected, and may not represent past conditions.

Most work and environmental exposures to chemicals fluctuate because of varying rates of emission and changes in ambient temperature, wind velocity, work activity, and the like. Unfortunately, advance notice may be given when measurements are to be taken in a plant or at a dump site, and either the work activity is cut back, or cleanup efforts are made prior to testing, so that measurements are not representative of previous actual or usual levels.

Reactions to chemicals frequently follow a dose–response curve; that is, the reaction is proportional to the dose. This is generally true for acute reactions to solvents, organophosphate pesticides, cyanide, asphyxiant gases, and most pharmaceuticals. Notable exceptions to the dose–response concept are chemicals such as PCBs, PBBs, and dioxins that result in greater toxicity from repeated small doses than from a single dose, because of their induction of enzyme systems, changes in immunological function, and long retention times within the body.[12]

Single doses of toxicological significance are usually associated with profound illness or death, as illustrated by a number of case reports in chapter 6 (see CNS case #9 and cardiac cases #1 and #2) as well as victims discussed in chapter 5 (cases #1 and #2 under "Waste Disposal Industry").

Even in these cases, the actual amount of the toxic chemical is unknown. One can only say that the dosage was "enough" to have either killed the person or made him or her profoundly sick.

In general industry, most doses of toxic chemicals are encountered on a intermittent basis, extending from days to decades. However, for a person living next to a toxic landfill, or in a home contaminated with toxic chemicals, the dosage can be essentially continuous. This factor is especially important in the young, the elderly, and persons who remain at home for whatever reason. A catch-22 situation has occurred with persons living in homes insulated with formaldehyde-releasing UFFI foam and persons living in mobile homes constructed of particle board and plywood manufactured with formaldehyde-emitting glues. Because of illnesses due to the formaldehyde emissions, and not realizing the cause, such persons have *stayed at home more*, thus increasing their exposures!

Less well recognized is the continuous exposure that results from chemicals that are retained within the body although ongoing external exposure has ceased. PBBs, PCBs, dioxins, asbestos, and many fat-soluble chlorinated pesticides are examples; when taken into the body, they are slowly, if ever, eliminated. Thus although the overt exposure may cease, the person remains exposed throughout a significant portion of his or her lifetime, and the adverse effects become cumulative.

By corollary, illness resulting from an acute exposure generally will be recognized, whereas the connection between illness resulting from a chronic, or even a subacute, exposure may go undetected. The death of screen actor Steve McQueen from mesothelioma at age 50 was traced back to his brief exposure to asbestos when he worked as a merchant seaman in his late teens.

It is generally considered especially critical when a toxic exposure occurs in the young. By definition, cells from multiple body systems are in a state of growth during childhood, and there is increased opportunity for chemicals to damage or intercalate into the dividing RNA and DNA of the cells. Additionally, many microsomal enzyme and excretory systems are poorly developed.[13]

An associated aspect of duration of exposure is the timing or sequence of exposure, such as that experienced by silicotics who developed an increased incidence of tuberculosis. A person with asbestos retained in the lungs who then becomes a welder, and thus is exposed to a number of pyrolysis products, will be at increased risk to develop lung disease. A person with a carbon tetrachloride–damaged liver is less likely to be able to metabolize the alcohol contained in a drink than one without the damage. Indeed, a clue to chemically induced

liver disease is the development of nausea, headache, or flushing
ously tolerated amount of alcoholic beverage, as, say, in a per
the dry cleaning industry and exposed to perchloroethylene.

The reparative processes of the body are remarkably eff
people tolerate repeated, small chemical insults without app
the frequency of an exposure to a chemical such as carbon
the body's ability to excrete it, then multiple doses can result in si
even death. Frequent, small exposures to a neurotoxic chemical may damage a
few cells with each exposure, but the cumulative effect may not be discerned
until a critical point is reached when there is notable interference with sensation
and/or function.

An analogy, easily understood, concerns what would happen if the mortar
holding a brick structure together were gradually chipped away. Admittedly,
the mortar could be removed in a number of areas, particularly if scattered,
with little noticeable effect. But a point would be reached when either so many
different areas would be devoid of mortar, or a single area would be so weakened
by the loss of the mortar, that the building would weaken and, with very little
force, collapse. This concept is particularly applicable when we consider damage
to the kidneys, liver, and central nervous system.

The amount of a chemical associated with a toxic effect in a human is rarely,
if ever, known. Only with controlled experimental conditions in animal and
human testing is it possible to determine the amount of a toxic agent. Even
then, the experimental route of dosing may not represent the conditions under
which humans are actually exposed.

Fortunately, most single exposures to most substances do not result in permanent damage unless the dosage is sufficient to alter or to paralyze a critical body
system. We cannot say, however, that repeated subclinical exposures will not
result in delayed, long-term effects. This problem is especially important where
there is exposure to chemicals with neurotoxic and immunotoxic effects.

An intermittent exposure may or may not cause permanent damage, depending
upon retention in the body and the body's ability to repair itself. Carbon tetrachloride and perchloroethylene may damage a "part" of the liver. Obviously
when enough "parts" have been damaged, the liver will fail. Failure of a part
of the body occurs secondarily to either repeated damage or unrepaired damage
that persists even after the exposure has ceased. The bronchospasm that occurs
from exposure to toluene diisocyanate (TDI) exposure can become permanent,
resulting in heart failure secondary to overload of the right side of the heart.
The worker who becomes sensitized to TDI while on the job handling plastics
may become so sensitized that otherwise unnoticeable levels of TDI emitted
from urethane-containing articles in the home become intolereable. These
unrecognized sources of TDI emissions may include foam rubber carpet
underlay, foam rubber cushions, pillows, and upholstery materials.

Duration of Exposure

Continuous exposure may involve single or multiple sources of contamination, such as one experiences by living adjacent to a chemical dump, drinking water contaminated by chemicals leached from underground tanks, or living downwind from such sources of pollution as a smelter, a municipal incinerator, a power-generating facility, a cement plant, a chemical plant, and/or other similar sources. Continuous exposure can also come from these same industrial chemicals when they are employed in home construction and home decorating or furnishing materials, with resultant offgassing as the materials cure and/or deteriorate. Products such as formaldehyde, urethane, vinyl chloride, acrylics, solvents, and so forth, have been associated with the "sick building syndrome," especially in poorly ventilated buildings where chemicals accumulate because of volatilization.[14, 15]

Exposure to a toxic chemical is by nature extrinsic. When the toxin is retained by the body, as are asbestos, silica, coal dust, halogenated biphenols, and dioxins, the exposure becomes intrinsic. This is not to be confused with an intrinsic source of chemical exposure. The source remains extrinsic, but, by retention, the exposure becomes intrinsic. The duration of exposure covers not only the time period when the person experienced exposure from an external source, but also the duration of time when the chemical is resident within the body. When intake exceeds output, then the concentration of a chemical can accumulate. A brief exposure to a chemical may cause illness and even death by virtue of its retention within the body. However, most exposures to toxic chemicals that result in illness are discontinuous and multiple.

Deposition, Storage, and Partitioning in the Body

Continuing the above discussion, the damage that a chemical does is related to its site of action within the body.

An example of an immediate and localized reaction is the migration of cells into and the production of fluid in the lungs, as a reaction to inhaling chlorine gas or acid fumes. A similar reaction is more readily observable on the skin when it is exposed to the same substances.

Solvents such as carbon tetrachloride, perchloroethylene, and trichloroethylene, although inhaled, become concentrated in the liver and other fat-containing tissues, including the brain, nerves, and bone marrow, and do their damage there.

When chemicals reach such organs as the liver, lungs, and/or kidneys, there can be interference with any of the normal functions. This includes interference with or induction of enzyme systems, disruption of cellular integrity, and cell damage with secondary scarring. When excretion is dependent upon hepatic conjugation, the foreign chemical can overwhelm the detoxification mechanism, resulting in increased concentration of the chemical with cellular damage and/or death.

Water-soluble chemicals gain entry into cells in a manner similar to that of the essential and ubiquitous elements, sodium, potassium, and chlorides. Most water-soluble xenobiotic chemicals are excreted in the urine, feces, or sweat, and exert their action by way of interference with cellular integrity and energy systems.

Protein-binding chemicals can change the surface of a cell, resulting in altered immunological function, and can alter the permeability of a cell or its integrity. Many chemicals that bind to proteins are bipolar, that is, have a water-soluble and a fat-soluble moiety, and thus can attach to various components, resulting in altered surface integrity.

Bone deposition was amply demonstrated in the leukemias that developed in radium-dial painters, and in residents living downwind from nuclear test sites who inhaled bone-seeking radioactive substances. Lead can be found in the teeth and bones of exposed persons, resulting not only in staining of these structures, but deranged function as well, reflected in anemia and altered red-cell structure. Tetracyclines administered to children during the tooth-formation period resulted in discolored permanent teeth. Bone is not only a depository for metals and inorganic elements, but for organic chemicals that are soluble in the fatty marrow. Thus, inhaled benzene is carried by the blood to the bone marrow where it interacts with the cells therein. If the resultant change is genetic and self-sustaining, causing a malignancy, it may be expressed as one of the blood dyscrasias, including leukemia. Other chemicals that are capable of similar deposition and are linked with blood dyscrasias include lindane and cyclopentadiene pesticides such as chlordane and heptachlor, as well as a number of solvents.

Fat solubility within a body not only leads to concentration of that chemical in a given volume of fat, but may result in storage and slow (or little) release with the passage of time. Anesthetic gases, solvents, and plastics can be recovered in the exhaled air hours to months following an exposure. Prolonged residence time within a body permits the chemical to exert its action for a period of time longer than the original extrinsic exposure time. When residence time is prolonged, and exposures are repeated, increasing concentrations of a chemical can accrue. Under these conditions, a reaction to a toxic chemical may not follow immediately, but may occur when the concentration of the chemical has reached a critical level.

A chemical can accumulate under the following conditions:

- Its intake exceeds its output.
- It is stored within the body with slow release.

- Its dosage exceeds the body's ability to metabolize or excrete it.
- Poisoning of the metabolic or excretory mechanism has occurred.

Transit or Route through the Body and Excretion

The route of a chemical through a body influences not only its site(s) of action, but excretion as well. Thinking as a toxicologist, ask, where does this chemical go, where is it concentrated, and how is it excreted?

Volatile chemicals, including solvents, are inhaled into the lungs, dissolved in the blood, and carried to various parts of the body, including the liver and the kidneys. There can be adverse effects all along the route; for example, a defatting action on the surface of the lungs, disruption of the integrity of the blood cells, and damage to the cells of the liver, kidney, and/or bladder.

Human, animal, and plant tissues are all rich in lipids. These lipid-containing cells and membranes are particularly important in the transport of xenobiotic chemicals.[16] This transit pattern is purposely utilized to induce anesthesia with such well-recognized solvents as ether, cyclopentane, and nitrous oxide. These chemicals can be detected in the breath of recovering patients. In the past, industrial chemicals including urethane and trichloroethylene were also tried as anesthetic agents until their toxicities were found to outweigh their usefulness. The glue-sniffing deviant is merely utilizing the lung–blood–brain route, well-documented in toxically caused illnesses, to achieve an altered sense of consciousness. Unfortunately, in many industrial "accidents" the altered state of consciousness may become permanent, and occasionally terminal.

Xenobiotics tend to do their greatest damage in places where the transit is either slowed or concentrated. Thus the urinary bladder, gallbladder, colon, and rectum are ready sites for the concentration and interaction of absorbed chemicals with the cells of the walls lining these structures.[17–22]

The current advertising rage, expressed in the media that urges more roughage in the diet, may be useful not for any intrinsic cancer-preventing properties of bran, but in speeding the transit of contaminated food along the alimentary canal. Obviously, as more and more pesticides and chemical contaminants find their way into our food supply, the likelihood will increase that one or more of them will interact with cells of the gut, damaging the cells and/or promoting a malignant change. Not only food contaminants but other chemicals such as asbestos, arsenic, and polycyclics that gain entry to the intestinal tract can cause damage there. Pharmaceuticals such as lincomycin and clindamycin, excreted in the bile and via the gut, alter the flora and the integrity of the gut and cause serious disease of that system.[23–25]

Many chemicals can be recovered from the urine. As an example of the speed of the transit of chemicals from intake to excretion, one can observe the odor of ketones in the urine shortly after eating asparagus.

Milk is a route of excretion for the lactating animal or human. Therefore, a significant source of exposure to xenobiotic chemicals in the newborn is its mother's milk. Fat-soluble contaminants such as PCBs, PBBs, hormones, and a number of pesticides such as chlordane and heptachlor can concentrate in the milk and deliver a significant dose to the infant. This dose is critical because the infant weighs very little (with the dose per body weight proportionally increased), nursing may be its only source of nutrition, the infant's cells are in a rapidly growing and dividing phase, and many of its enzyme and excretory systems are not fully functional.

Metabolism and Secondary Products and Biologic Activity

A foreign chemical can interact with the cells of a body in a limited number of ways. Following contact with a cell, there can be:

- No reaction.
- Interference with cellular structure and/or interference with cellular function: decreased, enhanced, additive, or synergistic effect(s).
- Cell death.

These mechanisms depend upon the toxicology of the chemical as well as the integrity and the reparative ability of the cell itself. Obviously, when there is no reaction, there is no cell damage; and if immediate cell death occurs, neither a malignant nor a mutagenic event can take place. However, if sufficient numbers of cells die, the structural integrity of an organ is affected. It may take the passage of time with accumulating cellular damage for the functional impairment to become apparent.

The kidneys have a reserve of approximately 75%, and so it is not until approximately 75% of the kidney tissue has been damaged that kidney failure becomes apparent.

Neurological damage from exposure to solvents that disrupt the cellular integrity may be manifest as tingling of the toes and feet, but may not be recognized by the patient until he or she develops weakness and the inability to walk.

Functional changes occur when a cell is damaged, and the reparative process is either incomplete or faulty. Damage to the DNA of the egg or sperm with incomplete repair can result in a child born with a genetic abnormality such as Down's syndrome or one of the trisomy defects. Incomplete or faulty repair of a damaged somatic cell can result in reduced or improper function. If the altered cell becomes self-replicating, a tumor, with benign or malignant potential, may result.

Interference with function may not always be reflected in measurable structural damage. This is especially true of behavioral and learning disabilities in children exposed to toxic chemicals during prenatal development.

The slowly developing dementia and personality changes found in persons exposed to such chemicals as solvents and carbon disulfide point up the importance of recognizing functional impairment. The structural changes frequently are not sought until after death.

Presence of Other Chemicals and/or Contaminants

The synergistic effect of carbon tetrachloride and ethanol on the liver is well known. Other chemicals stored or metabolized by the liver may influence the action, metabolism, and excretion of other chemicals introduced into the body by whatever route.

The reaction between disulfiram and ethanol was first described in rubber workers, and was then developed as a "treatment" for alcoholism. When we understand the neurological and liver effects of this reaction, it comes as no surprise that both patients treated with disulfiram (Antabuse) and rubber workers who are exposed to the chemical in the course of their work can develop serious effects from exposure to it, whether it is presented in the occupational or the "therapeutic" mode.

Status of the Host

In general, the young, the elderly, and the infirm are less able than other people to handle toxic chemicals. In the young there is the dual problem of immature body systems and rapidly growing cells; in the elderly, degenerative processes may have taken place that make the handling of xenobiotic chemicals less efficient.

The problem of chemical damage in young, developing individuals may in part explain the incidence of leukemia and brain tumors in children exposed to fat-soluble chemicals, such as solvents, chlorinated pesticides, and vinyl chloride, which concentrate in and damage cells of the bone marrow and brain.

At the turn of the century, prior to the passage of laws forbidding the employment of children, sickness and death in young people employed in the asbestos, textile, and mining industries was recognized. While industrial exposure of children is no longer allowed in the United States, they may receive exposures comparable to that of the workplace, due to the chronicity of exposure, as well as the relatively larger doses they receive, based on their smaller weight and proportionally greater surface area. A child living in a home contaminated with formaldehyde-containing construction materials, or wood preservatives such as pentachlorophenol, arsenicals, chlordane, or tributyl tin may receive essentially a continual exposure throughout his or her lifetime. Thus malignancies, chronic respiratory problems, kidney damage, skin disorders, and other medical problems that develop in children are a signal to investigate sources of exposure. The younger the age when an illness develops, the shorter the time span that needs investigation, and the more likely it is that cause and effect can be determined.

Categorically, it should be emphasized that every malignancy in a child younger than 18 years of age should be investigated thoroughly to rule out a chemically induced cause. Considering the potential that exists for improving public health and decreasing the costs and tragedy of these illnesses, it is a small and affordable effort.

The anatomical and hormonal differences between males and females influence the site and action of a number of chemicals, particularly those that involve hormonal and reproductive functions. Only females take birth control preparations with their attendant problems of blood clot formation, pituitary tumor formation, and birth defects. The DES once given to pregnant women influenced both male and female offspring, producing abnormalities in the genital tracts of each. Chemicals such as asbestos, benzene, various pesticides, lead, solvents, and so on, show no predilection toward either sex and can adversely affect whichever is exposed.

In general, poorly nourished individuals withstand insults, including chemical insults, less well than those who are well nourished;[26] and thermal and traumatic injuries interfere with the reparative processes.[27] An anachronistic situation exists when an obese person, whose tissues contain various fat-soluble chemicals, loses weight. As the chemicals are liberated from the fat depots, the person may experience symptoms of exposure. The classic situation of persons with lead stored in their bones is similar; they develop symptoms when they are in sunlight, the sunlight stimulating vitamin D formation and the mobilization of the lead.

The presence of a contaminant in the body may also render a person more susceptible to other disease, as has been recognized with the increased incidence of tuberculosis in silica-exposed persons.

Circadian rhythms are influenced by light–dark cycles with various effects upon the cytochrome oxidase systems, the metabolism of melanin, and hormonal release. The rhythms characteristic of a species (nocturnal in many rodents; diurnal in dogs, cattle, most birds, and humans) are dependent upon an integrated central nervous system with both sensory and motor systems working properly. A loss of characteristic circadian rhythm may reflect subtle damage to the central nervous system. Conversely, persons working the swing shift and those crossing multiple time zones experience fatigue and difficulty with sleep–wake cycles as a result of disruption of the diurnal rhythm. These stressful disruptions may interfere with the handling of xenobiotic chemicals.

As in most chemical reactions, the speed of a reaction is increased with increasing temperature. This is true also of the effects of chemicals upon living systems, where there is an increase in respiratory rate and blood flow in an effort to maintain temperature homeostasis. Workers in a hot environment will inhale more of a chemical because of their increased rate and depth of respiration.

Physical activity places demands for increased delivery of oxygen to the tissues. If a worker is performing physical activity in a hot environment, as, for instance, an agricultural field worker or a steel or foundry worker, the demands of the effort will require a further increase in the respiratory rate and blood flow, and thus lead to a further uptake of chemicals from the environment.

Where repeated exposures occur, the time interval of exposure is important if the person is to recover from the previous exposure. Thus, a person who works an 8-hour shift will have 16 hours to "recover" from an exposure, whereas a worker on a 12-hour shift will have not only a longer duration of exposure, but, perhaps more important, will have 50% less time to recover. This is especially important with exposures to an agent such as carbon monoxide, which combines with hemoglobin more rapidly than it dissociates; or for chemicals that require metabolism prior to excretion where rate-controlling enzyme systems are overwhelmed by either high concentrations or continued exposures.

This combination of increased exposure and decreased recovery time becomes especially apparent in overtime work situations. Repair and maintenance workers frequently experience greater exposures than production workers in the same department because of demands for speed in the corrective measures, as well as the prolonged periods required to complete the tasks. Those workers who clean and repair air filtering or hydraulic systems, clean solvent collection systems, and work in pits containing spent cutting oils and other waste materials generally experience greater exposures than workers on the jobs that create the wastes.

The timing of an exposure is especially important in the case of teratogens and mutagens. It is necessary for a chemical to reach a developing sperm, egg, or fetus at the critical time in development for interference in the normal sequential processes to occur.

This timing is also critical where sequential exposures occur. It is recognized that the solvents carbon tetrachloride and perchloroethylene are liver toxins. A person handling those chemicals either commercially or in a home situation would have less ability to metabolize alcohol than an unexposed person; a single bottle of beer could make that person sick, or even precipitate liver failure.

As must be appreciated from the foregoing discussion, many of these factors overlap and are interdependent. However, considering the factors individually makes the task of determining the interrelationships of chemicals and disease clearer.

REFERENCES

1. Chief Inspector of Factories. *Annual Report.* His Majesty's Stationery Office, London, 1945, p. 18.
2. Vorwald, A. J., Durkan, T. W., Pratt, P. C. *Experimental Studies of Asbestosis.* Edward T. Trudeau Foundation, Saranac, N.Y., 1950, pp. 1-47.
3. Estraderm$_R$. *Clinical Symposia Therapeutic Guide* 39(3): 1-7. Ciba Pharmaceutical Co., 1987.
4. Rabinowitch, I. M. *Canad. Med. Assoc. J.* 50: 199, 1944.
5. Hunter, D. *The Diseases of Occupation.* Little, Brown and Co., Boston, 1969, pp. 584-90.
6. Suskind, R. R. Percutaneous absorption and chemical carcinogenesis. *J. Dermatol.* 10: 97-107, 1983.
7. Stewart, R. D., Gay, H. H., Erley, D. S., Hake, C. L., Schaffer, A. W. Human exposure to 1,1,1-trichloroethane vapor: relationship of expired air and blood concentrations to exposure and toxicity. *Amer. Indust. Hyg. Assoc. J.* 22(4): 253-63, 1961.
8. Jaeger, R. J., Rubin, R. J. Plasticizers from plastic devices: extraction, metabolism, and accumulation by biological systems. *Science* 170: 460-62, 1970.
9. Autian, J. In *Casarett and Doull's Toxicology.* Macmillan, New York, 1980, pp. 534-50.
10. *Potency Ranking of Chemicals Based on Enhancement of Viral Transformation.* U.S. Environmental Protection Agency, Washington, D.C., 1981.
11. Thurman, G. B., Simms, B. G., Goldstein, A. L., Killian, D. J. The effects of organic compounds used in the manufacture of plastics on the responsivity of murine and human lymphocytes. *Toxicol. Appl. Pharmacol.* 44: 617-41, 1978.
12. Safe, S., Robertson, L. W., Safe, L., Parkinson, A, Bandiera, S., Sawyer, T., Campbell, M. A. Halogenated biphenyls: molecular toxicology. *Canad. J. Physiol. Pharmacol.* 60(7): 1057-64, 1982.
13. Fouts, J. R. Some studies and comments on hepatic and extrahepatic microsomal toxication-detoxication systems. *Environ. Health Perspect.* 55-66, 1972.
14. Anon. My office makes me sick. *Medical World News,* pp. 42-43, Sept. 9, 1985.
15. Pelletier, K. R. The hidden hazards of the modern office. *New York Times,* Sept. 8, 1986.
16. Bell, F. P. Lipid exchange and transfer between biological lipid-protein structures. *Progr. Lipid Res.* 17: 207-43, 1978.
17. Berg, J. W., Howell, M. A. Occupation and bowel cancer. *J. Toxicol. Environ. Health* 1: 75-89, 1975.
18. Tokudome, S., Kuratsune, M. A cohort study on mortality from cancer and other causes among workers at a metal refinery. *Int. J. Cancer* 17: 310-17, 1976.
19. Vobecky, J., Devroede, G., Lacaille, J., Watier, A. An occupational group with a high risk of large bowel cancer. *Gastroenterology* 75: 221-23, 1978.
20. Claude, J., Kunze, E., Frentzel-Beyme, R., Paczkowski, K., Schneider, J., Schubert, H. Lifestyle and occupational risk factors in cancer of the lower urinary tract. *Amer. J. Epidemiol.* 124(4): 578-89, 1986.
21. Crabbe, J. G. S., Cresdee, W. C., Scott, T. S., Williams, M. H. C. The cytological diagnosis of bladder tumors amongst dyestuff workers. *Brit. J. Indust. Med.* 13: 270-76, 1956.
22. Mancuso, T. F., Coulter, E. J. Methods of studying the relation of employment and long-term illness—cohort analysis. *Amer. J. Pub. Health* 49(11): 1525-36, 1959.
23. Nelson, Senator G., Chrmn., Hearings before the Subcommittee on Monopoly, of the Select Committee on Small Business, United States Senate. *Safety, Efficacy, and Use of Antibiotics—*

Clindamycin and Lincomycin. U.S. Government Printing Office, Jan. 28, 29 and July 8, 1975. Part 27, pp. 12397–642.

24. *Physician's Desk Reference*, Medical Economics Co., Oradell, N.J., 1986, pp. 1564–66.
25. Ibid., 1981, pp. 1805–11.
26. Wade, A. E., Bellows, J. Role of dietary corn oil in the function of hepatic drug and carcinogen metabolizing enzymes of starved–refed rats: response to the mixed function oxidase inducer, 3-methylcholanthrene. *J. Environ. Pathol. Toxicol. Oncol.* 7(3): 11–19, 1987.
27. Blazar, B. A., Rodrick, M. L., O'Mahony, J. B., Wood, J. J., Bessey, P. Q., Wilmore, D. W., Mannick, J. A. Suppression of natural killer-cell function in humans following thermal and traumatic injury. *J. Clin. Immunol.* 6(1): 26–36, 1986.

CHAPTER

5

INHERENTLY DANGEROUS INDUSTRY APPROACH

"Everyone looks forward to retirement, but there's a lot of 'em not makin' it. . . .
A lot of 'em, they're countin' the months instead of the years—and pass away. . . .
maybe I won't be able to take it any more. It's gettin' tougher. I'm not like a machine.
Well, a machine wears out too sometimes."

—S. Turkel, *Working*[1]

Throughout history, some industries have imposed greater burdens of sickness and death upon workers than others. The occupational hazards of mining continue from the days of Agricola, who lived from 1494 until 1555 and described the effects of cobalt–arsenic ore, which ate away the hands of the men handling it. Paracelsus, who traveled throughout his lifetime in middle Europe, described the phthisis characteristic of work in this dusty field. During his life, from 1493 to 1541, Paracelsus, active in philosophy, metallurgy, and surgery, is probably best remembered for having introduced mercury as a treatment for syphilis.

As noted in Chapter 1, probably the first to undertake the careful study of the diseases of workers was the Italian Bernardino Ramazzini (1633–1714), whose treatise *De Morbisi Artificium Diatriba* was based upon his direct observations of workers and their workplaces. He, like Paracelsus, described the phthisis in stone masons and miners, as well as vertigo and neuropathies in potters, which today we understand as secondary to lead poisoning. Ramazzini extended his studies to encompass the diseases found in such disparate occupational groups as blacksmiths, tanners, laundresses, privy cleaners, printers, farmers, welldiggers, weavers, and coppersmiths, among many others. He implored physicians to inquire into a person's occupation. Two hundred fifty years later, I urge the same thing.

Ramazzini's influence extended beyond his lifetime, when his work was translated by Robert James in 1746, and published in England as *Treatise on the Diseases of Tradesmen*. In discussing this book, the historian, Warbasse wrote in 1935:[2]

> The wise Ramazzini visited the workshops and studied the effects of occupation upon the workers. He saw diseases specifically related to certain industries. He saw deaths resulting from certain trades. He discovered the toll exacted by the quest for profits. It was opportune that this book came into England just before its industrial revolution began. But it made little impression on the conduct of industry; for more than a century and a half later, when England's competitive commerce took her into the Great War, she found that every other man she needed for military service had been seriously damaged by her industries, and one out of every four was physically unfit. Medicine had collected the information and showed the commercial destruction of men; it could do no more. There remains no sadder chapter in the story of human greed than that of the stunted lives and hastened deaths for which society has paid by its misuse of human beings in industry.

More than a half-century later, the worldwide spread of chemicals continues to threaten us all.

Since the 1970s in the United States, we have become more aware of the benefits of laws passed in recognition of the disease and misery wrought by the workplace and its products. An alphabet soup of agencies and acts, such as the EPA (Environmental Protection Agency), OSHA (Occupational Safety and Health Administration), NIOSH (National Institutes of Occupational Safety and Health), CPSC (Consumer Product Safety Commission), FDA (Food and Drug Administration), TSCA (Toxic Substances Control Act), and FIFRA (Federal Insecticide Fungicide and Rodenticide Act), all bear witness to this. Lest we think that these regulatory agencies and laws arrived de novo, we must take note that they were worked out and legislated by Congress in response to documented problems and concerns. It may be assumed that they all have as their intent the prevention of sickness and disease, however they may be administered. Unfortunately, many problems require expensive remedial efforts, and others are not to be remedied at any cost. The Super Fund of EPA, addressing the problems of toxic wastes and their cleanup, provides just one example of such remediation.

In the ''cost–benefit'' world, it is an interesting exercise to compare the cost of prevention of toxic chemical contamination with the cost of remedial cleanup and repair, if the latter is at all possible. Because this exercise has rarely been undertaken, except in a few instances to bolster an argument against prevention or cleanup, we have few hard data to depend upon. The parallel between control of infectious disease and occupational/environmental disease needs emphasizing. Within six years of the introduction of polio vaccine, 154,000 cases of the disease were prevented and 12,500 lives saved. A $41 million research investment resulted in savings of $2 billion in hospital costs and averted income losses of $6.3 billion.[3] If we could be half that successful in our efforts and expenditures to prevent illness and death from toxic chemicals, we would have reason to celebrate.

Compared to much of the world outside of North America, Scandinavia, Japan, and such countries as Britian, Australia, and New Zealand, our citizens (workers and nonworkers alike) have considerable protection against chemically caused disease in terms of legislation and enforcement. Most Third World citicizens have few of these protections. With the increasing exportation of manufacturing processes and products, it can be predicted that the environment and the health status of people living in these countries will be in jeopardy.[4,5] Many pesticides that have been banned from use in the United States are freely available to people who cannot read any language, and, in turn, contaminate foodstuffs imported back into the United States.[6] Many fabricating jobs, employing toxic chemicals such as asbestos, solvents, and plastics, are being carried out in less developed countries without any protection from EPA, OSHA, FDA, or a comparable regulatory agency. A plethora of drugs, without labels providing warnings and side effects, are sold outside the United States without the minimal protections afforded to U.S. users.

We must understand the comment that "the only thing that history teaches us is that we don't learn from history." Public and worker health concerns are not recent inventions. Much information about toxic chemicals and, certainly, many methods of control have been known for decades. It appears we need only the *intent* to carry out programs of preventive care.

Before 1900, the Chief Inspector of Factories in England had noted lung disease from the dusty conditions of the asbestos workers, and had recommended the use of respirators.[7] By 1910, multiple cases of pulmonary disease had been noted in asbestos-exposed workers, and by 1930, Merewether had defined asbestos-caused diasease as a class of illness.[8] Without an historical perspective, these reports appear commonplace. It must be remembered, however, that asbestos did not come into widespread industrial use until 1890, and that a decade later worldwide production was only 500 tons, rising to nearly 200,000 tons by 1920. Had the well-described findings been heeded and preventive efforts undertaken, how different the lives of hundreds of thousands of workers would have been.

In spite of the existing laws, regulations, and methods for control of asbestos contamination, there is still reason for concern. Tons and tons of asbestos-containing materials have been put in place in buildings, factories, and consumer products, as well as landfills, water conduits, and road surfaces. These products will continue to emit friable asbestos whenever they are disturbed, and will continue to add to the respiratory disease and carcinogenic burden of all exposed to them.

Reports and warnings, similar to those made by the Chief Inspector of Factories in England at the turn of the century, are being made today regarding workers who handle pesticides, plastics, solvents, dyes, welding materials, and other chemicals and products; and, once again, all too often opportunities to prevent illness and death are not being heeded, or are heeded only after harm has been done. One of the intentions of this book is to provide a method to anticipate future harm from chemicals as well as to investigate possible past damage, particularly in the industrial setting.

Some industries employ many chemicals, some of them toxic, and some augmenting or accelerating the effects of other chemicals. Industries such as rubber processing, welding, metal plating, coking, foundry work, and plastics manufacturing have complex and mixed processes. It is difficult to separate out a single toxic agent as the cause of disease where there are mixed exposures, and when certain job classifications bear heavier burdens of illness than others.

In some occupations, workers develop diseases characteristic of the substances to which they are exposed. Although it is uncommon to find one specific disease in a group of workers, occasionally a physician, an epidemiologist, or the workers themselves may be alerted to an abnormal condition and note the connection. This may be termed a **sentinel** disease, that is, standing alone, providing a warning. Notable examples of sentinel diseases in an industry are the finding of hepatosarcomas in vinyl chloride workers, bladder cancers in benzidine dye workers, male sterility in dibromochloropropane (DBCP) workers, and increased lung cancers in men who worked atop coke ovens. The most clearly recognized sentinel disease is mesothelioma, seen in asbestos-exposed persons.

More common, and less easily discerned, is an overall increase in the general level of sickness in a population exposed to toxic chemicals. There can be multiple effects, as, for instance, both fibrotic lung disease and cancer in asbestos-exposed workers; lung disease and cancer in welders and foundry and coke-oven workers; immunological and liver dysfunction in chlordane-exposed persons; and liver and neurological damage in solvent-exposed workers. Most chemicals do not cause a single abnormality, but some organ systems may be affected more often than others, as with leukemia in benzene-exposed persons and nasal cancer in those exposed to formaldehyde and nickel compounds.

When the effect of a chemical is known (by way of either animal experiments, case reports, or predictions based on structural–activity relationships), and where exposure can occur, it is necessary to look at the overall health and illness patterns of workers with these common exposures. This can be done easily in this day of computerized records, and it is one method recommended for monitoring exposed persons.

For instance, patterns can be developed by:

1. Linking workers' compensation claims with the various industries that exist within a geographical area.
2. Linking death certificate data with various industries.
3. Comparing morbidity and mortality data from various zip code areas, taking note of those areas known to have industries associated with toxic chemicals, and vice versa, looking at areas with increased sickness and death rates and looking for possible associations.
4. Investigating specific job classifications within an industry; for example, the health status of welders who are exposed not only to the emissions from welding rods but to the pyrolysis products emitted from the surfaces being welded.
5. Linking union benefit and retirement records with various job classifications and industrial sites.[9]

Where a variety of chemicals are used, or interact, it is difficult to pinpoint a single cause of illness. It helps to approach the problem with the thought: What are the most likely effects of the various chemicals(s), and are the illnesses found in this/these person(s) comparable with exposure to the agents?

In the field of chemical toxicology, one does not need to operate from a position of random ignorance. Considerable data are available on high-volume chemicals used in industry and commerce, from both animal and human reports. Most chemicals and processes used in industry have been developed and/or marketed because the chemicals possess characteristics that are deemed desirable in commerce. In this context, practically every "new" chemical has a "cousin" displaying similar characteristics and similar toxicology. In order to patent a "new" chemical or process, a minor change may be made in the structure, thus producing a "new" product that still possesses the action of the chemical that was copied. Many pharmaceutical agents have been created this way. They are called "me-too" products. If the structure, action, and/or application is similar to the original, then it can be assumed that the toxicology will be similar also. There are few surprises in chemistry—it is quite predictable.

Figure 5-1 shows six chemicals with similar structures. These chemicals share similar immunological, carcinogenic, and neurotoxic properties, as well as similar industrial uses, based upon structure. They are employed in the plastics, textiles, paint, and medical fields where exposures can occur to workers and consumers alike.

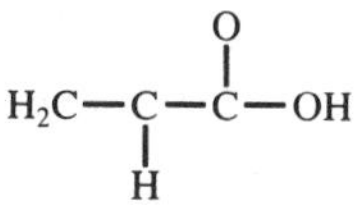

$$H_2C-\underset{\underset{H}{|}}{C}-\overset{\overset{O}{\|}}{C}-OH$$

Acrylic Acid

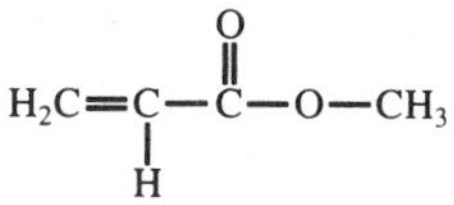

$$H_2C=\underset{\underset{H}{|}}{C}-\overset{\overset{O}{\|}}{C}-O-CH_3$$

Methyl Acrylate

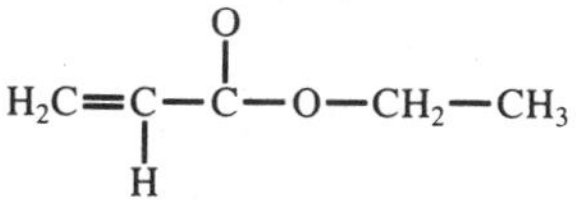

$$H_2C=\underset{\underset{H}{|}}{C}-\overset{\overset{O}{\|}}{C}-O-CH_2-CH_3$$

Ethyl Acrylate

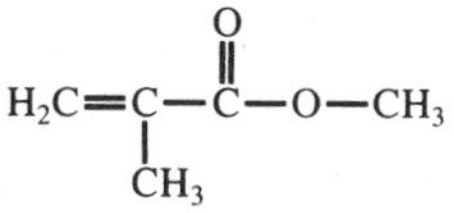

$$H_2C=\underset{\underset{CH_3}{|}}{C}-\overset{\overset{O}{\|}}{C}-O-CH_3$$

Methyl Methacrylate

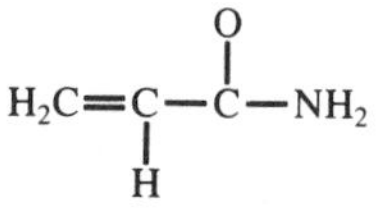

$$H_2C=\underset{\underset{H}{|}}{C}-\overset{\overset{O}{\|}}{C}-NH_2$$

Acrylamide

$$H_2C=\underset{\underset{H}{|}}{C}-C\equiv N$$

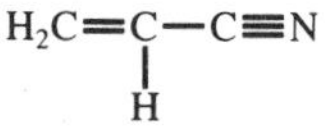

Acrylonitrile

Figure 5-1.

Figure 5-2 shows three commonly used, high-volume chemicals that share not only industrial properties, but toxicological effects as well. These chemicals are all employed as solvents, and benzene is a component of engine fuel. Benzene's adverse effect upon the bone marrow and its leukemogenic properties have been known since the turn of the century.[10-12] The carcinogenic properties of benzene were confirmed in rats where it was known to cause tumors in multiple sites.[13,14] It is critical to understand the importance of this finding: once again, it is demonstrated that a carcinogenic chemical is not site-specific. Additionally, as can be predicted by the structure of these chemicals, not only benzene but xylene and toluene as well have been shown to be carcinogenic.[15]

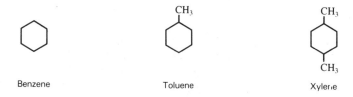

Benzene   Toluene   Xylene

Figure 5-2.

Because of benzene's known leukemic effect, xylene and toluene are substituted in many processes. The importance of investigating the thousands of workers exposed to these common chemicals that are utilized as fuels and solvents cannot be overemphasized.[16]

Figure 5-3 shows the chemical structures of three alcohols. Like their ethyl cousin, the form found in alcoholic beverages, the methyl and propyl forms too can cause central nervous system impairment and liver damage. Additionally, methyl alcohol is toxic to the optic nerve, causing blindness, whereas propyl alcohol is a carcinogen—properties that point up the importance of looking at structural–activity relationships.

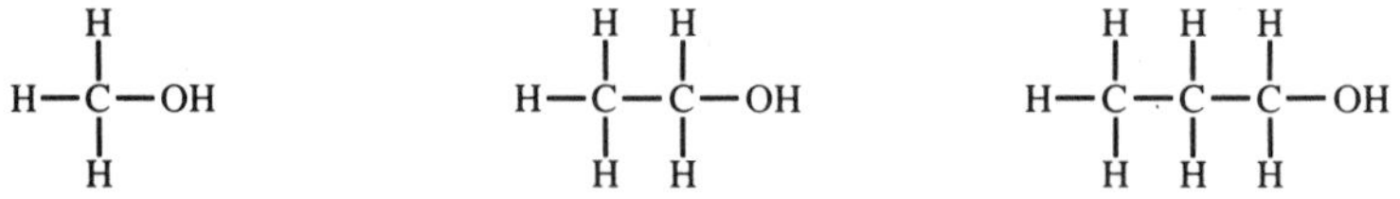

$$\begin{array}{ccc}
\overset{\displaystyle H}{\underset{\displaystyle H}{H-C-OH}} & \overset{\displaystyle H\ \ H}{\underset{\displaystyle H\ \ H}{H-C-C-OH}} & \overset{\displaystyle H\ \ H\ \ H}{\underset{\displaystyle H\ \ H\ \ H}{H-C-C-C-OH}} \\[1em]
\text{Methyl Alcohol} & \text{Ethyl Alcohol} & \text{Propyl Alcohol}
\end{array}$$

Figure 5-3.

MONITORING FOR EFFECTS

If a person is to monitor a population of workers or persons exposed to a known toxic agent or process, it is time-effective and cost-effective to follow "Sutton's law." (Willy Sutton, when asked why he robbed banks, said: "That's where the money is!") Likewise, with limited time, funds, and personnel, it is effective to monitor persons *known* to be exposed to toxic chemicals. In an otherwise perfect world, it would be desirable to monitor all workers, but this simply is not feasible. It is essential to begin monitoring those with known exposures, and to extend the monitoring process to all workers as time, funds, and personnel allow.

Certainly, as a minimum, workers in the following industries and workplaces should be monitored for chemically associated diseases:

- Heavy industry
    Foundry workers
    Steelworkers
    Coking industry workers
    Oil and gas refining workers
    Auto and machinery manufacturing workers
    Miners

- Petrochemical industry
    Oil well drillers and operators
    Handlers and formulators of petroleum products
    Workers with all solvents from manufacture to final user
    Workers with all plastics from manufacture to final user
    Gas station workers
- Agriculture
    Pesticide: handlers, manufacturers, applicators
    Fruit and vegetable pickers and handlers
    Dairy and livestock users of pesticides, growth-promoting agents, and
        hormones
    Meat handlers
    Grain handlers
- Trades
    Construction workers in general
    Carpenters
    Welders
    Cement workers
    Plumbers
    Heating and cooling workers
- Service industries
    Waste, trash, and garbage handlers
    Dry cleaning workers
    Auto repair (engine and body) workers
    Gas, electric, and sewer workers
    Beauticians
    Janitorial workers
- Maritime industry
    Ship builders and repairers
    Shipboard workers
- Health care industry
    Anesthetists
    Persons handling sterilizing agents
    Persons handling drugs and chemotherapeutic agents
- Office: workers exposed to:
    Chemicals used in duplicating materials
    Sprays, inks, and solvents
    Building materials in closed, self-contained buildings
- Electronics: assemblers and repairers of equipment, exposed to:
    Silicones
    Solvents

Rare metals
Soldering agents
Plastics

Most workers' exposures are not investigated until after the fact, when they become sick. I recommend that every worker maintain an "exposure diary," and if the worker is unable to do so for any reason, that another person be delegated to keep such records. This record-keeping function may be assumed by a union representative or coworker. Most workers, however, are not covered by union contracts of any kind.

OUT-SOURCING

"Out-sourcing," a new term, implies the manufacture or assembly of parts by an independent contractor. The resultant components are then utilized by a larger company to complete its product. Examples of parts produced in this way are the dash, seat, body, and trim components of an automobile made by injection molding and assembly of plastic parts.

The increasing practice of out-sourcing manufacturing will not only leave workers unrepresented in management, but raises doubt about their ability to monitor their job sites at all in regard to chemical exposures. The major company benefits economically from out-sourcing by being relieved of having full-time employees on the payroll. With out-sourcing, not only are there no knowledgeable personnel responsible for health or safety, but a principal disadvantage to the worker is that his or her job is dependent upon the vagaries of the major user. The person running a small shop rarely has the background, the interest, or even the time to be able to make informed decisions about whether or not to use a process that may involve toxic chemicals or by-products. The realities of economic and production pressures may mitigate against the decision. Because of such dispersal of manufacturing efforts throughout a community, it is especially necessary for each user and handler to become aware of potential problems from exposure to chemicals, and to prevent them.

Every company that uses chemicals must have either bought or manufactured them. Data concerning each chemical, reported on Material Safety Data Sheets (MSDS's) as required by the Department of Labor[17] should be maintained and linked to worker health records. Data from the U.S. Environmental Protection Agency's Toxic Substances Control Act (EPA—TSCA) and the Department of Commerce could be utilized to pinpoint areas where toxic chemicals are manufactured. Unfortunately, an adversarial mentality persists in much of industry, so that those in control of such information are reluctant to give it to those whom it would help most. Although it is cost-effective to maintain a healthy work force in the long run, there is generally a lag in efforts to prospectively monitor workers who are exposed to toxic chemicals.

If each worker kept a work-exposure record, the results of cumulative exposures could be analyzed, health effects of similarly exposed workers could be monitored, and if it developed that there were similar problems in similarly exposed workers, steps could be taken to eliminate causes of illness. This type of record could make workers' compensation claims easier to document, and could minimize such claims if the information gained by such detailed collection of information were utilized in a preventive fashion.

Given the burgeoning use of chemicals today, this collection of information may be the only way that certain diseases in exposed persons can be documented. With the widespread use of computers, there is no reason why information concerning the kinds and amounts of chemicals in use in industry cannot be linked with illness, death, and workers' compensation records, to provide information as to occurrence and prevalence of diseases. The main stumbling block on the part of governmental agencies appears to be the will to do so. Industry is reluctant to generate such data; for if increased sickness and death are shown, there may be increased liability. Policymakers need to realize the gains achievable through the preservation of public health and of trained workers, and the lessening of insurance and litigation costs that would result from enlightened use of this information. Then this failure to assemble meaningful information would be offset, and the health of the workers protected.

WORKER RECORD OF CHEMICAL EXPOSURES

Each worker record of exposures should contain the following information:

Chemical or product

Manufacturer

Method of use

Amount and frequency of use or exposure

Dates of use

Whether used by self or by other workers in the vicinity

Spills, etc.

As the person's job changes, or new products are introduced, the record should be updated. Examples of worker exposure are shown in Table 5-1.

Table 5-1.   Sample Worker Exposure Form.

Name _____ Birth date _____ Soc. sec. # _____

Job	Product	Used From To	Other Data
Welder	Cadmium rods	1973–75	Also did silver soldering
Painter	Vinyl paint	1975–76	Solvent for cleanup
	Urethane paint	1975–76	
	Acrylic paint	1975–81	
Cleanup	Solvents	1982–83	Lacquer thinner, kerosene, methyl ethyl ketone
Auto body	Plastics	1977–79	Vinyl chloride, acrylonitrile–styrene–butadiene plastics
Brakes	Asbestos	1970–72	
Mason	Cement sealers	1965–70	Urethanes and acrylics, caustics
Darkroom	Photo chems	1983–86	Color processing
Nurse	Formaldehyde	1974–81	Sterilized kidney machines
	Cyclophosphamide		Chemotherapeutic drugs
	Halothane		Anesthetic gases
Farm hand	Pesticides	1970–85	Seasonal, varied products
Machine operator	Plastics	1978–84	Injection molding
Battery repair	Manufacture	1977–81	Lead, acids, cadmium, and arsenic
Carpenter	Wood products	1971–85	Glues, arsenical and pentachlorophenol, paints, insulation

When the general chemical makeup of an industry is determined, it is possible to make a link between known effects of various chemicals and illnesses that they can cause. Conversely, when one is confronted with a sick worker, evaluation of the illness and the chemicals with which the person worked may aid the determination of the cause of that person's illness.

CASE REPORTS: INHERENTLY DANGEROUS INDUSTRIES

What follows are abstracts and comments concerning actual patients whom I have evaluated over the years. These abstracts are necessarily brief and do not include all of the information that was gathered in each case. Each was chosen to demonstrate a point, stimulate thought, and/or point out areas of concern and controversy. It will become apparent that there is an overlap of exposure between various industries. Chapter 6 contains additional case reports, arranged by the review-of-systems approach and addresses specific chemical exposures.

Rubber Industry

RUBBER WORKER #1

J. V. worked for a rubber manufacturing company from 1954 through 1970. In 1978, when he was 47 years old, he was found to have colo-rectal cancer, which spread despite medical care. He died of his illness in 1983 when he was 52 years old.

Mr. J. V. was exposed to at least four chemicals, known to be toxic, in the course of his work in the rubber industry. Review of the depositions of fellow workers revealed that the odors of the various chemicals were apparent in the work area, and that chemical spills were common. In addition to work practices that must be described as less than optimum, a coworker testified that the amount of at least one chemical was measured through the open top of a tank with a dipstick! (Aside from the inaccuracy of this method, this practice is condemned because of its unnecessary exposure.) The plant manufactured styrene–butadiene–acrylonitrile rubber, to which was added carbon black. Review of animal-testing data shows that the individual chemicals comprising the tri-polymer, as well as components of the carbon black, are carcinogens.

This particular worker had a history of hemorrhoids, a common condition if we are to judge from the number of hemorrhoidal remedies touted on the market. The roughened skin, fissures, and crevices associated with hemorrhoids that can trap particulates and chemicals may be compared with the roughened skin of the scrota of chimney sweeps who also developed malignancies as a result of contact with carcinogenic chemicals.

The fate of these unfortunate scrotal cancer sufferers, who often began working as chimney sweeps while they were children and small enough to fit into confined spaces, has been known since 1776, when first described by Sir Percival Pott. The plight of the sweeps was followed into the twentieth century by Butlin and others.[18] The offending agents, the polycyclics, were identified in the sooty particles that adhered to their skins. Three hundred years later, countless workers in the rubber, foundry, and chemical industries continue to be exposed to these same substances in a variety of forms.

It has been suggested that Mr. V. developed his malignancy on the basis of his "life-style." Much has appeared in the lay press about a possible connection between a fat-rich diet and the development of cancer of the gut. There are no animal data showing that fats per se are carcinogenic. The numerous chemical agents that are fat-soluble and carcinogenic include the industrial chemicals styrene and butadiene. It is entirely logical to postulate a mechanism of exposure to industrial chemicals via dietary and body fats, where the chemicals are dissolved in the fats and achieved contact with cells.

RUBBER WORKER #2

R. H. worked in the Banbury department of a tire building plant for 24 years. A Banbury, for the uninitiated, can be likened to a giant cake mixer whereby hundreds of pounds of various "ingredients," such as resins, oils, solvents, additives, and rubbers, are blended and heated to yield the "raw" rubber to build tires. The process gives off smoke, dust, and fumes, not only from the original chemicals but from the reactions of the various components.

By the time R. H. was 56 years old, he had severe pulmonary emphysema, bronchitis, and obstructive lung disease. He developed cor pulmonale (enlargement of the right side of his heart) and heart failure, secondary to his disabling lung disease. About a year later, he developed a pleural effusion, and an infiltrating squamous cell carcinoma of his lung was diagnosed. He died 8 months later, when he was 58 years old.

The rubber industry, with its various chemical sources of illness, is so well studied that the International Agency for Research in Cancer devoted an entire volume to this subject.[19] A variety of cancers is only part of the problem that workers face. Fibrogenic dusts, solvents, and other chemicals contribute to lung, liver, kidney, skin, and other diseases. A state-of-the-art bibliography on the subject is included in the bibliographies gathered at the end of this book.

Asbestos Industry

ASBESTOS WORKER #1

H. M. was 44 years old at the time he was found to have an abnormal chest X ray. He had begun to work for a brake manufacturer 19 years previously. He did a number of jobs over the years—grinding, riveting, and handling of brakes. A lung biopsy revealed poorly differentiated adenocarcinoma in *both* lungs. Despite radiation and chemotherapy, H. M. was dead 13 months after he learned of his abnormal X ray.

Mr. H. M.'s death is one of the thousands related to asbestos exposure. It has been clearly shown that, once taken into the body, asbestos, in any form, can cause cancer. It does not matter if the exposure setting is an asbestos mine, an asbestos textile mill, a construction insulation site, or a brake manufacturing or repair facility. The important connecting factor is the asbestos, as was well known by 1964, when H. M. started to work with the material.

Obvious examples of asbestos exposure include not only those indicated above, but exposure of persons working on ships and in shipyards, mines, and railroad repair shops, and with construction materials. Less obvious, but important, sources include thousands of schools, homes, and offices constructed with asbestos-containing insulation, floor tile, roofing, and dry wall materials. Asbestos may be handled safely if it is encapsulated and remains encapsulated, but once wear and tear, fire, or demolition breaks up the building materials, there is opportunity for exposure. Additionally, asbestos sprayed on structural members and existing in air plenum spaces is a source of airborne exposure. Workers, students, residents, and repair and custodial workers in these buildings may be at risk for exposure.

Asbestos does not break down, nor is it metabolized. Once it is mined from the earth and formed into various products, asbestos remains a source of exposure for as long as it can be disrupted. The argument that levels of asbestos in a building or an industry are comparable to outside air in an urban setting is fallacious. The asbestos in that air, emanating from brakes, clutches, buildings, and the like, came from those same asbestos mines, and is also unsafe. It will take decades and untold amounts of money to abate asbestos pollution. In the meantime, the prudent thing to do would be seek alternative technologies and to ban asbestos from as many uses as possible. Essentially, asbestos is a problem forever!

Heavy Construction

CONSTRUCTION WORKER #1

E. D. was 20 years old in 1948 when he began to work as a laborer in industrial construction. He worked at a number of major factories where steel, iron, and coke were made. He used power and hand tools to remove brick and insulation materials from blast furnaces, stoves, and coke batteries. He said his work was hot, noisy, dusty, and dirty. He said he wore a paper mask, but could still spit up dust after work.

Although E. D. went as far as the sixth grade, he could neither read nor write, and did not know the nature of the materials that he had worked around. He described a greenish insulation material between the firebrick and the steel walls of the structures.

He said he had developed a cough in about 1976, productive of about one cup of yellow phlegm a day. He had complained of increasing shortness of breath over the two years prior to when I saw him in 1981.

He told me he has smoked a package of cigarettes for 8 years only, and that for the 7 to 8 years before I examined him that he had smoked no more than 2 to 3 cigarettes a day.

E. D. had a rattling cought throughout my examination, and crackling sounds in both lungs, with both obstructive and restrictive components on pulmonary function testing.*

Mr. E. D.'s lung disease was undoubtedly due to his mixed exposures. In the coking

---

Obstructive lung disease means that the person has difficulty expelling air from his/her lungs. This may be caused by narrowing of the lumen of the bronchi or bronchioles, and/or emphysema, where the alveoli (air sacs) are torn or loose their elasticity. *Restrictive* lung disease occurs when the person cannot take air into the lungs, because of stiffening of the lungs, as from fibrosis. Obviously, either a blockage of the air passages or poor movement of the chest, as from the pain of fractured ribs, can result in restriction. In the context of chemically caused lung disease, we refer to diseases that are intrinsic to the lungs.

and steel industry, particularly in the tearing out and relining of the processing structures, workers are in contact with many fibrogenic products, such as asbestos, fiberglass,
silica from firebrick, and the components of mortar, which in high-temperature applications have a high content of chromium compounds. Considering the time span when
Mr. E. D. was engaged in this work, it is more than likely that asbestos exposure was
also a significant factor in the development of his lung disease. His symptoms of progressive shortness of breath and cough, combined with the restrictive component of his
pulmonary testing, are in keeping with the effects of asbestos exposure. Additionally,
the inside of coking batteries, as well as the inside of blast furnaces, contains many
polycyclic hydrocarbons, associated with irritation of the lungs as well as with cancer.
It is unlikely that smoking was a major cause of Mr. E. D.'s lung disease.

CONSTRUCTION WORKER #2

A. S. said that in 1981 he had became progressively short of breath over a three-week
period. He related that he had coughed up blood off and on for two previous years. He
was diagnosed as having pneumonia and treated with antibiotics. A follow-up chest X
ray indicated a "spot on his lung." The diagnosis was poorly differentiated squamous
cell carcinoma with involvement of both the upper and lower lobes of his right lung.
There were metastases to a lymph node and chronic reactive lymphadenitis. A. S. was
40 years old at the time his entire right lung was removed. After surgery he had severe
restrictive and obstructive lung disease. A. S. said he had smoked 3/4 package of
cigarettes for about 21 years, stopping in 1980.

From 1973 to 1979, Mr. A. S. had worked as a furnace repairman in the
steel industry, performing repairs on 13 furnaces and two cupolas. His job
involved using a chipping hammer to remove damaged firebrick, and then
replacing the brick using special mortar. He related that although he wore a
mask off and on, it was dusty, dirty work with black soot, and he inhaled the
material.

Mr. A. S.'s industrial exposure to carcinogens was by description a mixed
one, with contact with the components used to construct the refractory structures, as well as the substances processed in the furnaces and cupolas. A patent
search was conducted for the product used to mortar the firebricks together. The
patent revealed that the product was made up of approximately 50% chromite
ores, with particles that could pass through a mesh of 100 holes/inch, and 50%
of the particles able to pass through a 325 mesh, indicative of even smaller
particles, easily within the respirable range.

The patent noted that mixed chromium could be employed, including
"chromic acid, chromium salts and chromium-containing salts, for example
chromous sulfate, chromic sulfate, chromous acetate, chromic acetate, and the
chromous and chromic halides, chromium nitrate, chromium oxalate, the

chromates and dichromates of the alkali metals, e.g. sodium and potassium, and of ammonium and magnesium chromate and dichromate.''

Considering the composition of the refractory brick mortar, it should come as no surprise that it would burn a worker's skin, nor should it come as any surprise that a person could develop cancer as a result of inhaling the material. The bibliographic section at the end of this book documents the development of information concerning the relationship between exposure to chromium-containing compounds and cancer.

Automotive Service Industry

AUTO SERVICE WORKER #1

G. C. began to work in the auto repair industry when he was 14 years old. In the summer of 1985, when he was 47 years old, he was hired as the service manager of an auto dealership. His office was adjacent to the repair shop, where between 10 and 25 cars and trucks were under repair at a time. He said that although there was a ventilation system in place for gasoline engines, it was not used properly, and there was no ventilation for diesel engines. He said that the walls of his office had been painted brown so the soot would not show.

Over the course of a week in December 1985, G. C. became sick, complaining of severe headaches and shortness of breath. At the emergency room, he said that he was "flushed and red all over."

A diagnosis of carbon monoxide poisoning was made, compatible with Mr. G. C.'s history and physical findings.

Carbon monoxide perferentially combines with the hemoglobin of the red cells, displacing oxygen, and at a faster rate than it is given up from the cells. As a result, continual exposure to carbon monoxide will gradually result in increasing levels of it in the body. The symptoms of headache, nausea, and shortness of breath (from lack of oxygen) and the sign of redness of the skin are all compatible with carbon monoxide poisoning. If a person is not removed from the source of the carbon monoxide, unconsciousness and death can ensue.

Unfortunately, like many workers, Mr. G. C. was a smoker, using up to three packages of cigarettes a day for a number of years, before cutting back to about one pack a day, which he currently smokes. Without question, his cigarette smoking has had an adverse effect upon his heart and lungs. His pulmonary function tests showed both obstructive and restrictive lung disease. Cigarette smoking undoubtedly contributed in part to the obstructive component. The restrictive component is of concern because of exposure to asbestos, a pollutant from brake and clutch repair. His exposure to carbon monoxide, superimposed upon his lung disease, was the straw that essentially "broke the camel's back" and sent him to an emergency room.

Paint Industry

PAINT WORKER #1

P. A. developed a pain in his posterior upper chest when he was 52 years old. A biopsy revealed cancer, and he was treated with radiation. Over the next three years, he became increasingly short of breath, so much so that he had to sleep with his head elevated and with oxygen.

P. A. began to work as a paint mixer when he was 23 years old and did that for 29 years, until he got sick. In the course of his work, he said he handled 50-pound bags of pigments, which leaked, worked with urethane and vinyl paints, and handled such chemicals as toluene, xylene, methyl ethyl ketone, methyl isobutyl keton, methyl ethyl benzene, butanol, and other chemicals. He said that his shoes and socks were frequently soaked from chemicals spilled on the floor, and that he used a squeegee to push the spills into the drain.

Mr. P. A. applied for worker's compensation. His attorney said his case was settled for around $5000.00. Mr. P. A. is now 55 years old and permanently and totally disabled.

PAINT WORKER #2

A. B. was 18 years old and rode a bicycle up to 45 miles a day without problem, when he began to work as a paint sprayer. He sprayed paint for 2 to 8 hours a day, sometimes wearing a "particle mask," and did the job for about 11 months. He said the paints were delivered in 55-gallon drums, and then dumped into 5-gallon containers where other chemicals were added. He said he worked with toluene, xylene, acetate, urethanes, toluene diisocyanate, hydrochloric and chromic acids, as well as other chemicals. He said the paint pots held from 5 to 60 gallons of paint and that he used methyl ethyl ketone to clean out the pots, and to clean up spills on the floor.

On one of his jobs, he said he sat on a wooden swing and rode a monorail that ran a zigzag course through a paint curing oven to clean it. He used a chipper and an air hose to blow the waste material out of the oven. He said that the oven was about 60 to 65 feet long, and that he spray coated the oven, using about 30 gallons of a heated material, the composition of which he did not know.

A. B. was 21 years old when I examined him, and had clear findings of obstructive lung disease.

The paint industry has been identified as a source of sickness and death for decades.[20]

Waste Disposal Industry

WASTE WORKER #1

V. W. said he was a well person until he began to work for a liquid waste disposal company in 1979. In February 1981, he said that a truckload of waste materials was mixed with other chemicals, producing a cloud of poisonous gas. Eight workers were overcome by the chemical fumes, two of them dying.

At the time I examined V. W., he was 42 years old and told me that he had developed wheezing at night, severe enough to require him to sleep partially upright with three pillows. In 1979, he had developed numbness in his hands, which persisted for about 6 to 12 months, and he had severe nosebleeds in 1980. He had no history of heart disease, but by the time I examined him, his EKG showed cardiac ischemia and an old infarction. He complained of an immediate onset of headache after the gas exposure. The headaches, with some dizziness, have persisted on a nearly weekly basis since then. He had findings of obstructive lung disease on both physical examination and pulmonary function testing.

Mr. V. W. described the waste disposal facility as occupying 10 acres, with two incinerators, measuring about 50 feet long by 10 feet wide and 5 to 20 feet high, with stacks about 70 feet tall. He said that both solid and liquid waste chemicals were delivered in metal and fiber drums and in tank cars. He reported that the chemicals were fed into the gas-fired incinerator, and the sludge was mixed with dry cement, and hauled away while in the mud stage. He told me that the incinerator was cooled by waste water from an open pond that held about one million gallons, which, he said, smelled very bad. He said some chemicals were pumped into underground tanks before processing. Mr. V. W. and a coworker, Mr. K. S., said the facility handled wastes such as formaldehyde, greases, oils, paints, silicone materials, blood, solvents, pesticides, and so on.

He stated that, in addition to moving drums and barrels with a hi-low truck, for about 18 months he operated 60- to 95-pound jackhammers to tear as much as 5 feet of residue from the inside of the incinerators. He said there was a "cottony" insulation material back of the firebrick. He did not know if it was asbestos or not.

WASTE WORKER #2

K. S. was 28 years old when he worked at the same facility as V. W. His job, tearing out the waste, firebrick, and insulation from the inside of the incinerator, was similar to V. W.'s.

He described the smell of the poison gas as "awful" and "similar to rotten eggs." At the time of the incident described above, he said he saw his foreman and two coworkers on the ground, and was one of those taken to the hospital for treatment of toxic inhalation.

K. S. complained of a productive cought of two years' duration, as well as pleuritic chest pain since the gas exposures. He, too, complained of the onset of headaches since the exposure.

Although Mr. K. S. went to high school, he was in a special education class because of a learning disability, and cannot read or write well. He is married with two children, ages 6 and 3. Since he was overcome with chemicals, he hopes to get training in the butchering field. What the future holds for him and his family cannot be guessed.

After these workers were overcome, and two others killed, the facility was closed by an order from state authorities. Mr. V. W. is now working as a security guard, at considerably less pay.

A reasonable assessment of the situation would lead one to conclude that there are multiple sources of pollution from a facility such as this. The sources of pollution in a waste treatment facility include:

1. The raw chemicals.
2. The products of mixing various chemicals together.
3. The products of combustion from the chemicals given off in the stacks.
4. The sludge.
5. Liquid contamination from spills and the settling pond.
6. The scrap drums and barrels.
7. The construction materials used in the incinerator.

These pollutants have effects on:

1. The employees.
2. The neighboring residents.
3. The soil and water table in the adjacent area.
4. The soil and water table where the sludge is taken.
5. The air surrounding the facility.

It is not possible to assess the effects of the various chemicals upon Mr. V. W.'s or his coworkers' health. No systematic study was undertaken of either the employees or the community. Because of exposures to complex mixtures of solid, liquid, and gaseous chemicals, one must include the waste disposal industry in those considered inherently dangerous.

It also can be reasonably concluded that until a safe and effective way is devised to control the production and disposal of industrial wastes, these hazards to workers and the community will continue.

The Dry Cleaning Industry

DRY CLEANING WORKER #1

J. P. was a high school student, getting A's and B's, when she worked for a dry cleaning establishment. One day a drum of dry cleaning solvent containing about 20 to 30 gallons leaked out onto the floor over a period of about 7 hours. J. P. mopped up the fluid, using a manually operated wringer to squeeze out the mop. Two days later, she was seen in the emergency room of a hospital, complaining of headache, shortness of breath, and dizziness. Five days later, she was again seen in the emergency room with persistent complaints, but, in addition, had chills, fever, nausea, and vomiting. A year later, she had a neurological evaluation because of continuing headaches, visual disturbances, and "declining memory and lack of social activity." This evaluation showed impaired hand grip bilaterally, impared rapid alternating motions on the left, and impaired mentation, speech, and ability to learn new verbal information. The examiner described her "affect as incongruent with the topic of conversation," and reported that she was "giddy." A second examination, 11 months after the exposure, stated: "throughout the interview the patient holds he head down and keep her eyes closed. When she does open her eyes she giggles in a immature manner." This evaluation was made one month short of her 18th birthday. J. P. has been unable to return to school.

The product that she worked with has been described as "perc," most likely perchlorethylene, the most commonly used solvent in the dry cleaning industry, although one examiner thought that she had been exposed to 1,1,1-trichloroethane. In either event, J. P.'s exposure to solvents has resulted in brain damage, which is undoubtedly irreversible. The bibliographic section at the end of this book contains citations concerning these chemicals.

DRY CLEANING WORKER #2

R. N. began to work in the dry cleaning industry in 1948 when he was 25 years old. He worked as a spotter, cleaner, and presser, up to 12 hours a day, 6 days a week, for approximately 25 years. He developed chronic granulocytic leukemia, and died of multiple complications when he was 56 years old.

A number of chemicals have been employed in the dry cleaning industry and are available for either commercial or home use. "Perc" (perchlorethylene, also known as tetrachloroethylene) accounts for about 75%. Additional chemicals include 1,1,1-trichloroethane, trichlorotrifluorethane, benzene, used for spotting, and carbon tetrachloride, which was used in the past. These are the same chemicals used to degrease metals. With about 225,000 workers in the dry cleaning industry,[21] the potential for widespread harm is great. With the detection of "perc" in expired air extending over a two-week period after exposure,[22] there is ample time for exposure to continue and cells to be harmed.

Welding Industry

WELDER #1

C. B. was 56 years old in 1985, when he developed enlarged lymph nodes. He was diagnosed as having diffuse large cell lymphoma and was started on chemotherapy. He was dead two months later as a result of his lymphoma which transformed to preleukemia with a blastic crisis.

C. B. had worked as a maintenance welder for approximately 38 years, and in the course of his work had been exposed to smoke and fumes. He did not wear any respiratory protection, which is not uncommon considering the size and shape of a most welding hoods.

Review of the MSDS's of some of the products that Mr. C. B. used in the course of his work indicated that they contained coal tar, lead, cadmium, cadmium oxide, chromites and chromates, thorium dioxide, diethanolamine, mineral oil, and welding fume.

Considering the toxicity of these chemicals and the fact that Mr. C. B. spent approximately 68% of his lifetime in various welding activities, it must be concluded that these exposures contributed significantly to his disease and death. A steward in Mr. C. B.'s department expressed concern that five of nine welders in the department had developed malignancies.

WELDER #2

R. W. was 31 years old and working as a repair welder for a mining company. After about 3 hours of welding on a backing material that contained, among other components, an epoxy resin formed by condensation of epichlorhydrin with bisphenyl-A (an amine hardener) and asbestos, he "became sick, dizzy, developed chest pains, generalized nausea and malaise." R. W. drank approximately one can of beer a day and did not smoke. Blood tests showed elevated liver enzymes and elevated blood lipids (including a cholesterol and triglycerides), which persisted for a month.

WELDER #3

R. K. was 29 years old and worked at the same job site as R. W. (Welder #2). He also applied and did welding operations on the plastics-containing backing material. After about two weeks, he said he developed dizziness, soreness in this throat and chest, and weakness. Since then, he has developed shortness of breath, a persistent skin rash, and persistently raised blood lipids and liver enzymes. About 8 months after his initial reaction, he developed a gastrointestinal bleed, requiring blood transfusions, which has complicated his findings.

The composition of the original product and Mr. R. K.'s medical records have been evaluated by several scientists. A variety of thermal degradation products, released in the process of heating this epoxy resin, have been identified and associated with adverse effects to the central nervous and respiratory systems and the liver, and, not incidentally, with carcinogenic and mutagenic effects.

These two cases (#2 and #3) are presented as examples of a situation where the source of toxic chemicals is not so much a factor of welding on a metal surface, but is related to the composition and the pyrolysis products of the associated coating. The release of asbestos from the product as the epoxy is burned away must not be disregarded as a source of carcinogenic and fibrogenic exposure.

As in many welding jobs, where metal surfaces have been coated with paints, oils, and other substances, the pyrolysis products may present as great a hazard as the welding itself.

Firefighting

FIREFIGHTER #1

D. R. was 53 years old in 1981, when he underwent radical surgery in an attempt to control a moderately well-differentiated invasive squamous cell carcinoma of his throat. He had smoked for about 7 years, from the time he was 19 until he was 26 years old. D. R. died less than 9 months after his surgery, approximately 20 years short of his expected lifetime.

Mr. D. R. had worked as a municipal firefighter for 24 years, and had been exposed to the smoke, fumes, and toxins inherent in such situations. He had used asbestos-containing gloves and fire blankets until about 1981. Asbestos-containing building products have been an integral part of commercial and industrial buildings, as well as many homes, so that destruction of a building would release asbestos into the air.

Firefighters are exposed to multiple toxic agents including carcinogens, released as pyrolysis products from burning building materials such as insulation materials, fabrics, plastics, wall and floor coverings, paints, and so forth. Additional sources of toxic pyrolysis products result from the burning of materials stored within the buildings such as pesticides, oils, industrial chemicals, and the like. Most injuries to inhabitants of burning structures and to firefighters occur as a result of inhalation of toxic materials, rather than from thermal burns themselves. This factor has been of critical importance in hotel, commercial, and airplane fires.

Attempts to formulate fire-retardant products have given some hope of decreasing the morbidity and mortality of fire injuries. Considering the chemistry of these products, one must not assume that the retardants themselves are without hazard. This concern extends from those who formulate the fire retardants to those who incorporate them into various commercial products (fabrics, plastics, and so on), to those who use, live with, and handle the products on a daily basis, and lastly to firefighters. The retardants have been represented by a number of chemicals, including: the PBBs, marketed as Firemaster®, formulated to be used in plastics, and which resulted in a poisoning episode in Michigan; the PCBs, which were employed in high-temperature electrical systems because of fire retardancy and now are banned by order of the EPA; and Tris, or 2,3-dibromopropyl phosphate,[23] which was used in children's clothing and is also banned because of positive mutagenic and cancer findings. Other fire retardants in use include various brominated compounds such as brominated polymeric epoxy resins, decabrombiphenyl oxide, and various phosphorus and antimony compounds, as well as melamine compounds. As we look at the chemical structure and solubility characteristics of a number of these products, concern must be raised about their biological activity, including possible carcinogenicity, teratogenicity, and mutagenicity. At this writing, test results concerning these components have not been available. One would urge caution in regard to exposure to fire retardants until test data indicating safety are available.

NOTIFICATION OF HAZARDS AND RIGHT TO KNOW

Many industries, including blue-, pink-, and white-collar work, have chemical exposures inherent in them, with the greatest likelihood of exposure occurring in heavy and service industries. In an effort to indentify individuals exposed to hazardous substances in the workplace, the U.S. Congress considered the "High Risk Occupational Disease Notification and Prevention Act of 1987."[24] It is hoped that notifying workers at high risk will help prevent some of the 100,000 deaths and 340,000 disabilities that occur yearly as a result of occupational disease.[25-26] We cannot think that notification alone will solve these problems; we must also take action to minimize and/or eliminate the risks from hazardous chemicals, and better administer the regulations that are on the books.[27] This cannot be attempted unless the worker, the physician, and the public health worker are given adequate information about the identity and nature of chemical risks.[28] The passage of right-to-know laws on both federal and state levels may help end some of the proprietary witholding of information.[29-31] Only an educated and informed populace can make intelligent and responsible decisions about these exposures.

There is sufficient information in the scientific literature to protect workers from hazardous chemicals now. Any reduction in exposure to toxic chemicals will benefit those exposed, by decreasing their risk. This can be accomplished with knowledge.

A nation that is afraid to let its people judge truth and falsehood in an open market is a nation that is afraid of its people.

—John F. Kennedy

REFERENCES

1. Terkel, S. *Working: People Talk about What They Do All Day and How They Feel about What They Do.* Ballantine Books, New York, 1972, p. 715.
2. Warbasse, J. P. *The Doctor and the Public; a Study of the Sociology, Economics, Ethics and Philosophy of Medicine, Based on Medical History.* P. B. Hoeber, New York, 1935.
3. Annon. *Amer. Med. Assoc. News*, p. 8, Dec. 26, 1981.
4. Postel, S. Stabilizing chemical cycles. In *State of the World*, Brown, L. R., Ed. W. W. Norton and Co., New York, 1987, pp. 156–76.
5. Rifkin, J. *Declaration of a Heretic.* Routledge and Kegan Paul, Boston, 1985, pp. 1–140.
6. Weir, D., Schapiro, M. *Circle of Poison: Pesticides and People in a Hungry World.* Institute for Food and Development Policy, San Francisco, 1981, pp. 1–100.
7. Chief Inspector of Factories. *Factories and Workshops.* Her Majesty's Stationery Office, London, 1899, p. 264.
8. Merewether, E. R. A., Price, C. W. *Report on the Effects of Asbestos Dust on the Lungs and Dust Suppression in the Asbestos Industry.* His Majesty's Stationery Office, London, 1930, pp. 1–34.
9. Wolfe, J., Education director, International Molders and Allied Workers Union, Cincinnati, Ohio. Personal communication.
10. LeNoir, C. Sur un cas de purpura attribue a l'intoxication par la benzine. *Bull. Mem. Soc. Med. Hop. Paris* 3: 1251–60, 1897.
11. Santesson, C. G. Chronic poisoning with coal tar benzene; four deaths. Clinical and pathological–anatomical observations of several colleagues and illustrating animal experiments. *Arch. Hyg.* (Munchen) 31: 336–76, 1897.
12. Delore, P., Borgomano, C. Acute leukemia following benzene poisoning. On the toxic origin of certain acute leukemias and their relation to serious anemias. *J. Med. Lyon.* 9: 227–33, 1928.
13. Maltoni, C. Scarnato, C. Le prime prove sperimentali della azione cancerogena del benzene. *Osp. Vita.* 4(6): 111–13, 1977.
14. Maltoni, C., Scarnato, C. First experimental demonstration of the carcinogenic effects of benzene. Long-term bioassays on Sprague-Dawley rats by oral administration. *Med. Lav.* 70: 352–57, 1979.
15. Maltoni, C., Conti, B., Cotti, G., Belpoggi, F. Experimental studies on benzene carcinogenicity at the Bologna Institute of Oncology: current results and ongoing research. *Amer. J. Indust. Med.* 7: 415–46, 1985.
16. Mehlman, M. A., Ed. *Benzene: Scientific Update.* Alan R. Liss, New York, 1985, pp. 1–499.
17. Hazard Communication Standard 29 CFR 1910.120 (1983).

18. Butlin, H. T. Cancer of the scrotum in chimney-sweeps and others. *Brit. Med. J.*, 1-6, 66-71, 1341-47, 1892.
19. *IARC Monographs on the Evaluation of the Carcinogenic Risk of Chemicals to Humans: The Rubber Industry*, Vol. 28. International Agency for Research in Cancer, Lyon, France, 1982, pp. 1-454.
20. Rosner, D., Markowitz, G., Eds. *Dying for Work: Worker's Safety and Health in Twentieth Century America*. Indiana University Press, Bloomington, 1987.
21. Blair, A., Decoufle, P., Grauman, D. Causes of death among laundry and dry cleaning workers. *Amer. J. Pub. Health* 69(5): 508-11, 1979.
22. Stewart, R. D., Erley, D. S., Schaffer, A. W., Gay, H. H. Accidental vapor exposure to anesthetic concentrations of a solvent containing tetrachloroethylene. *Indust. Med. Surg.*, 327-30, 1961.
23. *IRAC Monographs on the Evaluation of the Carcinogenic Risk of Chemicals to Humans: Some Halogenated Hydrocarbons*, Vol. 20. International Agency for Reseach on Cancer, Lyon, France, 1979, pp. 575-85.
24. U.S. Senate bill S. 79, and U.S. House of Representatives bill H.R. 162.
25. Bradford, H., Garland, S. Warning: your job may be hazardous to your health. *Business Week*, F-128, June 8, 1987.
26. Hoffer, W. What should the government tell workers in risky businesses? *Amer. Med. News*, pp. 17-20, Apr. 4, 1986.
27. Glaberson, W. Is OSHA falling down on the job? *New York Times*, 3: 1-6, Aug. 2, 1987.
28. Williams, T. R. A review of Michigan's right to know law. *Mich. Med.*, 306-7, May 1987.
29. Bingham, E. The "right-to-know" movement. *Amer. J. Pub. Health* 73(11): 1302, 1983.
30. Himmelstein, J. S., Frumkin, H. The right to know about toxic exposures. *New Engl. J. Med.* 312(11): 687-690, 1985.
31. Ashford, N. A., Caldart, C. C. The right to know: toxics information transfer to the workplace. *Ann. Rev. Pub. Health.* 6: 383-401, 1985.

6

CASE REPORTS BASED ON THE REVIEW-OF-SYSTEMS APPROACH

Trying to control carcinogenic substances on a case-by-case basis is like trying to put out a forest fire, one tree at a time.

—Secretary of Labor Ray Marshall[1]

After reading the following case reports, we may well ask ourselves if our approach to toxic chemicals in general is any more efficient than such a piecemeal one. This chapter contains a number of actual patient reports, taken from files of some 8000 cases that I reviewed since 1970. What follows are accounts of what happened to men, women, and children who developed reactions as a result of exposure to a variety of chemicals. These chemicals include industrial chemicals and drugs, encountered through work, medical care, and the general environment.

There is a heavy emphasis on cancer cases because, as explained earlier, malignancy represents a degree of illness over which there is little dispute. Many more persons, who became "a little sick," became seriously ill, or died, have not been cited. The cases that are presented were selected because they demonstrate the usefulness of the review-of-systems approach in determining whether a condition is related to chemical exposure or not. Additionally, these cases give examples of sentinel health events,[2] which are useful in alerting others similarly exposed, as well as providing a stimulus for preventive efforts and evaluation of exposed populations.

Most persons "enjoy" good health, and have a tendency to dismiss "minor" symptoms, attributing them to a "cold," the "flu," or some other temporary inconvenience. Therefore, much illness is established, even far advanced, before most persons seek medical help. Most patients ignore early symptoms, such as a change in bladder, bowel, or sexual function; shortness of breath when performing tasks that were previously easy; or even confusion and loss of memory. Typically, it is the patient, not the physician, who becomes suspicious of a possible connection between an exposure to a chemical and his or her illness, and who then seeks medical help for either relief of the malady or confirmation of the connection.

In taking a medical history, it is imperative to try to determine a time when the person was symptom-free, and then to trace the historical development of his or her complaints. A period free of illness may be documented by using the time of a physical examination, an X ray, and/or a laboratory test done in the course of a preemployment evaluation. In general, it is safe to assume that anyone serving in the armed forces was healthy at the time of induction, which may also serve as a point in time when good health prevailed.

The following cases are recounted in review-of-systems order, starting at the head and proceeding downward through the body, as outlined in Chapter 3. Literature citations, arranged in chronological, state-of-the-art form may be found in the bibliography at the end of the book.

CENTRAL NERVOUS SYSTEM

CNS CASE #1

A. P., a 68-year-old retired woman, worked part-time to supplement her social security by doing lawn maintenance, and in the spring of 1986, she had been a healthy, vigorous person. On May 1, she mowed a lawn, bagged the grass, and did some weeding. On May 10, she began to be aware of a persistent headache and aching and a "picky" sensation in ler legs. She returned to the same job on May 17, and had difficulty doing the job because of the symptoms in her legs. By May 23, she noticed difficulty climbing stairs because of weakness, and by May 29, she began to notice a tendency to fall backward. She returned to the job on June 27, working very slowly, when she noticed a white residue on the grass and saw that the weeds next to the drive were dead. Her grandson, with whom she had been working that day, found a bill from a lawn application company, indicating that the lawn had been sprayed that morning. By July 3, A. P. could no longer work. An EMG (electromyelographic) study done on July 15 revealed an acute axon polyneuropathy. She has had intense pain and numbness in her legs, and subsequently developed a chronic rash on her upper body.

Investigation of the lawn maintenance company revealed that a triple combination of phenoxy herbicides (plant killers) had been sprayed on the lawn to control the weeds. These chemicals included: 2,4-D(2,4-dichlorophenoxy acetic acid), Dicamba (3,6-dichloro-*o*-anisic acid, and MCPP (2-[4-chloro, 2-methyl phenoxy] propionic acid). Their formulas are shown in Figure 6-1. Their similarity to other chlorinated phenoxy compounds was demonstrated in Figure 2-2. An earlier application of pendimethalin had also been used.

The compound 2,4-D is best remembered as half of the formula, called Agent Orange, utilized in vietnam to kill acres of forest. The other part of the formula is a close "relative," called 2,4,5-T(2,4,5-trichlorophenoxy acetic acid). These

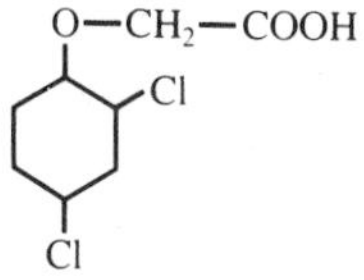

2,4-dichlorophenoxy acetic acid (2,4-D)

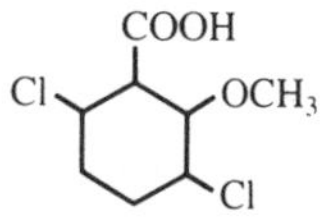

3,6-dichloro-o-anisic acid (Dicamba®)

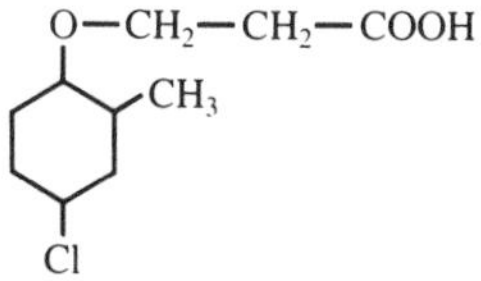

2-(4-chloro, 2-methyl phenoxy) propionic acid (MCPP)

Figure 6-1

chemicals are members of the family of chemicals called phenoxy acetic acids that kill plant life by stimulating growth beyond the plant's ability to keep up with it.

These chemicals have been associated with neurotoxicity, cancer, skin disorders, and birth defects. This case points up the necessity to obtain a work history, even in a "retired" person. Without this history, the etiologic factor in her illness would not have been found.

CNS Case #2

M. K. was 30 years old and working as a hospital laboratory technician when she lost the vision in her left eye in August 1982. She was told she had "idiopathic optic neuritis" and was out of work until October, when she returned half-time. A year earlier, in the spring of 1981, she had developed "intense" headaches that came on quickly, persisted for about three days at a time, and were not relieved with any of the common medications. About the same time, she said she was aware of increasing tiredness, and in February 1983, she awoke with her hands and feet numb, with the symptoms progressing throughout the day. She was hospitalized for a week, whereupon a diagnosis of multiple sclerosis was made. She suffered a miscarriage in April 1983. She again returned to work, but because of loss of ability to do fine work, she stopped her laboratory job in September 1983. She attempted a number of other jobs after she left hospital work, but because of her weakness and clumsiness, was unable to persist in any of them.

After stopping work, she indicated that her numbness took months to decrease, not resolving completely, and that her muscle weakness had become persistent by the time she was examined in September 1985.

A suspicion that M. K.'s neurological dysfunction was not multiple sclerosis was suggested by the fact that, unlike the usual case of multiple sclerosis, her symptoms have not progressed with time. Additionally, she did not have the classical findings of nystagmus (rapid lateral eye movements) or spastic paralysis with hyperreflexia, or the other classical changes seen in patients with multiple sclerosis.

Because two other technicians also complained of similar symptoms, M. K. became concerned, and investigated the chemicals she handled in the course of her work, where she embedded bone specimens in plastic. To do this, she mixed methyl methacrylate monomer with polymethacrylate in the open air, in a hood, into which she put the bone specimens to harden. She said she frequently had skin contact with the monomer as she repositioned specimens, or cleaned up spills. Additionally, she handled small quantities of dibutyl phthalate, ethyl alcohol, and xylene. She indicated she worked by herself in the small laboratory and had not been given any instruction as to the toxicology of the chemicals.

The acrylics belong to a family of chemicals widely used in the plastics, coating, paint, and textile industries. Various reports have been made that connect exposure to methyl methacrylate, acrylamide, and acrylics in general with both peripheral and central nervous system damage. These reports extend back to the 1940s and include both controlled animal studies and human case reports.

The widespread use of this family of chemicals with their attendant neurotoxicity is cause for concern, as plastics increasingly replace metals in the automotive and other industries. If warnings about the toxicity of this chemical were not available to a health care worker in a hospital, one wonders if they are available to workers in general industry, whose exposure may be even greater, and whose knowledge of chemistry is less than that of M. K.!

Like many other chemicals employed in the plastics industry, the acrylic compounds have a double-bonded portion (carbon-to-carbon linkage) in their structure that allows them to react with one another and with other chemicals. This structure also allows them to react with living materials, resulting in neurological damage, cancer, and reproductive effects. The chemical structures of several of the acrylics were given in Figure 5-1. One can see a similarity in their structures, which influences not only their commercial uses but their toxicology as well.

CNS CASE #3

E. S. is a 63-year-old man who worked for a company that made specialty papers. A resin was impregnated in the paper by using a solvent. He noticed that he began to fall, had difficulty climbing stairs, and had pains in his legs in about 1982. His wife became concerned for his mental state, in that he became forgetful and withdrawn, a departure from his previous even-tempered self. E. S.'s work was done in the basement of the plant where the solvents were recovered. He oversaw this operation, freed up paper caught in the machinery, and repaired machinery, work that brought him into close proximity to the solvent, by both inhalation and skin contact. At the time of my examination in 1985, he was obviously confused, and displayed clear signs of peripheral neuropathy.

Solvents are useful in industry because they dissolve other chemicals; that is, they dissolve greases, oils, and other non–water-soluble chemicals. This property allows them to combine with the oil and fatty portions of the body, resulting in damage to organs that are rich in lipids, such as the brain, peripheral nerves, bone marrow, liver, and fat stores.

CNS Case #4

B. M. was a 36-year old woman with a dependency on alcohol. Her physician prescribed the well-known "alcohol aversion" drug Anabuse®. After taking the drug for about six months, B. M. became totally paralysed. Additionally, her voice and speech was impaired and she was diagnosed as having severe encaphalopathy, or organic brain syndrome. She gradually regained the use of her arms and legs, although her legs have remained weak with a loss of sensation and a permanent unilateral foot-drop.

Anabuse, whose generic name is disulfiram, was originally employed in the rubber industry as an antioxidant. The idea of utilizing this reaction in a therapeutic fashion came from observations of rubber workers who found that even a small drink of alcohol after work would result in flushing, nausea, vomiting, and headache. Soon it was promoted as a "cure" for alcoholism. Early reports suggested that it was the combination of disulfiram and alcohol that was responsible for the profound effects and neurological damage, but other investigations showed that disulfiram alone could result in neurological damage.[3-6]

The medical literature frequently cites neuropathy as a complication of alcoholism, perhaps on the basis of poor nutrition and/or direct neurotoxicity, but there is no study in the literature that has surveyed an alcoholic population with neuropathy to determine the extent of disulfiram usage. Nonetheless, in every person with a history of alcohol abuse and neuropathy, a complete history of medications, specifically Antabuse, as well as occupation must be sought. Its purported benefits notwithstanding, it is questionable whether a drug with known neurotoxicity should be given to any patient, much less a person with an alcohol dependency, who may also develop neurological damage on the basis of the alcohol use and poor nutrition alone.

CNS Case #5

D. Q. was seen by his family physician in 1979, complaining of headaches over his right eye, associated with ghost images, occurring daily for the previous six months. He said the headaches were never present the first thing in the morning. He had been hospitalized in 1976 for investigation of the onset of his headaches, and underwent a number of tests including echoencephalogram, cerebral angiogram, CT scan, EKG, and chest X ray, all reported as normal. Subsequently, he was examined by a neurologist who noted that D. Q.'s headaches "have been absolutely daily events for the past two years, irrespective of season, climate conditions, day of week, time of day, etc." In May 1980, he was again hospitalized, and underwent a second arteriogram, which, not surprisingly, was still normal. While hospitalized, he became "virtually pain free." In November

1980, he was examined by another neurologist who noted D. Q.'s right-sided headaches as "being quite severe and disabling, with nausea and repeated vomiting, and lacrimation of the right eye as minor symptoms." This examiner noted no cardiac disease but observed that D. Q.'s hands showed "rather severe involvement of a reddened, scaly, cracked plaque." In October 1980, he underwent a third cerebral arteriogram (still normal!). Various diagnoses of "vascular" and "tension" headaches were made.

In the interim, D. Q. had been prescribed Lithium (a drug specific for manic–depressive illness); Valium (diazepam), a tranquilizer; Elavil (amitriptyline), a mood elevator; steroidal creams, shampoos, and ultraviolet radiation for his skin eruption; Inderal (propranolol), a beta-blocking agent given for a heart condition; and five pain preparations: Tylenol (acetamenaphen), codeine, Norgesic (orphenadrine), Percocet (a combination of the narcotic oxycodone with acetaminophen), and Indocin (indomethacin), one of the nonsteroidal anti-inflammatory agents. In about September 1981, after five years of headache complaints, he was prescribed ergotamine, an agent that constricts the blood vessels, and has been suggested for migraine relief. Three months later, Elavil was again prescribed. Two weeks later, D. Q. was admitted to a hospital with a heart attack, with second and third heart attacks occuring a month and one year later. D. Q. was 45 years old at the time of his first myocardial infarction.

Mr. Q. was employed by a company that made coffee filters. This process uses what is called "wet-strength" paper, which is achieved by impregnating the paper with a formaldehyde-containing resin. Rolls of the impregnated paper were passed through a machine where the paper was cut, heated, and pressed into the desired shape, thus liberating formaldehyde gas in the process. He had additional contact by handling the formaldehyde resin–impregnated paper.

The onset of his headaches occurred soon after the urea-formaldehyde and melamine-formaldehyde impregnated paper was introduced. Unfortunately, persons who investigated Mr. Q.'s persistent headaches obtained no work or environmental history, and it was not until he complained about the loss of his job, because of his continuing illness, that the subject was mentioned. In summary, Mr. Q. became a walking pharmacopoeia, and suffered an irreversible side effect from his medication, while the *cause* of his illness went unquestioned and unrecognized. Although the immediate cause of his inability to work was his heart ˌattack, his exposure to formaldehyde—which caused the headaches, and in turn resulted in the use of a number of medications, including a vasoconstrictive drug—remained the primary cause of his illnesses. If any case is representative of the need to obtain a work history, Mr. Q.'s case clearly is. Failure to determine the cause of his headaches led to his being treated with a number of drugs that, like all medications, carry side effects. Because he was not removed from the cause of his headaches, and the medications were continued, he suffered a significant side effect: a heart attack.

CNS Case #6

V. S. was an active 55-year-old man when he used a degreasing product to clean the engine room of the tug on which he was employed as a marine engineer. Over a period of about eight weeks, he noticed pain in his feet at night, numbness, tingling, and finally weakness. An examination showed wasting of the anterior muscles of his legs, and decreased sensation to vibration and pinprick. Additionally, his liver enzymes and blood urea nitrogen (BUN, a measure of kidney function) were elevated, as were his blood lipids (including triglycerides and cholesterol). He sought medical help, and a number of causes were considered, including diabetes and exposure to a toxic substance.

In an investigation of V. S.'s apparent exposure to a toxic agent, the makeup of the degreasing product was sought. Importantly, the label of the product stated only: "contains petroleum distillates." The Material Safety Data Sheet (MSDS) was equally unrevealing, indicating that the formula was "proprietary" and under the Hazardous Ingredients section, that it contained a "high boiling aliphatic."

It appeared highly likely that the permanent peripheral neuropathy that Mr. S. developed resulted from exposure to a solvent whose most likely component, consistent with the rapid onset of his problems, was n-hexane. After litigation was instituted, the true identity of the product was learned, and it was n-hexane, well established as a neurotoxin in the scientific literature.[7,8] The permanent damage done to this man, combined with the failure to identify the product and failure to warn him of the danger of exposure, resulted in a $250,000 settlement. The product is a commonly used solvent, not only in the shipping industry but in other industries as well.

This case report demonstrates another phenomenon, the elevation of serum lipids in persons exposed to a number of chemicals, including solvents, PCBs, PBBs, epichlorhydrin, and a number of plastic monomers. The condition is poorly studied and not completely understood, although atherosclerosis has been produced in experimental animals via exposure to kerosene.[9] The elevation of lipids may be on the basis of solubility of the foreign chemicals in the fat stores of the body, with an attempt made by the body to transport and dilute those chemicals by producing more lipids. Obviously there is a need to rethink our concepts about lipids, which emphasize diet and hereditary factors while ignoring the role played by exposure to bioactive chemicals. This field of inquiry is ripe for study, and it is hoped that such investigations will be undertaken. As a start, all persons with elevations of blood lipids should be questioned about chemical exposure.

CNS Case #7

K. McC. was 42 years old in 1981 when he had the sensation that "everything went blank." The next day, he said, he developed numbness and tingling in his hands, arms, and neck, followed by difficulty with walking and with his balance. He was hospitalized, and underwent a number of neurological tests, including an EEG, a CT scan, and electromyelography; and he saw a number of physicians as well as a psychologist. He was found to have both neurological and cognitive dysfunction, the latter expressed as reduced mental capacity, and problems with memory, attention, and concentration.

Unfortunately, throughout the various examinations, factors referable to his work or environment were not addressed, or were addressed only in a cursory manner.

Mr. McC. operated a degreasing tank up to $1\frac{1}{2}$ hours/day, for a period of about seven years. The tank measured approximately 4 feet by 2 feet and contained approximately 20 gallons of 1,1,1-trichloroethane (TCE), which was heated. Mr. McC. dipped metal parts into the tank, used a spray of solvent to clean the parts, and worked with an open 5-gallon pail of the solvent next to his workbench, where he then welded the parts. He said the ventilation was poor, especially in the winter.

At the time when Mr. McC. was examined in 1985, he continued to display severe peripheral neuropathy as well as loss of mentation and loss of balance. In addition, he had elevated blood triglycerides, and obstructive lung disease. At age 46, with a wife and three children, ages 14, 17, and 21, Mr. McC. is permanently and totally disabled. Neither he nor his family can expect to enjoy the fruits of his 22 years of work.

Bibliographic references for 1,1,1-trichloroethane are included in the reference section at the end of the book.

CNS CASE #8

C. A. was a healthy woman when she began to work for a thermometer company in 1980. She was 29 years old. Approximately 12 months later, she developed headaches, muscle tremors, irritability, nausea, and tiredness, and would fall into a "dopey sleep" as soon as she got home from work. By March 1984 she had developed nosebleeds and frequent sore throat.

In the course of making thermometers, she used a vacuum pump to draw mercury from an open dish into glass tubes, which she then heated in order to seal the ends of the thermometers. She said there was spilled mercury on the floor, tabletop, and work surfaces from thermometers that broke. She reported that she emptied incorrectly calibrated thermometers by heating them over a burner and collecting the mercury as it ran out into an open pan.

In addition to her exposure to mercury, she also worked with hydrofluoric and nitric acids, acetone, alcohol, toluene, "spirits," propane gas, and wax that was heated.

It was before the end of the nineteenth century that Lewis Carroll immortalized the Mad Hatter in *Alice in Wonderland*. It was customary to use mercury in preparing felt hats, and the "mad" symptoms that the Hatter displayed were classic for mercury poisoning. Nearly a century later, this young woman was made sick by exposure to the same substance. There is little reason for complacency; after Ms. A. and a coworker developed signs of mercury poisoning, the company closed its plant in Michigan, and moved to North Carolina. No details as to job conditions for workers at the new location have become available.

CNS Case #9

L. G.'s death certificate indicated his immediate cause of death as a cerebrovascular accident, "due to or as a consequence of hypertension." Under "other significant conditions" it noted "diabetes mellitus." When L. G. died on May 2, 1979, he was 50 years old. He had been assigned to spray parathion on crops all day on April 4, 5, and 6. The last day, he visited a physician because of weakness, dizziness, headache, nausea, difficulty in breathing, and a "strange cough," and the physician treated him for an infection. On April 9, he was found on the ground behind a pesticide sprayer, and was taken to a hospital. Hospital personnel noted the odor of parathion on him, and he was treated with atropine to counteract the effects of the cholinesterase inhibition due to parathion. His symptoms were primarily of labored breathing and pulmonary edema. While hospitalized, L. G. developed a right hemiparesis. On April 12, the nursing notes indicate that the odor of parathion was still present. He continued to have congestion of his lungs, and dark urine with proteinuria. A CT scan of his head on April 10 showed "a large area of decreased density in the left temporal and parietal areas," compatible with his right-sided stroke. A search of the medical records did not reveal high blood pressure readings except for the initial event while he was in the emergency room. He was discharged from the hospital on April 28. Four days later, on May 2, after he had returned home from his first physical therapy session, L. G. collapsed and died.

Mr. G. spoke little English, and it is questionable if warnings concerning the use of parathion were given in his native Spanish, or given at all.

Much rides on the information given on death certificates; for instance, insurance benefits, funding for medical care and research, governmental programs, and efforts toward prevention (or lack of any of the above) are frequently pegged to death certificate data. Is it accurate to say that Mr. G. simply had a stroke as a result of high blood pressure (which was undocumented except for his initial findings in the emergency room)? Stroke and heart attack are currently considered leading causes of death, but did this man die of a stroke, or from organophosphate pesticide poisoning, as a reading of the hospital record indicates? If the precipitating cause of Mr. G.'s death was clearly parathion poisoning, and not recorded on this death certificate, then one may question how many other cases of pesticide and other toxically related deaths go unreported.

CNS Case #10

During C. K.'s examination, his wife said that he had been an "A-1 mechanic before he got sick." Two years previously, Mrs. K. said that her husband had begun acting "peculiar" and "talking irrationally," and that he had become forgetful, and had lost about 30 pounds of weight. In 1983, C. K. was hospitalized for 12 days, and treated with a major tranquilizer, haloperidol. During this time, his wife said that he developed "chewing motions of his mouth," and she commented that he "pads his feet on the floor" all of the time. Significantly, a review-of-systems inquiry revealed that he had been found to be sterile in 1961.

C. K. had begun to work as a metal finisher in 1952, and for 30 years he had used hand tools to file and smooth the leaded seams on auto bodies. He said that molten lead was applied by a coworker who was stationed next to him. He said that he never wore a mask, and that his work clothes, including his underwear, were dusty and stained from the lead that got on them. Mrs. K. said she shook the dust and metal particles out of his work clothes before she put them in the family laundry. He said that the last five years that he worked, he was supplied with coveralls.

Data from worker lead testing were obtained for the time span from June 1977 until June 1983, when testing was conducted every two to three months. C. K.'s values ranged from 23 to 55 micrograms, with 46% of the tests showing over 40 micrograms, and two of the tests 53 and 55 micrograms. The "normal" lead concentration for adults is considered to be from 0 to 40 micrograms, with an OSHA limit of less than 50 micrograms.

Mr. C. K.'s original history is compatible with organic brain damage from chronic exposure to lead (see Chapter 1 references). His neurological symptoms of tardive dyskinesia are secondary to his having been treated with haloperidol. In this case, we see that a single neurological condition, when treated with a major tranquilizer, can lead to even more disability. Perhaps, because of his exposure to lead his brain had an increased sensitivity to the known effect of this major tranquilizer, and the combination resulted in further damage. Unfortunately, both agents apparently have produced permanent, irreversible damage to Mr. K.'s brain.

It is essential to determine whether there are any prior exposures to neurotoxic agents in persons who develop tardive dyskinesia, especially if the abnormality develops early in a treatment regime.

Neurotoxic Chemicals

Below is a list of chemicals associated with nervous system disorders. Some of the chemicals cause temporary problems, while others cause permanent damage to the brain, the spinal cord, and/or the peripheral nerves. In general, those chemicals that are fat-soluble, and thus can be retained in the lipid-containing nerve cells, are more likely to cause long-term, permanent damage than those that are water-soluble. Some chemicals cause damage by accumulation: each individual dosage may be below a threshold value, but because of fat-solubility and slow excretion, the chemical gradually accumulates in the body until a critical point is reached when the person becomes aware of his or her loss of function. Many of these chemicals cause toxic effects not only to the nervous system but to other organs as well, including the liver, kidneys, heart, and lungs.

NEUROTOXIC CHEMICALS

acetyl ethyl tetramethyl tetralin (AETT)

acrylamide

benzyl sulfonyl fluoride

2-*t*-butylazo-1-hydroxy-5-methylenehexane   (BHMH)   (azo-substituted aliphatic, used in polyester plastics)

carbon disulfide

carbon monoxide

carbon tetrachloride

chlordecone

cyanide

dichloro-diphenyl ethane (DDT)

2,4-dichloro phenoxy acetic acid (2,4-D)

dimethyl amino propionitrile (DMAPN)

dipentyl-dichloro-vinyl phosph*ate*

disulfiram (Antabuse)

n-hexane

(Kepone)

lead

gamma benzene hexachloride (Lindane)

mercury

methyl bromide

methyl mercury

methyl methacrylate

methyl *n*-butyl ketone

nitroimidazole

organophosphates

  parathion

  leptophos

  tri-ortho cresyl phosphate (TOCP)

pentachlorophenol

pentachlorophenolate

phenyl benzyl carbamate

phenyl-phenyl acetate (PPA)

polychlorinated biphenols (PCBs)

styrene

thallium

1,1,1-trichloroethane

trichlorethylene

CNS Discussion

Of all of our faculties, our intellectual processes are perhaps the most important, being what sustain us if we loose locomotion through arthritis, suffer kidney or intestinal disease, or become deaf or blind. The loss of neuromuscular control and loss of mentation brought about by exposure to toxic chemicals represents as serious a situation as cancer or heart disease. All patients with complaints referable to the nervous system should be questioned as to occupational and environmental factors.[10] Little investigation has been done, and until recently little information had been required by regulatory agencies in regard to neurotoxins used in the workplace and the home, and/or released into the general enviroment.

The specter of senility and Alzheimer's disease haunts anyone who has known one of these victims. Frequently, loss of mentation secondary to toxic chemical exposure is labeled as Alzheimer's. The term is merely a descriptive label, indicating nothing as to etiology. A systematic investigation of chemical and drug exposure in each patient suffering with Alzheimer's dementia may prove valuable. One neurological disorder, previously thought to be hereditary, has been found to be caused by a plant extract.[11] How many other etiological agents there are remains to be investigated.

Multiple studies have confirmed uptake by the brain of a variety of chemicals taken into the body via inhalation, ingestion, skin exposure, and penetration via the naso-olfactory pathway. The latter is particularly interesting in that lesions of the brain have been produced by this method in animals. The olfactory tract (referred to as the first cranial "nerve") pierces the cribiform plate at the base of the skull and extends along the base of the frontal lobe. The posterior connections are in close proximity to the brain, including hypothalamus and pituitary. Lesions in the brain, produced via the nasal route, have been found after exposure to aluminum[12, 13] and formaldehyde.[14–16] Impairment of mental function due to formaldehyde exposure has been found in humans.[17]

With the promotion of pesticide use in and around homes and on foodstuffs, the burgeoning plastics industry, the pollution of air and water supplies with solvents, lead, and mercury, and the storage of foods in containers made of leachable plastics, we may be heading toward an epidemic of neurological cripples. There would not be simply problems of incoordination, loss of balance, and memory, but more subtle changes such as persistent headache, inattention, malaise, and, most ominous of all, loss of mentation with failure of creativity, curiosity, and productivity. It is serious when these symptoms and signs appear

in older age groups, but no less than a tragedy when they occur in the young
or the newborn.

EARS, NOSE, AND THROAT

ENT Case #1

M. C. worked as an electro-polisher in the electronic components industry from 1976
until 1981. He used sodium dichromate, chromic acid, and sulfuric acid to do his job.
In August 1978, he noticed a crack in his lower lip, which progressed to an ulcer. In
1982, at which time he was 32 years old, a biopsy showed a carcinoma. Despite three
radical surgeries, he has developed metastatic disease.

The link between exposure to chromium compounds and cancer has been
known for many years. Hence, it is imperative that employers protect workers
from coming into contact with chromium-containing compounds and monitor
them in an *expectant* fashion. Certainly, an employee who develops a sore
anywhere on his or her skin should receive immediate medical attention.

Chromium compounds are encountered in the manufacture of steel alloys,
the plating industry, and the tanning of leather. Less well recognized but equally
important sources of exposure are the chromate pigments used to color paints,
linoleum, rubber, ceramics, and textiles, as well as in the bricks, cements, and
plasters found in high-temperature furnaces. The men who use air hammers to
tear out the insides of refractory furnaces are engaged in an occupation fraught
with carcinogen contact, being exposed not only to chromium compounds but
also to asbestos and soot.

Bibilographic sources concerning chromium compounds are listed in the
references at the end of this book.

ENT Case #2

H. P. and her husband moved into a new home in October 1977. Almost immediately,
she began to have symptoms of "sinus trouble," attributed to "allergies" and "sinusitis"
by her internist and an ENT specialist. She was prescribed antibiotics and antihistamines, to little avail, and finally required nasal suction to remove mucus. After H. P.
developed nosebleeds and coughing, she was evaluated by a second ENT specialist,
who ordered special X-ray examinations. Quite by accident, as the first ENT specialist
was aspirating her nose, he removed a piece of tissue. The pathology report was read
as hyperplasia. The tissue was submitted to another facility for confirmation, and an
undifferentiated squamous cell carcinoma was diagnosed. Still a third opinion was
"inverted papilloma and hyperplasia and atypia." In March 1978, H. P. underwent five
hours of surgery to remove the cancer that had spread inside her head. She returned to
her home, but continued to have bleeding and did not heal well. She lost her senses of
smell and taste, had difficulty talking and hearing, had little saliva so could not swallow

with ease, and lost weight, down to 82 pounds. She required radiation therapy to control the remaining tumor, and this resulted in damage to her pituitary, thus interfering with her reproductive system. She had noted a change from her ordinarily regular menstrual cycle after she moved into the house, and before her nasal cancer was diagnosed.

Ms. P.'s husband did some library research concerning nasal cancer (an uncommon malignancy, especially in a young woman), and came upon data describing nasal cancer that had been produced in rats exposed to formaldehyde. The P. home had been insulated with UFFI (urea formaldehyde foam insulation). Ms. P. had been present when the UFFI was installed, and was at the home site on a daily basis. After she and her husband moved into the house in October, and because she did not feel well, she stayed inside much of the time.

In June 1982, five years after installation of the UFFI, the home was tested by an engineering consultant who found levels of formaldehyde varying from 0.063 to 0.078 mg/m^3 in the basement and attic, respectively. Bulk samples of the insulation materials averaged from 83.6 μg/g in the fiberglass in the basement to 7485 μg/g formaldehyde in the UFFI sample in the attic. The engineering consultant pointed out that the levels found in the house were below the NIOSH recommended TLV standard of 3 mg/m^3, as well as below the then-proposed standard of 0.6 mg/m^3.

The NIOSH TLV standards are based on a 40-hour workplace exposure, but a home exposure can extend up to 168 hours a week. Taking this into consideration, the recalculated equivalent work exposure would have been between 0.75 and 0.14 mg/m^3. Many persons can detect the odor of formaldehyde at 0.1 mg/m^3; additionally, there is a marked difference in sensitivity between different persons.

In addition to UFFI as a source of exposure, formaldehyde can be present in a conventional home or mobile home as a result of offgassing from glues utilized in the manufacture of particle board and plywood used in the construction of the home or furnishings. Mobile home exposure is particularly significant, as many of these structures are constructed virtually entirely with particle board and plywood materials, and they are usually well insulated, thus not permitting ventilation of the formaldehyde gas. Concerning building exposure, the offgassing of the formaldehyde can be expected to be greater in warm weather, when a house is closed up, or when a heating unit is in operation. Moreover, the levels of formaldehyde will decrease with time, although any formaldehyde in the remaining insulation or building material will continue to offgas as long as it is in place. The final concentration in a structure is dependent upon concentration in the building materials, temperature, humidity, ventilation, and the passage of time.

ENT Case #3

E. H. was 28 years old when he and his wife moved into their home in 1976. He owned his own excavating business and was in vigorous health. Both Mr. and Mrs. H. developed runny noses, itchy eyes, headaches, and bronchitis. In 1982, E. H. developed soreness and bleeding of his nose. He was referred to a major clinic where a nasopharyngeal squamous cell cancer was found. Despite surgery and radiation, E. H. died in November 1984, when he was 38 years old.

Mr. H. developed cancer at the same site as test animals that had been exposed to formaldehyde gas. His case was the second of three human nasal cancer–UFFI cases that I have evaluated. A fourth patient, a 49-year-old teacher, developed vocal cord cancer as a result of exposure to formaldehyde used to preserve biological specimens for his classes.

Mr. and Mrs. H. had their home insulated with UFFI (urea formaldehyde foam insulation), which was installed by blowing a mixture, similar in consistency to shaving cream, into the spaces of their house, between the inner and outer walls. They noticed a strong odor, but were unaware of the toxicity of the formaldehyde that was being slowly released from the product as it was in place.

Their house was tested in March 1985, nine years after the UFFI had been installed, and the formaldehyde levels ranged from 1.98 ppm near the downstairs mantle to 2.30 ppm in the upstairs bedroom, to 2.47 ppm in the living room.

Despite years of knowledge about the irritative and mutational effects of exposure to formaldehyde in animals and humans, this product was promoted as a "solution" to the energy crisis, and installed in thousands of homes in the United States and Canada.

When asked about a connection between E. H.'s exposure to formaldehyde and nasal cancer (in 1985), his treating physician wrote: "I do not know whether the possible exposure to the formaldehyde would be related to his carcinoma. I did tell his wife that it would have been an unusually short time from exposure in 1977 to a grossly evident tumor in 1984. The usual delay times required for carcinogenesis in man are measured in decades with the exception of the more rapidly-induced leukemias."

This misconception concerning latency, even by a member of the medical profession, must be addressed. The oft-cited cancers, involving exposure to asbestos, for instance, have taken up to four decades to develop; however, others have developed cancer as a result of exposure to a variety of chemicals in much shorter time spans. We need only consider the high rate of cancer among children, whose latency period is obviously no longer than their lifetimes. Chemicals having greater degrees of biologic activity have shorter latency times. Notably among these is formaldehyde.

ENT Case #4

C. L. was 53 years old in 1981 when he suddenly became aware of shortness of breath. After an X-ray was taken, he said he was told he had interstitial fibrosis. About eight months later, he said he awoke with a severe burning sore throat and could feel a lump in the left side of his neck. He continued to see physicians at a clinic over the next year, complaining of a sensation like a "cotton ball" stuck in his throat. A biopsy, done 14 months after that complaint, revealed poorly differentiated epidermoid cancer of the tonsil with metastases. C. L. underwent surgery and radiation and, understandably enough, was also hospitalized for treatment of depression. He had smoked about one pack of cigarettes a day for approximately 20 years, stopping in 1971, ten years before his diagnosis of cancer.

C. L. related to me that he had been employed by a large automobile manufacturing company from 1957 until 1983, beginning work there when he was 26 years old. He handled stock and supplies on the truck and railroad docks, driving both gasoline-powered and battery-powered lift and tow trucks. He stated that he worked around fiberglass, foam rubber, roofing materials, and chemicals. He said there were strong fumes from diesel-powered trucks at the loading dock, with two or three of the trucks running their engines as the men handled the stock.

The only work history in his medical records was obtained in February 1981 when his pulmonary fibrosis was diagnosed. It said: "His work history is that he worked for Company X as a hi-low operator, where his only industrial exposure to fumes consists of the diesel fumes from the hi-low and from the trucks that unload at his dock. There has been no recent change in what he does at his job. He has never worked in a mine or foundary, but about a year ago, he had a six month exposure to asbestos on the production line. Twenty years ago he also spent one year as a spray painter for Company Y."

A biopsy at that time confirmed the diagnosis of fibrosis that had been seen on his chest X ray, but there was no mention of any attempt to identify fibrogenic material in the lung specimen. Despite this diagnosis, at age 53, Mr. L. was not cautioned against returning to work where he would be further exposed to respiratory irritants and pollutants.

With this dual information at hand, about his fibrotic lung disease and his exposure to asbestos and diesel emissions, the appearance of additional symptoms (in his throat) should have signaled a red alert to his physicians, compelling them to search for a malignancy. The delay in the diagnosis of his cancer allowed it to spread, requiring more extensive treatment and resulting in additional disability.

ENT Case #5

C. B. worked three years as a "racker" on a line where auto parts were chrome-plated. She attached parts onto a rack that were then dipped into a solution containing chromium salts. She described her work area as filled with fumes. She developed a sore and bleeding nose, and by the time I examined her, when she was 44 years old, she had developed a perforation in her nasal septum. Over the subsequent 6 years she demonstrated decreased pulmonary function.

ENT Case #6

J. F. worked at the same chrome-plating plant as C. B., for a period of four years. She developed "colds," frequent sore throat, and "chest flu" several times a year. She too had developed perforation of her nasal septum and a marked decrease in her pulmonary function when I examined her a second time six years later.

These two women (cases #5 and #6) demonstrate an uncommon finding: perforation of the nasal septum. It has been described in persons exposed to chrome ore and salts for many years. Accompanying the nose deformity—which itself is not a minor affliction, considering the accompanying irritation, change in voice, and increase in infections—is pulmonary disease, which I found in nearly all the plating workers that I examined.

CHEST

Chest Case #1

N. S. worked as a meat wrapper for a supermarket company from 1958 until 1979. She used a clear plastic wrap, cut the plastic with a hot wire and then sealed the package on a hot pad. She noticed the onset of shortness of breath in about 1971, and it became progressively worse with time. In addition, N. S. experienced chronic tiredness and weight loss, slept after work as though she had been "drugged," and developed a rash on the palms of her hands. When I examined her in 1980, she had chronic obstructive lung disease and associated pulmonary hypertension.

The development of lung disease following exposure to the pyrolysis products of vinyl chloride plastic film was first labeled "meat wrappers' asthma," although the handling of meat was not what led to the development of this disease. The critical factor is the vinyl chloride film and the breakdown products give off when it is heated. One can develop the disease whether meat, T-shirts, or dishes are being wrapped in the film, whenever the film is heated to be either cut or sealed. The same breakdown products are give off when vinyl chloride plastic film and articles are burned, either in a building fire or in an incinerator. Therefore, the exposure problem is one of degree and intensity.

When vinyl plastic is heated above 135°C, irritating and sensitizing chemicals are emitted. When the temperature increases above 150°C, both particulate and nonparticulate gases are given off, including hydrochloric acid, chlorobutane, benzene, toluene, benzyl chloride, phthalic anhydride, dioctyl adipate, and other compounds generated from the plastic components. These products are taken into the body by the respiratory system, causing damage not only locally in the lungs but elsewhere in the body, via immunological reactions. The toxicity of vinyl chloride encompasses respiratory, hepatotoxic, immunotoxic, and carcinogenic effects.

CHEST CASE #2

R. H. was 46 years old in September 1982 when he saw his physician because of anterior chest pain, night sweats, and fatigue. A lymph node biopsy revealed poorly differentiated bronchogenic carcinoma with metastases. Despite aggressive therapy, he died in December 1983. R. H.'s medical records are representative of those of many patients. His histories, obtained by his various physicians, related that he had smoked 1 to 2 and up to $3\frac{1}{2}$ packages of cigarettes a day for a number of years, and had quit smoking three years prior to his diagnosis of lung cancer. No occupational history was obtained by any of his physicians.

When R. H. began his job in May 1977, he underwent a physical examination, including a chest X ray. The films were interpreted as showing linear strands of increasing density of both lungs, compatible with old inflammatory disease. Over the next year, he was treated in the company medical department because of eye inflammation secondary to exposure to fiberglass and plastics. Two years later, he was treated for redness and puffiness of his face after mixing epoxy plastics.

Skin reactions, including skin cancer, are common in workers who subsequently develop internal cancers. It follows logically that chemicals that irritate and interact with cells covering the body will also irritate and react with those on the inside.

In November 1981, when R. H. transferred to a new company, another chest X ray was made and read as "essentially normal." He continued to do similar work, mixing and grinding polyester resins and working with dibutyl phthalate, peroxides, glues, solvents, fillers, urethanes, fiberglass, impregnated wood, and amine hardeners, as he fabricated full-scale auto models.

Review of the chemicals listed on the company's MSDSs revealed that a number were considered positive on cancer tests. Review of the scientific literature confirmed that there were many reports of cancer produced in animal systems by a number of the chemicals.

Mr. H. worked in an area approximately 100 by 250 by 18 feet with about 60 other men. Although there was a filter system on a back wall, the men wore only paper masks when they sanded the plastic materials. He indicated that it took five men approximately 28 days to fabricate a full-scale model of an automobile.

If an occupational history had not been obtained, this man's death might have been chalked up as just another smoking-related statistic. Although his smoking history cannot be dismissed, it is important that he stopped smoking three years prior to his diagnosis of cancer. It is also important to note that cigarette smoke paralyzes the respiratory cilia, thus making inhaled materials difficult to remove from the lungs and trapping them there. There is little question that there is a synergistic effect between the chemicals in cigarette smoke and the chemicals inhaled in the course of work. Most of the chemicals with which Mr. H. worked were very reactive, a property that accounts for their usefulness in industry! Considering his exposure, it must be concluded that these chemicals were a significant factor in the development of his lung cancer and his death at age 47.

One must ask, how could this death have been prevented, and what of the health of other similary exposed workers? Because the reactivity of these chemicals is known, it is essential to prevent exposure to them. This requires the use of respirators to filter the air, not masks, which only strain out the larger particles and do nothing to stop fumes and gases. Respirators and masks are uncomfortable, and wearing one makes it difficult to do one's job. Also, although one worker may not be mixing plastics or grinding chemically impregnated surfaces, his coworker may be doing so, making it necessary for all workers in an area to be protected. A more acceptable alternative is to employ engineering methods to enclose contaminated work areas and supply a strong laminar air flow to carry the fumes and particulates away from the workers. With this method, one must also understand where the exhaust materials are going; they should not be allowed to either contaminate the neighborhood environment or be sucked back into the plant at another entry site.

We must not become complacent and think that only model fabricators are exposed to these chemicals. With the recent decline in the use of metals in autos and other products, there has been a concomitant increase in the use of plastics and the very chemical systems that contributed to Mr. H.'s death. The average U.S. auto contains about 300 pounds of plastic,[18] including not only trim, but structural components as well. There is little monitoring being done of workers who handle plastics, and virtually no monitoring of the populations living in the vicinity of ventilation from these plants. With out-sourcing and the production of plastics that is being done in small plants, it is more than likely that there will be an increase in illness and death from exposure to these products. It is imperative that all workers handling plastics systems be monitored on a regular basis, and that their exposure be minimized.

Data for various plastics components can be found in references cited in the bibliographic section at the end of the book.

CHEST CASE #3

D. G.'s death certificate states as the immediate cause of his death "cardiopulmonary arrest, etiology unknown." Farther down on the certificate, under "other significant conditions," it states: "bronchial alveolar cancer of the lung." D. G. was 30 years old when he died. He did work similar to that of R. H. (chest case #2), and was exposed for four years to the same type of chemicals, including epoxies, solvents, resins, paints, lacquers, urethanes, polyesters, styrene, and so forth. D. G. had previously smoked two packages of cigarettes a day for five years, and then one pack a day for about five years prior to stopping smoking two months before his diagnosis of lung cancer.

While not ignoring a smoking history, one must go further in investigating an illness or death. Intellectual curiosity alone should drive a person to wonder why a 30-year-old man developed cancer and died. It is highly unlikely that smoking was the sole cause. Mr. G.'s treating physician did document that he handled "fiberglass, woods, plastics and paint," and that he wore a "particle mask only when he handled the above." While that is a start in obtaining information about exposure to chemicals, it is necessary to proceed further and determine the identity of the actual products. It becomes apparent that there is a pattern, a tragic pattern of sickness, in many workers who handle plastics on a regular basis.

CHEST CASE #4

S. R. was 39 years old when he developed a severe cough and noticed shortness of breath and chest pain, especially when he took a deep breath.

He related that he was 11 years old and in the sixth grade, in 1955, when he left school and took his first job as a dishwasher. After that, he drove a truck, worked in a furniture factory, and performed street repairs until 1976, when he began to work as a laborer, tearing out the inside of blast furnaces. He said the furnaces were about 100 feet tall and 52 feet across, and that he and the other men used jackhammers to break out the old firebrick, carbon blocks, and insulation materials.

S. R. said he usually wore a mask, either paper or rubber, but that even the best of the rubber masks did not keep the dust out. He said that on one job the men could work for only about 10 minutes at a time, and then had to wait for about 20 minutes for the dust to settle so that they could see to work. He said 10 to 15 men operated jackhammers while standing on a scaffold that was slowly lowered down the inside of the structure, as the dusty insulation material fell to the bottom. He said he handled fiberglass, firebrick, mortar, and asbestos, the latter as recently as 1983.

S. R. told me that he had smoked a pack of cigarettes a day for approximately 25 years. His medical tests revealed small cell undifferentiated cancer of the lung with metastases. He underwent radiation and chemotherapy to no avail, and died when he was 41 years old, on November 16, 1985, leaving a wife and five children, three younger than 18 years old.

Considering the job conditions of workers who build and maintain heavy structures for the foundry, steel, and coking industries, as well as the materials that go into them, it is little wonder that workers in these industries get sick, develop cancer, and die at young ages.

Chest Case #5

H. M. was 25 years old in 1964 when he went to work for a brake shoe manufacturer. He ground and riveted asbestos-containing brake shoes. H. M. smoked cigarettes from about age 16 until he learned that he had an abnormal chest X ray in September 1983. He was diagnosed as having poorly differentiated adenocarcinoma of both lungs. Despite chemotherapy and radiation therapy, he developed metastases to his brain, and died of his widespread cancer when he was 44 years old, leaving a 15-year-old daughter.

The company that employed Mr. M. required an annual chest X ray for employees who worked with asbestos, beginning in 1976. Despite this rule and despite the fact that he worked in an asbestos-contaminated area, his most recent chest X ray, prior to his diagnosis of cancer, was done by the company physician in 1979. It was reported as normal.

X rays are often promoted as a part of "cancer prevention" programs. Unfortunately, however, X rays, like many other diagnostic tests, provide only early detection, and not prevention. Indeed, upwards of 90% of persons diagnosed by X ray as having lung cancer are dead within a year, despite radiation, surgery, and chemotherapeutic efforts. In the case of Mr. M. and others like him, only control of respirable asbestos or its substitution with a nonhazardous material could have prevented his sickness and death.

Chest Case #6

A. C. began to operate industrial sewing machines in 1943, and, with the exception of a few years of layoff, performed that job until 1984. She stitched seats for autos, using vinyl plastic, velvet, various fabrics, and foam rubber. Ms. C. never smoked, and had been in good health until 1973. That year, she related that she suddenly became short of breath while at work, and that her breathing difficulties became worse when she handled foam rubber and plastics. In 1978, the plant physician restricted her from working in the TDI (toluene diisocyanate) area, but she said that although her work station was moved, the chemical odors permeated the plant.

When examined two years after she left work, she had chronic severe obstructive lung disease, documented by pulmonary function testing.

TDI is one of the components employed in the manufacture of urethane products, including foam rubber. It is extremely sensitizing, and once a person has become sensitized to TDI, even minute amounts can cause severe pulmonary symptoms. Occasionally, a person sensitized to TDI in the workplace will then react to the small amounts of TDI given off from such home products as foam-padded furniture and carpet underlay. For this reason, it is advised that persons sensitive to TDI avoid using foam rubber articles in their homes. Conversely, persons who develop asthma, or similar breathing problems, without demonstrable cause are urged to search their homes and workplaces for possible contamination from offgassing of TDI from urethane-containing materials.

Chest Case #7

W. E. worked as a beautician, approximately ten hours a day, $5\frac{1}{2}$ days a week, from 1954 until 1969. In styling clients' hair, he handled hair dyes, styling solutions, and aerosol sprays. The aerosols were largely powered with vinyl chloride, which was a common and cheap propellant employed during that time period. He smoked about one pack of cigarettes a day for 30 years. In 1971, when he was 43 years old, W. E. was found to have poorly differentiated adenocarcinoma of the lung, and his right lung was removed. Following the surgery, he developed overload on the right side of his heart as a result of his pulmonary insufficiency. He attempted to work as a real estate salesperson, but had problems walking upstairs, and related that because he felt so poorly he had difficulty being "cheerful" with his clients.

Vinyl chloride is a small, reactive molecule that can interact with living cells to alter the genetic material. It is similar in structure to other small molecules that are also associated with causing cancer. (See Figure 2-5 for chemical structures.)

Vinyl chloride originally was shown to cause a rare form of liver cancer, called hemangiosarcoma. Since then, vinyl chloride exposure has been implicated in cancers of the brain and lung. Exposure limits have been severely limited, but the wide use of this material still affords the opportunity for exposure.

Chest Case #8

S. S. was a healthy 33-year-old man who enjoyed swimming and playing tennis two or three times a week. He was a nonsmoker, and worked as a clerk in a federal agency. A pesticide containing a synthetic pyrethrin mixed with aromatic petroleum hydrocarbons and petroleum distillate (not further defined) was sprayed in the building where he worked prior to the weekend, and the ventilation system was turned off. He said that when he came to work on Monday morning he noticed a bad odor, and developed dizziness and a burning sensation in his chest. He continued to work and felt "listless." The next day, he felt worse so went to the first aid department and then to an emergency room. After x rays and blood tests, he said, he was told that he had a "respiratory infection" and was prescribed antibiotics and an anti-inflammatory drug (for the pain in his chest). He continued to feel sick, and after an additional visit to his own physician, was hospitalized for 11 days.

Nine months later, the building was again sprayed, and S. S. stated that within three hours he developed "cold symptoms." He stayed at home for two days. When S. S. returned to work, he developed nausea and chills, and passed out. He was taken by ambulance to a hospital, where his physician found wheezing in his chest, and spirometry confirmed obstruction. He was out of work for about six weeks; and when he again tried to work, he developed fever, chills, and shortness of breath while in the building. His physician restricted him from working in the building that had been sprayed with the pesticide.

The semisynthetic pyrethroid pesticide to which Mr. S. was exposed (along with the petroleum products) is a sensitizing and allergenic chemical. Once a person becomes sensitized to it, symptoms will result after smaller and smaller doses. The person may also develop symptoms following exposure to related compounds. The latter include such plants as chrysanthemums, which are employed by gardeners as "natural pesticides" to keep insects away from neighboring plants. The petroleum portion of the pesticidal mixture has its own ill effects, in addition to carrying the pyrethroid deep into the lungs and increasing penetration.

The case demonstrates additional factors pertinent to the review-of-systems approach, and emphasizes the need for education of emergency room personnel to recognize chemically related diseases. It is noteworthy that:

1. Mr. S. was well and active before he had an acute onset of illness.
2. He was diagnosed as having a "common" condition (respiratory infection), without an adequate work history having been taken.
3. He was treated for a respiratory condition, without a determination of the cause of his illness.
4. When he was reexposed, he became sick almost immediately.
5. Like many persons with induced chemical sensitivities, he became symptomatic after exposure to levels of agents that previously would not have bothered him.

This case also illustrates additional points:

1. The inappropriateness of the use of many pesticidal products within closed buildings. This occurs not only in offices and factories, but in hotels and apartment complexes.
2. The spread of these chemicals via the ventilation system.
3. The need to warn personnel in workplaces, and residents of apartments and hotels, of the intended use of these products and to allow them to vacate, be assigned to a work area free of the chemicals, and/or have the right of refusal to use the products.

This case represents an increasing problem, one that could be characterized as "chemical assault," involving persons who are exposed to pesticides and other agents without their knowledge or consent. Many of the effects of exposure are not minor, and, as illustrated in the case of Mr. S., are not temporary. These assaults on the body can result in life-long adverse health effects. The use of many pesticidal agents is promoted on the basis of cost; but the cost of medical care and time lost from work, as well as the intangible cost of poor health, must also be weighed before the use of these products is considered.

CARDIAC

CARDIAC CASE #1

L. B., a 51-year-old cement finisher, was hired to apply the finish coat to the concrete surface of a municipal parking structure. For approximately two months, he applied a urethane coating by both spraying and mopping the material onto the concrete. He developed a rash on his legs and quit the job because of concern over his health. Two days later, L. B., with no known history of heart disease, was found dead by his wife. The coroner's report indicated his cause of death as cardiopulmonary arrest, with probable acute myocardial infarction.

A coworker indicated that he himself had developed a rash while working with the substance and so quit the job, and stated that at least one other employee had been hospitalized because of illness due to exposure to the coating substance.

The MSDSs for the two part mixture (which were not at the job site) listed only "polyols" (with no further identification or percentage), "solvents" 15% (the other 85% not identified) with "un-reacted methylene bis-4 cyclohexyl-isocyancate (1%), as well as a mixture of xylene 20%, 2-ethoxyethyl acetate 20%, and un-reacted toluene diisocyanate monomer (1%)." The remaining 60% of the second part of the product was not identified.)

After Mr. B's death, air sampling was performed by OSHA, U.S. Department of Labor, and it revealed TDI levels ranging from 0.2 to 1.04 mg/m^3. The contracting company was cited by OSHA for failure to supply respiratory protective devices, protective footwear, or sleeves, and for allowing exposure to TDI at approximately 7.4 times the allowable limit.

Urethane paints and coatings are a source of exposure to TDI that is both irritating and sensitizing to the respiratory system and the skin. Reaction may result from subsequent exposures to lower and lower levels because of the sensitizing effect. Some workers who have become sensitized to TDI can react to minute levels of the substance, below the odor threshold. Because of its severe sensitizing effect, which can remain permanent, engineering controls and/or product substitution are probably the only way to prevent exposure to TDI and its associated isocyanates, such as MDI (methylene diisocyanate). Lest any reader fail to recognize the name isocyanate, it was methyl isocyanate, an intermediate in a pesticide manufacturing process, that was responsible for the deaths of over 2000 persons, and injury to countless others, in Bhopal, India. Considering the toxicity and sensitizing effects, it is reasonable to conclude that in some situations the use of isocyanate-containing materials is inappropriate. In the case of Mr. B., the large surface area available for volatilization of the components permitted significant exposure.

Once again, a death certificate indicated heart disease when the true cause of death was chemical poisoning.

Cardiac Case #2

L. C.'s death certificate lists "presumed cardiac arrest" as his immediate cause of death, with "presumed arteriosclerotic coronary disease" as an associated causative factor. He was 46 years old when he died on February 10, 1983. An autopsy report stated: "Absence of gross and microscopic evidence of ischemic heart disease with coronary vessel arteriosclerosis of moderate severity." Significantly, the autopsy noted "acute necrotizing bronchopneumonia, left and right lungs, with edema and congestion." Cultures of his lungs and trachea revealed no growth of any disease organisms. The microscopic examination of his liver was reported as normal.

L. C. had returned home from work shortly after noon, four days previously, complaining of a severe headache and chest pain. His wife said that he slept most of the next day, and all of the next night, which was an uncommon thing for him to do, as he had usually worked the night shift, and had difficulty accommodating to sleeping during the daytime. He ate little and slept a great deal the next two days, still complaining of chest pain. A physician, whom he saw on February 9, gave him some antibiotics, which he took after he got home. On the 10th, his wife urged him to go to the hospital, but he was reluctant to do so. He spoke with his children after school at about 3:30 P.M., and at about 4:00 P.M. his wife found him dead.

L. C. had recently used a large quantity of 1,1,1-trichloroethane in the course of his work.

Although 1,1,1-trichloroethane (TCE) is useful in industry as a degreasing agent and solvent for oils and fats, it also essentially "degreases" the cells and tissues of the body, breaking down the lipid and fat barriers. The normal glycolipid barrier of the lungs can be broken down by exposure to solvents, resulting in destruction of tissue and an outpouring of fluid. Additionally, the narcotic properties of many solvents, including TCE, can anesthetize a person. Mr. C.'s somnambulance was in keeping with a solvent exposure.

This case demonstrates two points: (1) the importance of taking a work history, to determine if there are any exposures to toxic substances; and (2) the importance of obtaining the information that only an autopsy can provide in cases of sudden death, especially in relatively young individuals. The autopsy findings on Mr. C. are compatible with solvent-induced destruction of his lungs.

Once again a person's death was attributed to that great wastebasket category of coronary artery disease, when it reality it was due to a chemically induced pulmonary reaction. The death certificate should be amended to reflect the correct diagnosis.

BREAST

BREAST CASE #1

D. K. began to take the Enovid brand of birth control pills in about 1963, when she was 35 years old. In September 1974, her medical record noted "breasts small, 2 to 3 pea sized nodules along scar on right breast—right breast larger." She discontinued the birth control pills in 1975, when she was 47 years old. The nodules were confirmed again in January 1977. By May of that year, she had lost 15 pounds and sought medical advice. An examination in May, by a nurse clinician, stated: "Many different sized masses in right breast, different shapes, all movable, all firm and hard. Large mass under scar at 10 o'clock. Tender." A surgical consult was recommended. A week later, a physician's examination noted "Birth control pills for 12 years . . . small nodules 1–2 cm., firm, nontender in upper quadrant of both breasts." In October 1977, both the history of birth control pill use for 12 years and masses in both breasts were confirmed. By June 1978, three masses were described in each breast. A biopsy of her right breast mass was done in May 1979, revealing papillary intraductal carcinoma with infiltrating component. A mastectomy was done in June, with metastases found in several axillary nodes. She was referred to a university center for chemotherapy.

Ms. D. K. had taken a birth control preparation containing mestranol 0.075 mg and norethynodrel 5.0 mg, and, later, a similar formulation having a mestranol level of 0.1 mg (the estrogenic portion) and a norethynodrel level of 2.5 mg (the progestational portion).

Most birth control pills are variations on this same theme, the estrogen portion supplied as mestranol, ethinyl estradiol, and quinestrol, with the progestational part supplied as norethynodrel, norethindrone, megestrol, ethynodiol, or medroxyprogesterone. In addition to synthetic chemicals with a steroidal configuration, several other estrogenic substances have been marketed for human use: DES (diethylstilbestrol), chlorotrianisene, and methallenestril.

The American Cancer Society predicted that upwards of 130,000 women would develop breast cancer in 1987, and that 41,000 or so would die of the disease in the same year. It is the most frequent malignancy found in women and the second cause of cancer-related deaths in women. In addition to the deaths, the physical disability and mental anguish of many of these cases is incalculable.

Attention has been given to this disease, but mostly in the area of early detection—which appears to have had a minimal effect on incidence. Little has been done about the basic cause of breast cancer, although a great deal of information exists concerning chemicals that have induced breast cancers in animals.

During an investigation of the stilbene family of chemicals, of which DES (diethylstilbestrol) is the most infamous, in a search of the agricultural literature for information as to the development, promotion, and use of these chemicals, found that DES and a number of other hormonal agents had been proposed for use in farm animals to control times of breeding, milk production, and to increase body weight for marketing. DES use became widespread in the 1940s to "caponize" male chicks, that is, to suppress the male hormones, and make the birds fatter. Soon, poultry raisers were administering DES, in the form of a pellet embedded under the skin of the neck, to entire flocks of birds, both male and female. It did increase the fat of the animals and contributed to greater profits. Poultry raising was becoming more mechanized, with the birds killed and dressed at the source rather than at the butcher shop. When the problem of disposing of the offal became apparent, a use was sought for this protein waste. It was suggested to a number of mink raisers that chicken offal (heads, intestines, etc.) would be suitable food for mink, who are natural carnivores. Apparently, no one considered the effect of the DES that had been implanted in the chicken heads, which remained in the waste. Soon, a number of mink farmers noted that many of their animals failed to breed, the young that were born were not thriving, and the adults were losing their hair—a disaster for a mink farmer! Questions were asked, and considerable investigation followed, finally leading to the DES-contaminated feed as the cause of the hair loss and failure of the mink to breed.[19]

Several researchers gave graded doses of DES to laboratory animals, including rats, mice, and mink, and discovered the same or similar problems, including interference with development of the genital tracts in the newborns that did survive. This effect was clearly established in the 1940s. Still, producers continued to use this product in raising cattle and poultry because it increased the weight of the animals (mostly in the fat portion), thus increasing the profit to the raisers. Considering today's admonitions against animal fat in the diet, it is difficult to understand the benefit of this fat to the consumer. It was argued that the DES use was safe because it was administered in pellet form, implanted in the neck of the animal. However, for DES to promote its intended action, it had to be carried to other parts of the animals' body. Because of the fat solubility and long half-life of DES, measurable levels of it remained behind in the meat sold to consumers.

Exogenous hormonal agents, both natural and synthetic, have been shown to interfere with the natural hormones and thus prevent pregnancy.[20] Originally applied to farm animals, this treatment was used to control production.[21] In early tests of hormonal chemicals in animals, stimulation of the breasts was noted, and a number of malignancies resulted.[22-25] Still, these products were marketed and prescribed to many thousands of women. The distribution of these hormonal agents, employed in pharmaceuticals and in animals feed, has been widespread, and the effects have been nearly incalculable.

KIDNEY

KIDNEY CASE #1

J. C. was 47 years old when he noticed blood in his urine. Tests showed cancer of his right kidney, and it was removed in 1982. J. C. operated an antique car restoring business, and in the course of his work handled a number of plastics, paints, and solvents, including the following: toluol, xylol, acetone, isopropyl alcohol, methylene chloride, ethylene glycol monoethyl ether, methoxyethyl acetate, and a number of chemicals identified only as petroleum distillates, aromatic hydrocarbons, and aliphatic hydrocarbons. In addition he worked with acrylic paint systems. J. C. smoked cigarettes, beginning at age 17 and stopping at the time his kidney cancer was diagnosed.

Mr. C. obtained a number of the MSDSs; however, many of the ingredients listed did not add up to 100%, many others were not specific as to their identity, and others indicated only "solid" acrylic, "proprietary alcohol," "bodying agent."

These chemicals are taken into the body not only via inhalation but through the intact skin. Many of them are eliminated in the urine, where they come in contact with cells of the kidneys and bladder, and are able to cause damage as they pass along the urinary system. It is likely that the various reactive chemicals that he handled contributed to the development of his kidney cancer.

KIDNEY CASE #2

A. H. was admitted to the hospital because of kidney failure and hypertension when he was 58 years old. He had been admitted twice before because of lead poisoning, and within the previous month because of severe anemia. Six days later he was dead, his final diagnosis being chronic glomerulonephritis and uremia. A. H. had worked 24 years as a laborer at a company where lead was handled on a regular basis.

Lead is concentrated in the bones and excreted by the kidneys. As it passes through the kidneys, it damages the cells, resulting in decreased ability of the kidneys to eliminate fluid and solute loads. While lead is stored in the bones, it interferes with the body's ability to make blood cells; hence, the red blood cells are not only too few but abnormal. It must be concluded that the lead that Mr. A. H. handled for over 40% of his total lifetime contributed to his anemia and kidney failure.

KIDNEY CASE #3

In March 1984, D. L. was involved in the automobile chase of a suspect as he was working as a police officer. His patrol car crashed into a utility pole, and an electrical transformer exploded and burned. He was heavily exposed to the fumes and smoke and was taken to a hospital for treatment. By September or October of that year, he had noticed blood in his urine. In January 1985 he was diagnosed as having cancer of the bladder, and underwent surgical removal of the malignancy.

Significantly, D. L. had been thoroughly evaluated in January 1982 for an unrelated problem, and, among other tests, had an IVP (intravenous pyelogram) that was reported to show no abnormalities of his kidneys or bladder. D. L. was 39 years old at the time his cancer was found. He had never smoked cigarettes.

Many electrical transformers, especially older ones, contain PCBs as coolants. These chemicals have been associated with the development of malignancy at a number of sites, in both animals and humans. Additionally, it is highly likely that the transformer fluid was also contaminated with dioxins, the most powerful carcinogen of these being TCDD (tetrachloro dibenzo-p-dioxin).

It is uncommon to see bladder cancer in persons as young as 39 years. It is also uncommon to think of police duty as carrying a burden of exposure to toxic chemicals, aside from riot control gases. With a baseline of no disease in 1982, a known exposure to toxic chemicals in 1984, and the development of a malignancy ten months later, a cause-and-effect relationship is highly likely.

This sort of documentation is quite rare, in that few people have a body system evaluated twice within such a short time, with the appearance of cancer during the interval. This case, like the others cited in this book, underscores the importance of taking a work and environmental history from all cancer patients. Seemingly "minor" incidents must not be dismissed; for although an exposure may be brief, it can be both intense and significant. As PCBs are retained within the body, the duration of exposure becomes significant, essentially extending from the time of the original contact.

KIDNEY CASE #4

When I saw H. W., he was 60 years old. His blood pressure readings were 250/130 on the right, and 240/142 on the left, and he had significant protein in his urine. He first learned of high blood pressure and diabetes in 1967, and he took medication for each. H. W. worked 32 years, buffing chrome-plated bumpers. He used a dry buffing compound and said that although it was extremely dusty work, he did not wear any sort of mask until after 1978. He said the bumpers were plated in an area about 300 feet away from his work station, and that the fumes and the acids "cut his wind," especially when equipment broke down.

Exposure to chromium compounds is associated with lung disease (as described in the references at the end of this book) and with kidney damage as well. Chromium compounds are taken into the body via inhalation, by ingestion with the saliva, and through the skin, causing damage at a number of sites. Chromium is excreted in the urine, and as it passes through the kidneys, causes damage to the cells therein.

KIDNEY CASE #5

M. C. was 37 years old in 1981 when he applied for life insurance. A routine examination found that he had protein in his urine and significant kidney dysfunction. He underwent a biopsy of his kidney, which revealed a focal sclerosing glomerulosclerosis. This was interpreted by his physicians as "nonspecific." As an incidental finding, he had reported that he had essentially lost his sense of smell ten years previously.

M. C. began working as a welder when he finished high school. He did aluminum welding, and then silver brazing for the next 16 years, followed by electric welding for the last 2 years.

Review of the MSDS indicated that the brazing material contained the following: silver metal 50%, copper, 15.5%, zinc oxide 16.5%, and cadmium oxide 18%. The MSDS correctly notes the product was "dangerous mostly from cadmium fumes. Inhalation of dusts or fumes of cadmium affects the respiratory tract and may affect the kidneys." It states further that "cadmium and certain compounds are listed by the National Toxicology Program as substances 'that may be reasonably anticipated to be carcinogens' (evidence in humans is limited; evidence in experimental animals is sufficient)." The MSDS goes on to warn about the need for respiratory protection, not only for the person using the material, but for persons in the vicinity of its use. Additionally, the MSDS states that "studies indicate that there is an increased incidence of prostatic cancer, and possibly kidney and respiratory cancer in cadmium workers."

With a reasonable degree of scientific certainly, it is likely that Mr. C.'s kidney damage is a result of his exposure to the components of the brazing material, especially the cadmium portion.

GASTROINTESTINAL

GI CASE #1

E. H. had been a healthy, active person until he was 53 years old, when he developed diarrhea that persisted for about two years. An X ray revealed a malignancy, and E. H. underwent radical surgery for removal of his colon. The tumor had invaded his prostate and urinary bladder. Despite the surgery and radiation therapy, E. H. developed metastases to his lungs and liver, and died when he was 57 years old. He had never smoked.

Mr. H. worked for an industrial chemical company for a period of 24 years, and was exposed to benzene, ethylene dichloride, methylbromide, ethylbromide, and additional chemicals, including Tris (the fire retardant that was banned from children's clothing). In all likelihood, Mr. H. was also exposed to polybrominated biphenols (PBBs), as he worked at the plant where they were produced. PBBs are the fire-retardant chemicals that killed thousands of Michigan farm animals when bags of the chemical with labels similar to those of a mineral supplement were mistakenly incorporated into farm animal feed in 1973. PBBs, like PCBs, are similar to each other in structure, action, and toxicity. The chlorinated biphenols and their dioxin contaminants were studied extensively,[26] but it was not until after 1975 that thorough testing of the brominated congeners was undertaken. Based on similarity of structure and physical properties, it was predicted that the PBBs would be carcinogenic, as the PCBs already were known to be.

The chemicals that Mr. H. worked with have been associated with positive cancer studies in animals, as well as with other effects including interference with immunological and liver function. In all probability, his cancer was a result of his exposure to those industrial chemicals.

GI CASE #2

In July 1981, when D. W. was 52 years old, he became suddenly weak and tired, so much so that he was hospitalized seven times before 1983 to try to determine the cause of his profound weakness, arthralgias, abdominal tenderness, flushing and itching of his skin, darkening of his urine, and mild jaundice. Two liver biopsies as well as other tests were performed, with the result that the causes of his liver abnormalities and general sickness remained unknown.

When examined in 1985, he had elevated cholesterol and triglycerides, an elevated liver enzyme test, high blood pressure, and lesions on his hands, arms, and ears resembling keratoses. He had had a basal cell carcinoma removed from the side of his nose five months previously.

Mr. W. had begun to work as an electrical lineman when he was 18 years old, and did that kind of work, with the exception of a four-year period of employment as a barber, until he retired because of illness in 1983. He indicated that he had worked full days without any significant medical problems until July 1981.

From 1964 until 1976, Mr. W. rebuilt power distribution systems, stringing electrical lines, and about two or three times a year he installed transformers. From 1976 until he retired, he worked as a superintendent of utilities, and in the course of that work he helped dismantle and move electrical equipment from a substation to another site. He said that the job took from 30 to 45 days to complete. Among other tasks, the job required removing a regulator and a transformer, which he did in the summer, on July 8, 9, and 10, 1981. For two of those days he said he worked with his arms up to the elbows in transformer oil, the transformer holding approximately 500 gallons of the old oil. He said that neither the transformer nor the regulator was labeled as to content.

He said that his physician had questioned him about exposure to toxic chemicals, and at the time he could not identify any. Before he authorized the movement of the transformer, however, he had the fluid tested, and it was then that he learned that the regulator contained PCB levels of 1480 ppm, and the transformer contained 229 ppm of PCBs.

With that information, his physician decided that Mr. W.'s liver biopsy findings of chronic active hepatitis and his multiple medical problems were compatible with exposure to PCBs.

GI Case #3

H. K. was 56 years old when he developed dull upper abdominal pain, with intermittent cramping. Surgical exploration found an adenocarcinoma of his descending colon, with metastases to liver and spleen. Ascitic fluid removed from his abdomen showed atypical mesothelial cells. He was treated with chemotherapy, to little avail, and died within the year.

Mr. K. was interviewed shortly before he died, and said that he had begun working as a carpenter when he was 27 years old, and had worked with mostly untreated lumber for about ten years. In about 1959, he began to work with lumber that had been treated with either a copper–chrome–arsenate or pentachlorophenol solution. He said that the lumber was delivered from a commercial treating plant and was often wet from the chemical solutions. He reported that when he was working on double-wall construction, he sprayed the chemicals onto the inner portion of the wall, as well as dipping the ends of the cut lumber into the treating solutions. He said that in cutting the lumber with a saw, he became covered with sawdust. In 1970 he began to do carpentry work full-time in a shop, where he estimated he used about 5 gallons of the copper–chrome–arsenate solution a day for a period of two years. He then began to use a pentachlorophenol solution for the construction of windows, and continued using this preparation for a period of six years, until he was diagnosed as having cancer in 1978.

Mr. K. said that when he was working as a shop carpenter and handling the wood-treating chemicals on a daily basis, he never wore a mask or gloves, and he had not been cautioned about the hazards of the chemicals. Because these chemicals are associated with a number of adverse effects, including cancer, the EPA issued a Rebuttable Presumption Against Registration on October 18, 1978, and followed that up with additional information in 1981.[27,28] Considering the duration, method, and intensity of exposure, it is highly probable that Mr. K.'s cancer was related to his exposure to the wood-treating chemicals.

REPRODUCTIVE

Reproductive Case #1

J. S. was 20 years old when her first baby was born in January 1975, approximately $2\frac{1}{2}$ months premature. She had a spontaneous delivery, and the baby boy weighed 3 pounds 9 ounces. He was transferred to a specialty hospital for care of his problems due to prematurity. As he grew, he was described as having problems with breathing and coordination.

In February 1976, J. S. was five months pregnant with her second child when she began to have signs of "cervical incompetency." This was treated by suturing the cervical opening. Her second child, a girl weighing 6 pounds 5 ounces, was delivered in April 1976. Her physician noted that her "cervix was entirely amputated–$\frac{3}{4}$ amputated, except for a portion which was attached anteriorly." He also noted that "there seemed to be somewhat of a defect of the endocervical area, left of the posterior area." Two months later, in July 1976, she underwent a hysterectomy to stop her bleeding. The records stated that her "cervix sloughed off, the uterus come out through the vagina." The surgeon then noted absence of her cervix, "with the cervix flush with the vaginal vault." The pathology report said that "fragments of the cervix are polypoid in shape and are covered mostly by metaplastic stratified squamous epithelium."

In October 1982, J. S. noticed a small nodule in her vagina and went to see her physician, who biopsied the mass. The pathology report was clear cell adenocarcinoma. She was evaluated by a number of physicians because of her malignancy. Despite radiation to her pelvis and chemotherapy, her cancer spread, and she died when she was 29 years old.

Because of the unique character of Ms. J. S.'s cancer in so young a person, her physicians questioned her as to whether her mother might have taken a drug called DES (diethylstilbestrol) while she was pregnant with the patient. Unfortunately, the physician who cared for the patient's mother when she was pregnant with J. S. in 1953–54 had died, and his records were not available. The mother related, however, that she had experienced a miscarriage prior to the birth of her daughter and had been given pills to take while she was pregnant with her.

The unique finding of clear cell adenocarcinoma in a young woman is almost certainly due to DES. The rarity of the disease prior to the introduction of this chemical into human medicine, and the fortunate decline in its incidence since the relationship was reported in 1971,[29] have led to a decline in this tragic disease. Ms. S.'s connection to chemically caused cancer does not end with DES. After her second child was delivered in 1976, she was prescribed a drug to prevent breast engorgement. This was Tace® (chlorotrianisene), a congener of diethylstilbestrol. The chemical structures of the two chemicals are shown in Figure 6-2. It should be noted that the products are similar, each consisting of benzene rings connected by an ethylene bridge, and each possessing estrogenic activity.

The carcinogenicity of the family of chemicals called stilbenes, of which these two drugs are members, is well known; and it is in keeping with scientific precepts to deduce that Ms. S.'s cancer was initiated by the DES that she received when she was in utero, and that the further administration of chlorotrianisene, a member of the same carcinogenic family of chemicals, promoted and/or accelerated the final expression of her malignancy.

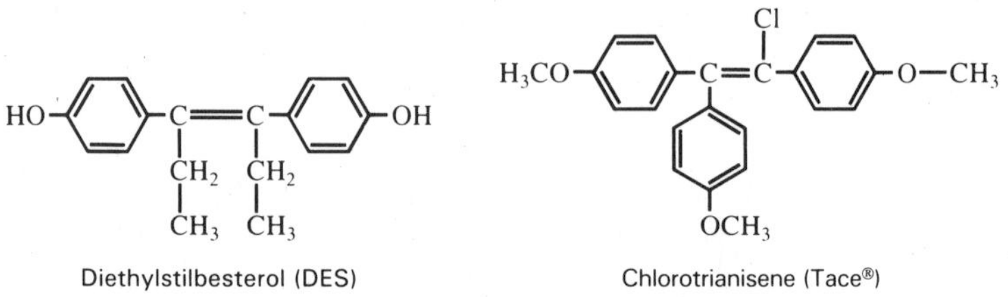

Diethylstilbesterol (DES)　　　　　Chlorotrianisene (Tace®)

Figure 6-2.

A further tragedy is that DES was without efficacy in preventing miscarriage, and that both DES and chlorotrianisene are not needed for postpartum milk suppression. The latter can be achieved by the absence of suckling and the use of a supportive brassiere.

REPRODUCTIVE CASE #2

Baby L. M. was the third child born to Mr. and Mrs. M. There were two older children, ages 9 and 11, both healthy. This third delivery was done by cesarean section because of fetal distress, and soon after birth baby L. M. was found to have an abnormal heart. After many hospitalizations for heart failure, and a variety of infections and lung problems secondary to her heart trouble, she underwent open-heart surgery to attempt correction of her defects. The surgeon found that instead of two atria she had a single atrium; that her inferior vena cava was absent; that the veins from her liver drained directly into the inferior portion of her atrium; and that she had a ventral septal defect, and a patent ductus. Additionally, because she had only one atrium, the mitral and tricuspid valves were fused, and were abnormal in both position and shape. L. M. was 11 months old at the time of this major reconstructive surgery.

Baby L. M. was hospitalized on numerous occasions because of her inadequate heart function, and when she was 28 months of age, her abnormal mitral valve was replaced with a prosthetic device.

L. M.'s parents wondered why she had been born with a deformed heart, and so began to search for possible causes. Mrs. M. had never smoked and had taken no medications during her pregnancy that could account for L.'s birth defects. Mr. M., however, had worked for a paint formulating company for approximately five years before L. was conceived. The company had installed a mixing room where Mr. M. worked, and he had direct contact with various chemicals. Among other components, the paints contained glycol ethers, which were also used for cleaning and other purposes at the plant. No protective devices, equipment, or clothing was issued to the workers, nor were they given information concerning possible hazards to health associated with handling these chemicals. There also was no laundry service, and so the workers took their clothes home to be laundered. Because his clothes frequently became contaminated, Mr. M. could not wear them home; so he changed and carried his contaminated garments home in a plastic bag. In doing the laundry, Mrs. M. was exposed to chemical vapors in the air, and as she handled the chemical-soaked clothing, she had skin contact with these substances as well. This exposure extended throughout the period of time prior to and during her pregnancy.

Mr. M. happened upon some information about the adverse effects of the glycol ethers at work, and then pursued the issue further. Many animal studies reporting the connection between exposure to this class of chemicals and birth defects appear in the medical literature.[30-32]

As was explained in Chapter 1, on cause-and-effect relationships, it is difficult to pinpoint beyond the shadow of a doubt an exact connection between exposure of the fetus to glycol ethers and her heart defects. However, given that Mr. and Mrs. M. were healthy, Mrs. M. took no medications during her pregnancy, the couple had two older children who were born without birth defects, and both parents were exposed to a known teratogen (Mr. M. directly in his work, and Mrs. M. directly, although secondarily, through a home contamination), we can say *with a reasonable degree of scientific certainty*, more likely than not, L.'s heart defects probably were due to her exposure to the paint chemicals, including the glycol ethers, while she was in her mother's womb.

Although L. is still alive at the time of this writing, she will never lead a normal life. A future potential problem may result from the considerable number of X-ray examinations that were required to diagnose and treat L., and the possibility that this amount of X ray may induce a malignancy in her. It goes without saying that her birth defect has had a profound effect upon her, and her parents, and siblings in worry and concern, disruption of a normal family pattern, and significant economic burdens.

When L. was still younger than 3 years old, her *direct* medical costs had been in the neighborhood of $138,000 ($115,000 paid by a medical insurance company, and $23,000 paid by her family). Considering that this is just the beginning of the costs for the life-long treatment of the child, we must ask: would an adequate ventilation system at the paint company have been more economical than the cost of treating these birth defects?

In many cases, the decision to install ventilation and other safety equipment is predicated upon cost. The cost decision is usually based on short-term considerations, and it is not made until the pressure of complaints and/or litigation forces a corrective action. Unfortunately, for those adversely affected, the action is rarely remedial for the damages incurred. To speak of ethics and morality in the current economic climate is to incur ridicule; but if we are to avoid the tragedies outlined in this chapter, something more than the "bottom line" will have to be taken into account. There is no free ride: costs deferred because of inaction or inadequate protection will surely be paid in the future.

REPRODUCTIVE CASE #3

C. W. was approximately five to six weeks pregnant when the area where she worked was under construction. In order to install some insulation materials, an adjacent area was sprayed with a material that was identified as containing the following: ethyl acrylate, acrylonitrile, itaconic acid, ethylene oxide–nonyl phenol adduct, and formaldehyde. C. W. developed a headache and felt nauseated at the time. She also developed some vaginal bleeding, and went to see her obstetrician. The bleeding stopped spontaneously. The following week, additional spraying was done, and again the fumes made her feel sick, so she went home early. C. W. had no further concern about the incident until her son, D. W., was born with the following defects: meningomyelocele (opening in the bones covering the spinal cord); imperforate anus, with his colon ending 3 cm short of his anus; hypoplasia of his lower limbs, and a short upper arm; small kidneys, lateral urethral opening in his penis, a hydrocele, and a rectovesical fistula, draining his urine and feces into the same opening; short neck and low-set ears; barrel chest; deformities of his head and face.

After D. W.'s birth, he was transferred to a specialty hospital for children, but despite multiple surgeries, multiple consultations, and heroic care, he lived for only three weeks.

Ms. W. sought to learn why her child had been born with so many defects. It was obvious, from the multiple organs involved, that the insult had come early in his intrauterine life. Of the 13 developing body systems outlined in Arey's *Developmental Anatomy*, D. W. had defects in seven: the urological, genital, skeletal, muscular, nervous, sense organs, and trunk. These combined injuries indicate a profound interruption of normal development at an early stage of development, which correlated closely with the time of Ms. W.'s exposure. Also, because of the multiple abnormalities it is likely that the insult was a result of exposure to a combination of chemicals, each having a slightly different effect.

Most of the chemicals in the product were fat-soluble and thus able to be taken into and retained by the body.

The baby's parents instituted a lawsuit against the applicator and the manufacturer of the product. During the lengthy pretrial period it was revealed that no premarket testing had been done on the product in order to determine whether there were teratogenic or any other adverse effects. Review of the toxicology of the various components showed multiple reports of toxic effects, including teratogenicity.

BLOOD

BLOOD CASE #1

W. M. discovered he had granulocytic leukemia in September 1981. At the time, he was 39 years old, and had been in good health. Between the time of his original diagnosis and the time that I examined him in October 1984, he had been hospitalized 14 times for treatment of his leukemia. He began to work for a chemical supply company in 1973 when he was 31 years old. He said that bulk chemicals were delivered by tank car and in 55-gallon drums, and that he formulated various products to be used as degreasers, carburetor and parts cleaners, and other solvent products. He said that he handled a variety of chemicals, including: chrysic acid, methylene chloride, 1,1,1-trichloroethane, ethylene glycol monobutyl ether, formaldehyde, isopropyl alcohol, cresols (not further identified on the MSDS), dichlorobenzene, a chemical identified on the MSDS only as "C.10–C.12 hydrocarbons," orthochlorobenzene, trichloroethylene, and methyl isobutyl ketone. He said he handled up to 8000 gallons of chemicals on an average workday, and wore only a face shield, but no respirator.

Two additional MSDSs listed products as "aromatic hydrocarbon" with the formula "complex mixture of petroleum hydrocarbons." These sheets obviously were an inadequate way to provide meaningful information to a worker.

After W. M. was diagnosed as having leukemia, *he continued to work* part-time until July 1983, when he had to stop because of weakness and complications of his disease and therapy.

Solvents are taken up in the fat portions of the body, which includes the bone marrow. Because cells are actively multiplying in the marrow, chemicals can interact with the DNA of these cells and interfere with the orderly process of blood formation. Mr. M.'s multiple exposures to chemicals that had been reported to cause cancer in laboratory animals and humans undoubtedly played a significant part in causation of his leukemia. Important also was the fact that he continued to work, and continued to be exposed to toxic chemicals, after his malignancy was diagnosed. Not only must causes be sought in cancer cases, but when exposure to chemical carcinogens is found, the exposed person absolutely must cease additional exposure.

BLOOD CASE #2

E. Y. was 39 years old when he was found to have reduced red cell, white cell, and platelet components of his peripheral blood, with abnormal forms. A bone marrow study showed increased cellularity with abnormalities of both the red cell and myeloid elements. A month later, giant polypoid segmented forms of white blood cells were found in his blood, with megaloblastic forms appearing the next year. At the time of his original diagnosis of pancytopenia, his physician inquired into E. Y.'s work and learned that he was a carpenter who regularly handled wood-treatment chemicals (copper chrome arsenates and pentachlorophenol), as well as solutions to treat ground termites (not identified, but undoubtedly of the cyclopentadiene family, including chlordane and heptachlor) and paints and solvents. His physician stated that "one is always fearful of leukemic transformation sooner or later in this patient, as many of these marrows damaged by chemicals later transform to acute leukemia." Despite superb supportive care, E. Y.'s blood picture never approached the normal, and he continued to have headache, fever, loss of energy, and stomach discomfort, and he developed bleeding. Seven years later, he was admitted to the hospital with acute blastic leukemia, diagnosed as most likely myeloblastic with erythroid components. His physician noted that his relapse was "probably because of reexposure to paint chemicals." He died when he was 46 years old, as a result of the complications of leukemia. Patients with aplastic anemia, like those with leukemia, usually succumb to infection, bleeding, or the effects of the severe anemia, related to the abnormal and ineffective blood cells.

The label of the wood-treating solution said that it contained chromium trioxide, copper oxide, and arsenic pentoxide, and warned of acute effects: "May be fatal if swallowed. Avoid contact with skin, eyes or clothing. Wash thoroughly after using. Avoid breathing spray mist." These warnings are not unlike those found on many household products—more or less commonsense concerns. But this product was not intended for an infrequent, household-type use. This is a product to which construction workers may be exposed in a significant manner. There was no mention on the label of the effects of chronic, long-term, or repeated exposures—nor, it must be emphasized, was there any information as to possible health effects on people living and working in dwellings constructed with such materials.

Considering the work that a carpenter does—handling, sawing, pounding wood no more than an arm's length away, with the production of sawdust—it is difficult to see how the carpenter could avoid contamination from wood-treating chemicals.

In 1975, in the state where Mr. Y. worked, wood preservatives used in the construction and repair of homes amounted to over 600,000 pounds, of which 97% were of the copper–chrome–arsenic type. The next year, over 1 million pounds of wood preservatives were used in the state.[33]

Both arsenic and chromium compounds are recognized as carcinogenic, and have been so known for a number of years. Arsenicals can bind to tissue proteins, and chromosomal breakage and aberrations have been found in the cells of arsenic-exposed workers. It is impossible to determine the contribution of each of these major components, arsenic and chromium, in formulations that combine them, but they must be at least additive, and quite possibly are synergistic.

It is highly likely that the chemicals to which Mr. Y. was exposed in the course of his work damaged his bone marrow. In addition to the copper–chrome–

arsenic product, one cannot ignore his additional exposure to paints and solvents, and the unidentified pesticides. A finding of aplastic anemia indicates that there has been severe insult to the bone marrow. Once the marrow has been damaged, it is unlikely to recover fully, or recover in a normal pattern. Further chemical insults can accelerate or precipitate a frankly malignant change.

Every patient who develops aplastic anemia or leukemia should be evaluated for exposure to chemicals that could have been encountered in either the workplace, the home, or the surrounding environment.

It is frequently stated that of all the carcinogens only arsenic has been found to be carcinogenic in humans only and not in animal species. However, it has been determined that arsenic follows the laws of other carcinogens, and data concerning cancers in animals are available. Interference with normal cell growth and subsequent malignancy is a property of *all* living cells, be they human, animal, plant, insect, or bacterium. (See the bibliographic section on arsenic and cancer at the end of the book.)

SUMMARY

There has been controversy in all the cases cited here. Still, the review-of-systems approach allows one the opportunity to narrow the search for causative agents. Although there is a plethora of chemical agents in the environment, most chemical exposures are relatively finite. Therefore, the historical approach is useful in delineating the most reasonable and likely cause for a disease process.

REFERENCES

1. Department of Labor/OSHA press release 77-866, Regulation of workplace cancer hazards, p. 1, Oct. 3, 1977.
2. Rutstein, D. D., Mullan, R. J., Frazier, T. M., Halperin, W. E., Milius, J. M., Sestito, J. P. Sentinel health events (occupational): a basis for physician recognition and public health surveillance. *Amer. J. Pub. Health* 73(9): 1054–62, 1983.
3. Grant, G. A., Shelton, F. R. Chronic toxicity studies of tetraethylthiuram (Antabuse). Ayerst Research Laboratories, Montreal, Canada, 1949.
4. Savolainen, K., Hervonen, H., Lehto, W. P., Mattila, M. D. Neurotoxicity of disulfiram in rat: effects on autonomic and peripheral nervous system. *Toxicol. Letters* 18 (Supp. 1): 34, 1983.
5. von Rossum, J., Roos, R. A. C., Bots, G. T. A. M. Disulfiram polyneuropathy. *Clin. Neurol. Neurosurg.* 86(2): 81–87, 1984.
6. Bilbao, J. M., Briggs, S. J., Gray, T. A. Filamentous axonopathy in disulfiram neuropathy. *Ultrastruct. Pathol.* 7: 295–300, 1984.
7. Takeuchi, Y., Ono, Y., Hisanaga, N., Kitoh, J., Sugiura, Y. A comparative study on the neurotoxicity of *n*-pentane, *n*-hexane and *n*-heptane in the rat. *Brit. J. Indust. Med.* 37(3): 241–47, 1980.
8. Graham, D. G., Abou-Donia, M. B. Studies of the molecular pathogenesis of hexane neuropathy. *J. Toxicol. Environ. Health* 6(3): 621–31, 1980.

9. Noa, M., Illnait, J. Induction of aortic plaques in guinea pigs by exposure to kerosene. *Arch. Environ. Health* 42(1): 31-36, 1987.

10. Matte, T. D., Baker, E. L. Dizziness: an occupational consequence. *J.A.M.A.* 257(23): 3283, 1987.

11. Spencer, P. S., Nunn, P. B., Hugon, J., Ludolph, A. C., Ross, S. M., Roy, D. N., Robertson, R. C. Guam amyotrophic lateral sclerosis–parkinsonism–dementia linked to a plant excitant neurotoxin. *Science* 237: 517-22, 1987.

12. Perl, D. P., Good, P. F. Uptake of aluminum into central nervous system along naso-olfactory pathways. *Lancet* 8540(1): 1028, 1987.

13. Perl, D. P., Brody, A. R. Alzheimer's disease: X-ray spectrometric evidence of aluminum accumulation in neurofibrillary tangle-bearing neurons. *Science* 208: 297-99, 1980.

14. Heck, H. d'A., Chin, T. Y., Schmutz, M. C. Distribution of C-14 formaldehyde in rats after inhalation exposure, in *Formaldehyde Toxicity*, Gibson, J. E., Ed. Hampshire Publishing Co., Washington, D.C., 1983, pp. 26-37.

15. Goldberg, L. Risk assessment with formaldehyde: introductory remarks, in *Formaldehyde Toxicity* (see ref. 14), pp. 275-78.

16. Zitting, A., Savolainen, H. Biochemical effects of subacute formic acid vapor exposure. *Res. Commun. Chem. Pathol. Pharmacol.* 27: 157-62, 1980.

17. Kilburn, K. H., Warshaw, R., Thornton, J. C. Formaldehyde impairs memory, equilibrium, and dexterity in histology technicians: effects which persist for days after exposure. *Arch. Environ. Health* 42(2): 117-20, 1987.

18. Roach, R. E. *The Detroit News*, p. 3-C, May 13, 1987.

19. Celler, E., Chairman. Mink claims. Hearings before the Committee on the Judiciary. U.S. House of Representatives. Mar. 2, 5, and 9, 1951.

20. Makepeace, A. W., Weinstein, G. L., Freidman, M. H. Inhibition of post-coital ovulation in rabbits with progesterone. *Amer. J. Physiol.* 119: 512-16, 1937.

21. Trimberger, G. W., Hansel, W. Inhibition of oestrus and ovulation in commercial animals with the use of progesterones. *J. Animal Sci.* 14: 224-32, 1955.

22. Burrows, H. Carcinoma mammae occurring in a male mouse under continued treatment with oestrin. *Amer. J. Cancer* 24: 613-16, 1935.

23. Cramer, W., Horning, E. S. Experimental production by oestrin of pituitary tumors with hypopituitarism and of mammary cancer. *Lancet* 247-49, 1936.

24. Kirkam, W. R., Turner, C. W. Induction of mammary growth in rats by estrogen and progesterone. *Proc. Soc. Exp. Biol. Med.* 87: 139-41, 1954.

25. Trentin, J. J. Effect of long term treatment with high levels of progesterone on the incidence of mammary tumour incidence. *Proc. Amer. Assoc. Cancer Res.* 1:50, 1954.

26. Baughman, R. W. Tetrachlorobenzo-*p*-dioxins in the environment, high resolution mass spectrometry at the picogram level. Ph.D. thesis, Department of Chemistry, Harvard University, Cambridge, Mass., Dec. 1974.

27. 40 CFR 162.11. 1978.

28. *Wood Preservative Pesticides: Creosote, Pentachlorophenol and the Inorganic Arsenicals (Wood Uses)*. Office of Pesticide Programs, U.S. EPA, Jan. 1981.

29. Herbst, A. L., Ulfeder, H., Poskanzer, D. C. Adenocarcinoma of the vagina. Association of maternal stilbesterol therapy with tumor appearance in young women. *New Engl. J. Med.* 284: 878-81, 1971.

30. Hardin, B. D., Niemeier, R. W., Smith, R. J., Kuczuk, M. H., Mathinos, P. R., Weaver, T. F. Teratogenicity of 2-ethoxyethanol by dermal application. *Drug Chem. Toxicol.* 5(3): 277-94, 1982.

31. ECETOC (European Chemical Industry Ecology and Toxicology Centre) Technical Report #4. *The Toxicology of Ethylene Glycol Monoalkyl Ethers and Its Relevance to Man.* Brussels, Belgium, July 1982.
32. Millar, J. D. NIOSH Current Intelligence Bull. #39, *Glycol Ethers.* DHHS (NIOSH) Pub. #83-112, 1983.
33. Wong, L., Budy, A. M. Annual Report #10 Hawaii Epidemiologic Studies Program. University of Hawaii, Feb. 2, 1977.

CHAPTER

7

PESTICIDAL CHEMICALS — ORGANOCHLORINES

So normal among us all is the acceptance of genocide whilst rejecting murder, the straining at gnats while swallowing camels, that we may well ask ourselves whether this double standard of behaviour is, as altruism is said to be, paradoxically an evolved characteristic favoring the survival of our own kind.

—J. E. Lovelock, *Gaia*[1]

If you want to buy any of the blood pressure or heart medications, a common antibiotic such as penicillin or tetracycline, or a pain medication such as codeine or Darvon, you must have a prescription from a person trained in the indications for, metabolism of, excretion of, and adverse effects of the medication. Additionally, the medication must be dispensed by a pharmacist, a person trained in the same areas of knowledge. Both the prescriber and the dispenser of the product must be licensed by the state where they operate. In the case of narcotics, considered "Dangerous Drugs" under the Federal Narcotics Act, persons prescribing and dispensing these chemicals must have federal licenses.[2]

Although FIFRA (the Federal Insecticide, Fungicide and Rodenticide Act[3,4]) does require licensing to handle certain pesticides, the training is minimal, any toxicology beyond the acute is seldom mentioned, and carcinogenic and chronic effects are not covered at all in the training for applicator licensing. The license to apply restricted pesticides may be held by a contractor, but the agents hired by the holder of the license to perform the actual application may have little or no understanding of the toxicology of these products, and may indeed be illiterate.

Although the FDA (Food and Drug Administration) does require that the consumer be advised of the adverse effects of some drugs upon dispensing (birth control pills being the most notable example), there is no provision under FIFRA for the consumer to receive advance information about a product intended for home (or workplace) use. Indeed, most homeowners who have developed problems as a result of the application or misapplication of these products have had difficulty determining what product was used, and then further difficulty obtaining information from the pesticide applicator or manufacturer as to likely effects.

Pesticides are biocides (''bio'' meaning life and ''cide'' meaning kill)— chemicals that kill life. Any prudent person must learn about and understand the toxicology of these products before their use. The products registered and controlled under this EPA regulation kill insects, fungi, and rodents, as its name indicates. Not included in the eponymic name FIFRA are the chemicals that kill plants, the herbicides.

There is no federal record-keeping repository to track the health status of persons living in homes treated with any of the various pesticidal chemicals approved for home use. Following the recognition of potential health problems of persons living in military housing that had been treated with chlordane, the Committee on Toxicology of the National Research Council recommended that an epidemiological study be carried out to correlate helath effects with blood levels.[5] That recommendation occurred in 1979. At the time of this writing, that study has not been undertaken.[6,7]

As of 1982, the time of its most recent recommendation, the Committee on Toxicology, National Research Council, recommended an ''interim'' exposure guideline for airborne concentrations of chlordane not to exceed 5 $\mu g / m^3$. This level was deemed ''interim'' in 1979, and pragmatically arrived at, based upon the military housing contamination information.[8] A spokesperson for Velsicol Corporation, the manufacturer, has insisted that chlordane ''has to be used in a safe manner'' and stated: ''there are occasions when we receive reports of misuse, occasionally there's a spill. But these can be managed.''[9] This statement appears contrary to the Velsicol Corporation's booklet ''Contain and correct application accidents,''[10] as well as much that is known of the persistence and biological activity of the chemical. We must seriously ask whether there is a ''safe'' method of application for a chemical with such persistence and such toxicity as chlordane—that is, safe for the applicator and for the recipient.

Although the chlordane/heptachlor problem is at issue in the following cases, other pesticides have also been associated with adverse health and environmental effects. Understanding the toxicology of these products, we can predict that still other pesticides will be implicated in similar fashion in the future. Few states, other than California, have requirements for reporting pesticide poisonings, and none maintains records on the use of pesticides by governmental, agricultural, or private applicators.

The following cases illustrate some of the concerns associated with the poorly controlled use of a number of pesticides, as well as a lack of monitoring for contamination and/or long-term effects. These concerns include methods of application, resultant medical problems, difficulties and cost of documenting contamination, lack of a central clearinghouse for monitoring, and, finally, the lack of any remedy short of legal action. With one exception, the families have been left with residences that, for health reasons, they should not continue to occupy and which they cannot ethically (and, in some cases, legally) sell.

It must be emphasized that all of these cases, with one exception, are in various stages of litigation, and it is unclear what the outcomes will be. These cases are offered as examples of medical problems associated with exposure to chemicals, and the various ways they were "diagnosed." These specific cases indicate ways that similar contamination problems can be recognized and documented. As with the hundreds of asbestos cases that came to litigation, the scientific data on the association between exposure and adverse effects are irrefutable. The slow grinding of the legal process also parallels the legal experience with asbestos; and, in all likelihood, there will be considerable delay before the legal, economic, and regulatory ramifications are settled to anyone's satisfaction.

These cases emphasize the point that, as in most other chemical-exposure diseases, the problems were borne by the victims. Most of the investigations were at the instigation of those affected, with confirmation from family physicians derived secondarily in most cases, and there has been no concerted national effort to document, remedy, or prevent these catastrophes.

Following my experience in these and similar cases, I recommend that the use of any and all pesticidal preparations in and around the home be thoroughly evaluated, and informed consent of the resident(s) be provided, *before* any use of these chemicals. This recommendation applies not only to "restricted pesticides" that require application by a licensed person, but to those available "over the counter" as well. In this case, protection begins at home.

EXAMPLES OF CHLORDANE CONTAMINATION CASES

CHLORDANE EXPOSURE #1

Mr. and Mrs. L. are a couple in their twenties, he a firefighter and she a nurse, who bought an older house in 1979. In the process of remodeling the house, they discovered evidence of termit damage and so sought a pest-control contractor. In June 1982, the basement and crawl spaces of the house were treated by rod injections, injection of the pesticide into the voids of the concrete blocks, and broadcast spraying of the soil. Because an infestation of live insects was found, the house was treated a second time in September 1982. At that time, the chlordane was applied to some of the interior walls.

In the summer of 1982, Mr. L. began to have headaches that lasted for 2 to 3 hours at a time, worse at night, which he said awakened him from his sleep. He also said that the headaches were worse in the fall and winter when the house was closed up. He said the odor of the pesticide, which caused him nausea, persisted and was worse in the basement. Mrs. L. also developed headaches, an uncommon symptom for her, but of lesser severity than suffered by her husband.

Mr. L. maintained a workshop in the basement, and did much of the remodeling work in the shop. He was also an avid hunter and stored his hunting clothes in the basement area.

On a number of occasions, when he was hunting, he said that he "had to pull over on the road because my headaches got so severe that I couldn't drive."

In the spring and summer of 1983, he said his headaches lessened, but in the fall and winter of 1983, his headaches, along with light-headedness, increased. He suffered a blackout spell, and was hospitalized. He underwent a number of tests including three EEGs (electroencephalograms), a computerized tomography of his head with contrast, a 24-hour EKG, an exercise treadmill test, and an echo cardiogram. The first two tests were done to rule out any brain abnormality such as a tumor, bleeding, malformation, or seizure condition. The latter three were to rule out a cardiac cause for his loss of consciousness. All of the tests were normal. Because Mr. L. worked as a firefighter, his physician also proposed a diagnosis of "due to toxin or carbon monoxide with occupation."

After his physician asked him about chemicals, Mr. L. began to perform some investigation and called the pest control company that had treated his house, finding that chlordane had been used. He contracted the local Poison Control Center and EPA to get information about chlordane. After he became aware of the toxic nature of chlordane, he contacted the State Department of Agriculture, Trade and Consumer Protection.

Several investigations followed, with the result that the house, much of its contents, and the soil around the house were found to be contaminated beyond repair, and had to be demolished. In September 1985, the applicator was charged with a criminal complaint for his unlawful use of the pesticide. He pleaded no contest and was fined $500.00.[11, 12]

The order from the State Department of Natural Resources noted: "Proper means of disposal (of the house) shall be demolition followed by disposal to a fully engineered landfill with leachate collection. In addition, soils adjacent to the foundation of the building should be excavated and disposed of with the building materials." The order continued. "Demolition may be done by a general contractor, however, the contractor(s) must be informed as to what the nature of the project is (i.e. demolition due to concentrations of toxic chemicals). I recommend, at the minimum, the use of half-face respirators with the appropriate cartridge by all personnel involved in the demolition and excavation project." It also stated: "All trucks hauling the demolition and excavated materials from the demolition site to the designated landfill must be wetted down with water and covered while in transit. The gate controller and landfill operator must be informed as the nature of the load upon its arrival at the landfill site."[13]

This case, like others, demonstrates the problem with "disposal" of persistent chemicals. In many cases, the contamination is only moved, not removed. "It is a mathematical certainty . . . that each time a pesticide is used, some form of environmental contamination will result."[14]

Chlordane Exposure #2

Mr. and Mrs. A.'s new home was treated with chlordane in February 1983. Soil testing done three months later showed levels of chlordane at 180 ppm, and those of its associated congener at 23 ppm. A week later, an electrician who had done work on the A. home filed a report with the California Pesticide Enforcement division concerning problems with muscle twitching and fatigue after he had been in contact with the contaminated soil.

State of California testing done in July 1983 showed levels of chlordane ranging from 550 ppm back of the house filter to 27.3 ppm in the soil 6 inches down in the same area. Additional testing, done three years later by several laboratories, detected persistent levels of chlordane and heptachlor inside the home, in the garden, and along the foundation.

In November 1983, Mrs. A. was examined by her physician because of cough and tightness in her chest when she smelled the chemical odor. In October 1984, she saw a second physician because of continuing chest complaints as well as problems with headaches, nausea, tremors, arthralgias, fatigue, and her eyes. In January 1986, she saw a third physician because of persistence of her symptoms, which also included nasal stuffiness. By September 1985, she had returned to her second physician, still complaining of the same symptoms, as well as increased respiratory sensitivity.

Mr. A. and his 25-year-old son each had problems with cough, headaches, fatigue, and weakness, which were severe enough for them to see a physician in April 1985.

Chlordane Exposure #3

Mr. and Mrs. W. contracted to have their home treated for termites in May 1982. During the process, Mr. W. heard what he described as a "shower sound" coming from the basement. He said that when he went to the basement to investigate, he found "the protruding nozzle where he (the pest control operator) was shooting the chlordane in and was spewing all over the place." Mr. W., the carpet, and the contents of the room were saturated, and the chemical permeated the rest of the house, and into the furnace duct system. Mr. W. called the contractor, and Mr. and Mrs. W. were told to remove the contaminated articles from the basement—which they did! They were not given any guidance as to precautions to take, and so they handled the articles with their bare hands, with no respiratory protection.

Mrs. W. described symptoms referable to her eyes and breathing. Mr. W. developed nausea, headaches, sore throat, and watery eyes after the incident. Their daughter, T. W., who was 11 years old at the time, developed pneumonia in the fall of that year, and was *at home in bed* for 4 weeks. Their son, S. W., 18 years old at the time, was treated throughout the summer for congestion, eye irritation, and headaches.

The applicator had obviously miscalculated when he drilled through the concrete floor of the porch, thinking it was above an area of soil. The W. family removed both the contaminated objects and the saturated carpet from the basement. They said that several weeks later a worker from the pest control company used scrubbing brushes and pails to complete the cleaning job.

In February 1987, nearly five years later, a sample of dust/soil was taken from the basement room, which showed chlordane levels of 447 ppm.

CHLORDANE EXPOSURE #4

D. P.'s house was treated with a chlordane preparation in July 1982 while she was present. Within about two weeks, she experienced headaches, tiredness, weakness, and marked bruising. Less than a month later, she was diagnosed as having aplastic anemia, characterized by markedly reduced white and red blood cells and platelets. Despite therapy, she died two months later. She was 24 years old at the time of her death.

Environmental tests obtained a year after the application of the pesticide showed 2.7 mg/kg of chlordane in the basement masonry and 120 mg/kg in dust from the cold air duct. The consultant indicated that the 120 mg/kg "corresponded to approximately 50% of lethal dose for chickens," the LD-50. D. P.'s body was exhumed, and tests were run on fat samples from her body. Her fat showed levels of less than 0.002 ppm of chlordane, 0.028 to 0.036 ppm of heptachlor epoxide, and 0.053 to 0.064 ppm of nonachlor. The latter two chemicals are breakdown products of chlordane.

D. P.'s case is not without concern. The association between chlordane exposure and blood changes is discussed below. Certainly, at a minimum, every person with a blood dyscrasia (aplastic anemia or one of the leukemias) should have a thorough environmental and occupational history taken, and fat samples should be assayed for environmental contaminants. This type of investigation is scientifically and medically prudent, considering the toxicology of these products and the relative ease and minor cost of the process.

CHLORDANE CASE #5

J. R. sprayed a chlordane preparation in and around his home for a period of about ten years. In August 1975, he was diagnosed as having myelomonocytic leukemia. Two months later, he was dead at age 37, leaving behind a wife and five children. An autopsy of his body confirmed the diagnosis of leukemia, but with additional involvement by carcinomatous cells in his liver, spleen, and lung. The characteristics of the second malignancy were comparable with poorly differentiated intrahepatic bile duct adenocarcinoma.

J. R.'s body had already been buried at the time that an investigation of the causation of his illness was undertaken. Fortunately the tissue blocks that had been obtained at the time of his autopsy were available. These autopsy tissues were analyzed for chlordane and showed levels of 26 ppm in his liver, 12 ppm in his spleen, and 4 ppm in his brain. Because the sampling had been done on fixed tissue, the chemist postulated a possible variation of +/−25%. In any case, there was significant contamination.

EVALUATION OF PESTICIDE CONTAMINATION AND TOXICOLOGY

Although the above examples are related to chlordane exposure specifically, the evaluation of any pesticide contamination case follows a similar method of inquiry. In cases of possible human exposure to pesticides, which by their nature are designed to be toxic, the following factors should be considered:

1. Possible cause-and-effect relationships between various medical problems that family members have developed and their chemical exposures.
2. General health risks that accrue from exposure to the chemicals in question.
3. Potential and future health problems that individual family members may develop as a result of exposure to these chemicals.
4. Reports regarding animal studies with these chemicals and adverse human case reports. (A neglected area of inquiry in contamination cases is the status of household pets. Frequently cats and dogs are the first to get sick, and not uncommonly the pesticidal poisoning of a family is diagnosed by a veterinarian. If contamination is severe enough, there may be no feral animals such as rats and mice. When these wild animals have not been killed off, they may be trapped, and their bodies analyzed for pesticide levels. Additionally, if domestic animals become sick, their tissues may be analyzed.[15, 16])
5. The state-of-the-art sequence of knowledge concerning the adverse effects of exposure to these chemicals in animals and humans. (That is, what information is known and when was it reported.)

PESTICIDES APPROVED BY EPA FOR TERMITE CONTROL

Evidently, as early as the 1940s studies determined that the cyclopentadiene family of pesticides, of which chlordane and heptachlor are members, carries significant health hazards following exposure.[17-19] The other members of this family are aldrin, dieldrin, and endrin. Their structures, along with those of the seven chemicals approved by EPA for subterranean termite control, are shown in Figure 7-1. Note that the similarity in structure portends similarity not only in pesticidal properties, but in toxicology. The structure of Dursban[R], an organophosphate, is responsible for anticholinesterase effects, in addition to the properties conferred upon it by its organochlorine structure. This pesticide, Dursban[R], is promoted by its manufacturer, Dow Chemical Co., as a substitute for chlordane. Despite an EPA finding of omissions and inadequacies in the company's studies,[20] it continues to be used both inside and outside of build-

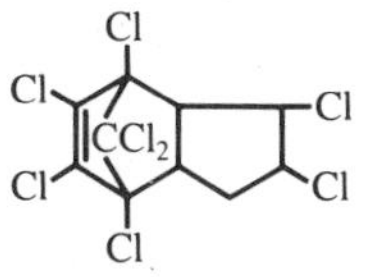

Chlordane

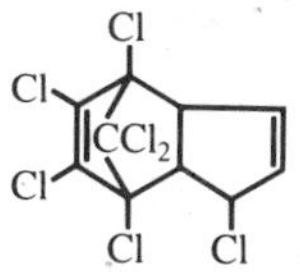

Heptachlor

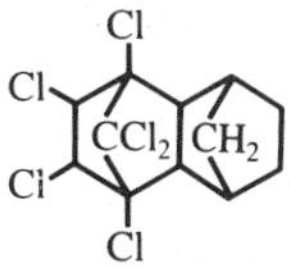

Aldrin

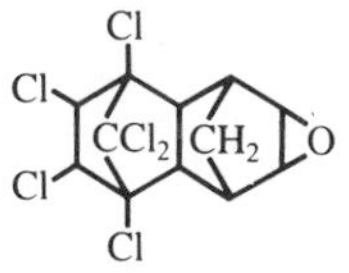

Dieldrin

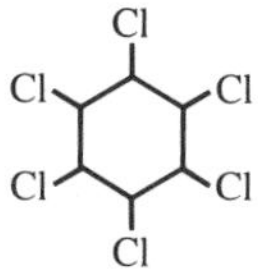

Lindane

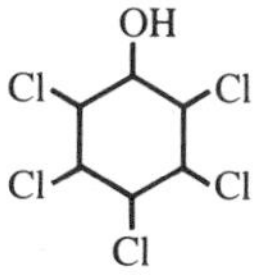

Pentachlorphenol

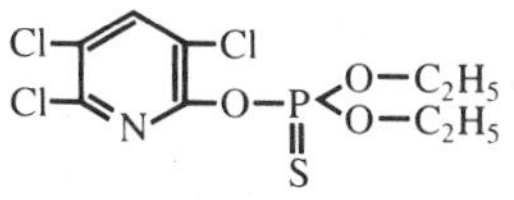

Chlorpyrifos
(Dursban®)

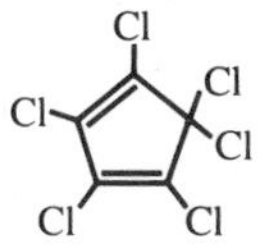

Hexachlorocyclopentadiene
Nucleus "C-56"
Starting Chemical

Figure 7-1.

ings. Like data submitted by many other manufacturers to EPA for registration of pesticides, many of the data submitted for Dursban^{R} are unpublished,[20] and access to the data is limited, negating a prime precept of science, that is, open critical review of research efforts.[21] One might argue, if a product is claimed to be safe, why can't the public have access to the data on which the opinion is based?

There is little question that these chemicals are effective against some insects. Most have been widely employed, even on foodstuffs, but many problems concerning human health and environmental effects of these agents remain undocumented and unstudied.[22,23] Also, the use of these products undoubtedly has lessened building structural damage from insects, but their ultimate costs, in terms of the environment and human health, remain to be assessed. Alternatives, not only in terms of pesticidal chemicals but in construction materials and techniques as well, must be developed to decrease our reliance on these petroleum-based chemicals. Despite gaps in knowledge, much is known, and sufficient data exist for us to make informed decisions about whether or not to use these chemicals.

Using chlordane as an example, a discussion of toxicology and the decision-making process follows.

TOXICOLOGY OF CHLORDANE/HEPTACHLOR

The effects of these two isomers are considered together because technical chlordane contains heptachlor as part of its formulation. Additionally, the two chemicals differ by only one chlorine atom (see Figure 7.1).

Adverse effects have been reported in the following animals: horses, rats, mice, dogs, chinese hamsters, cows, sheep, goats, chickens, rabbits, pigeons, sheepshead minnows, teleost fish, freshwater *Saccobranchus fossilis*, bluegills, *Daphnia*, cutthroat trout, newts, *Planaria*, and humans.

Adverse effects have been described in the following body systems:

brain	bone
stomach	bone marrow
small intestine	spleen
lung	skin
liver	thyroid
adrenal glands	kidney

Entry into the body can be obtained via intact skin, ingestion, and inhalation.

Significant biological properties include: liphophilicity, and thus storage in body fat; bioaccumulation; biomagnification, including in mother's milk; persistence; and volatilization from soil.

Adverse effects have been documented in the following metabolic systems:

DNA synthesis

Enzyme induction

Cyclic AMP

Pyruvate carboxylase

1,6-Diphosphatase

Glucose-6-phosphatase

Carbohydrate metabolism

Phosphenol pyruvate carboxylase fructose

Alpha-fetoprotein

Bile excretion

Porphyrin metabolism

Gamma-aminobutyric acid

Specific Adverse Effects with Documentation

Liver enlargement[24]

Stimulation of hepatic microsomal enzymes[25,26]

p-450 microsomal enzyme induction[27,28]

Potentiation of cirrhosis[29]

Mutagenicity[30-36]

Malformations in offspring[37]

Alterations of fertility[38,39]

Translocation via placenta in non–breast-fed infants[40]

Postnatal adrenal endocrine dysfunction[41]

Steroid interference[42-44]

Enhancement of xenobiotic metabolism of other chemical carcinogens[45]

Electrophilic metabolic products[34]

Porphyrin metabolic alteration[46,47]

Inhibition of protein synthesis[48]

Electrolyte imbalance in fish[49]

Immunological effects[50,51]

T-cell immune responses decreased[39,42]

Decreased reticuloendothelial activity[52]

Aplastic and hemolytic anemias[53,54]

Thrombocytopenia[55]

Megaloblastic anemia[56,57]

Bone marrow effect[58]

Leukemia[59]

Neuroblastoma in children[60]

Neurotoxicity[61-65]

Cancer[66-88]

As can be appreciated by a review of the above data, there is hardly an animal, a body system, or a process that cannot be adversely affected by exposure to chlordane and/or heptachlor, and their associated products.

HAZARD + EXPOSURE = RISK, AND RISK ASSESSMENT

Considering the foregoing information on chlordane/heptachlor, and using principles of decision making outlined in Chapter 9, one must conclude, in regard to the families discussed above, that:

1. The various symptoms (headache; pulmonary problems; arthralgias; eye, nose, and throat irritation; fatigue documented throughout the family members are compatible with exposure to chlordane/heptachlor and their metabolic products.
2. The presence of *significant* and *persistent* quantitites of chlordane/heptachlor in and surrounding the homes indicates a persistent and ongoing source of contamination.
3. The presence of chlordane, heptachlor, and their associated products in the bodies of the tested persons indicates that they have absorbed and retained these toxic chemicals. This retention provides for interaction with various body processes.
4. There is absolutely no doubt that exposure to chlordane and heptachlor carries significant adverse health effects, as amply documented in the scientific literature. These are hazardous substances, and:

HAZARD + EXPOSURE = RISK

5. Putting these factors into perspective, one may conclude, within a reasonable degree of scientific probability, more likely than not, that these family members are at increased risk of continuing to have adverse effects due to their ongoing exposure; and, most ominously, they are at increased risk of developing malignancies and immunological abnormalities.
6. Because of ongoing exposure from contamination of the house and surrounding soil, and because of the serious, documented adverse effects in these cases, it is urged that the residents move out of contaminated dwellings, as did the L. family, taking care not to carry contaminated articles with them.
7. There is ample documentation in the scientific literature, including governmental position papers, to have alerted the manufacturer and distributors of these chemicals to their persistence, as well as the serious health and environmental risks. In the current parlance, it hardly seems possible that the benefit derived from the use of chlordane/heptachlor could justify the risk of its use in a milieu occupied by humans.

The exposed family members will require life-long, thorough medical evaluations, including immunological testing. Because chlordane and heptachlor are fat-soluble, there exists a catch-22 situation: these chemicals are mobilized from fat stores and released when weight loss occurs, so that, as in PCB- and PBB-exposed persons, it is not advisable that these individuals purposely lose weight. Unfortunately, with unintended weight loss—as occurs with an infectious disease, diarrhea, or other problems—these chemicals may be mobilized and lead to symptoms.

LABELING

Under FIFRA regulations, labels of pesticidal products must give certain information, warnings, and precautions. A legal issue now being considered is the adequacy of labels of toxic products. There is considerable knowledge of the toxicology of chlordane and the potential for adverse health effects in persons living in structures containing measurable quantities of chlordane products, as well as in those persons who apply them. Thus, one must conclude that current warnings on the labels of the various chlordane products are *inadequate*, for the following reasons:

1. Failure to cite other cancer data (in addition to the single quote from the National Academy of Sciences).
2. Failure to cite other adverse effects.
3. Failure to indicate the chemicals' persistence.
4. Failure to discuss modes of entry into the body.
5. Failure to indicate bioaccumulation, biomagnification, and liphophilicity.
6. Failure to indicate volatilization from the soil.
7. Failure to discolse the "related compounds," which more than likely are heptachlor and/or heptachlor epoxide.
8. Failure to indicate the toxic effects of heptachlor and heptachlor epoxide.
9. Failure to indicate chronic effects.

Having reviewed the various labels from the products used in the homes (described above) that were contaminated with chlordane preparations, I believe that without full disclosure of the risks of exposure to these products, the label is defective, and that the use of the product is contrary to FIFRA regulations.[89,90] It is pertinent that EPA recognizes that the current label is not commensurate with the known hazards; this recognition has led to proposals for revised labeling, consistent with some of the anticipated hazards.[91]

CURRENT STATUS OF CHLORDANE

Cancellation of chlordane use has been considered by EPA for a prolonged period of time, but not without considerable pressure from the public.[92] On August 11, 1987, the only manufacturer, Velsicol Corporation, suspended sale of chlordane.[93] The use of existing supplies held by distributors and pest control companies will be allowed.[94,95] The end of sales of chlordane/heptachlor will not end the problems they cause, because of the persistence of these chemicals in the environment and within the body. EPA suspended sale of chlordane as of April 1988.

FUTURE PROBLEMS FROM PESTICIDES

Any product that replaces chlordane will also carry hazards, although perhaps not identical to those of chlordane. It is likely that lack of adequate testing and licensing procedures, and lack of a concerted effort to register and monitor all residences in which these pesticidal chemicals are applied, will lead to future illness and future lawsuits. Cancer is a major cause of illness and death, and various pesticides have wide-ranging effects that include cancer, neurological disorders, and birth defects; so it is imperative that a registration and monitoring system be put in place to track the use of these chemicals.

Unfortunately, only case-by-case methods are currently available for the approximately 30 millions homes treated with chlordane/heptachlor to date. Previously, these chemicals were widely used in agriculture, including food crops; so the extent of the problem is just beginning to surface. The 1980–81 heptachlor contamination of the milk supply in Hawaii, which resulted from feeding heptachlor-contaminated pineapple waste to cattle, and contamination of the milk supply in Arkansas are just two examples of the potential for adverse effects in humans. The further transmission of these chemicals to the suckling infant via its mother's milk and in commercial food has been a concern.[96-98]

Data concerning the toxicology of various pesticides, including the cyclopentadienes, have been available for the past four decades. This information is relatively easy to obtain, compared to data about some of the newer pesticides. Monitoring in the past has showed widespread contamination throughout the population,[99] and differences in levels have been stratified according to race, with blacks having greater evidence of DDT and Lindane contamination than whites.[100] The effect of contamination with biologically active chemicals on the black population in general can only be guessed. One must wonder, however, if the higher rate of many cancers in the black population is reflective of this contamination. At the least, pesticide contamination must be considered in the differential diagnosis of all sick persons, either as an inciting factor or as a contributory factor, the latter mediated by alterations in endocrine and immunological function.

The considerable detail of the above cases is given to provide guidance for the evaluation of any adverse-effect chemical situation. As can be appreciated, it is not necessary to "reinvent the wheel" here. There is considerable knowledge available concerning pesticidal products; or if a product is new to the marketplace, there are data available about related congeners. Considering the long history of the cyclopentadiene pesticides, this damage need not have occurred, had the regulatory officials and the manufacturers and distributors of these products only heeded the scientific findings.

To the pesticide applicator and the consumer goes this advice: before you use *any* pesticidal product be sure you obtain information about, and understand its potential for adverse effects. These are products designed to kill, and their harm may extend to non-target species. As Rachel Carson wrote so movingly in *Silent Spring* more than 25 years ago:

> The "control of nature" is a phase conceived in arrogance, born of the Neanderthal age of biology and philosophy, when it was supposed that nature exists for the convenience of man. The concepts and practices of applied entomology for the most part date from that Stone Age of science. It is our alarming misfortune that so primitive a science has armed itself with the most modern and terrible weapons, and that in turning them against the insects it has also turned them against the earth.[101]

INFORMATION SOURCES

The following sources are provided for persons wanting additional information on pesticides.

Book References, Specific for Pesticides

Carson, R. *Silent Spring*. Houghton Mifflin Co., Boston, 1962, pp. 1–368.

Hayes, W. J. *Pesticides Studied in Man*. Williams and Wilkins, Baltimore, 1982, pp. 1–672.

Horwitz, C., David, S. *Protecting Farmworkers from Pesticides: A Legal Services Corporation Attorneys' Pleading and Practice Manual*. Migrant Legal Action Program, 2001 S. Street, N.W., Washington, D.C.

Morgan, D. P. *Recognition and Management of Pesticide Poisonings*, 3rd ed. EPA-540/9-80-005, Jan. 1982. Available from the Superintendent of Documents, U.S. Govt. Printing Office, Washington, D.C.

Mrak, E. M., Chairman. *Report of the Secretary's Commission on Pesticides and Their Relationship to Environmental Health*. U.S. Dept. of Health, Education and Welfare, 1969, pp. 1–667.

Nader, R., Brownstein, R., Richard, J. *Who's Poisoning America*. Sierra Club Books, San Francisco, 1981, pp. 2–5, 8, 14, 28–31, 37–39, 48, 76, 313–24.

van den Bosch, R. *The Pesticide Conspiracy*. Doubleday, Garden City, N.Y., 1978, pp. 1–226.

Weir, D., Schapiro, M. *Circle of Poison*. Institute for Food and Development Policy, San Francisco, 1981, pp. 1–100.

Organizations

U.S. Environmental Protection Agency, Office of Pesticides and Toxic Substances, 401 M Street, S.W., Washington, D.C. 20460.

The National Coalition against the Misuse of Pesticides (NCAMP), 530 7th Street, S.E., Washington, D.C., 20003.

United Farm Workers, La Paz, Keene, Calif. 93570.

REFERENCES

1. Lovelock, J. E. *Gaia: A New Look at Life on Earth.* Oxford University Press, New York, 1979, p. 59.

2. Comprehensive Drug Abuse, Prevention and Control Act, 1970.

3. Pub. L. 92-516, 86 Stat. 973.

4. Pub. L. 95-203, 91 Stat. 1451.

5. National Research Council, Committee on Toxicology. *Chlordane in Military Housing.* National Academy of Sciences, Washington, D.C., 1979.

6. Letter to J. D. Sherman from Wendell Kilgore, Ph.D., dated Feb. 13, 1987.

7. All Things Considered. Termite chemical—chlordane, Parts I and II. National Public Radio, Washington, D.C., Sept. 14 and 15, 1982.

8. National Academy of Sciences, Committee on Toxicology. *An Assessment of the Health Risks of Seven Pesticides Used for Termite Control.* National Academy Press, Washington, D.C., 1982, p. 47.

9. Johnson, M. State is asked to ban chemical insecticide. *Detroit Free Press*, p. 14-B, Feb. 27, 1987.

10. Bergman, J. M. Contain and correct application accidents. Velsicol Chemical Corporation, 341 E. Ohio St., Chicago, Ill. 60611 (no date).

11. State of Wisconsin v. Douglas Semlar, Case #85-CM-389 (June 6, 1985).

12. State of Wisconsin v. Douglas Semlar, Case #85-CM-389 (Sept. 12, 1985).

13. Degen, Michael C., Solid Waste Specialist, Department of Natural Resources, State of Wisconsin. File #4410, Letter to S. Laufenberg, dated Jan. 21, 1986.

14. Valiulis, D. Groundwater contamination and the fate of agrichemicals (quoting D. E. Glotfelty), *Agrichemical Age*, pp. 10–13, 1986.

15. Hayes, H. M. The comparative epidemiology of selected neoplasms between dogs, cats and humans: a review. *Europ. J. Cancer* 14(12): 1299–1308, 1978.

16. Buck, W. B. Animals as monitors of environmental quality. *Vet. Human Toxicol.* 21(4): 277–84, 1979.

17. Ingle, L. Toxicity of chlordane to white rats. *J. Econ. Entomol.* 40: 264–68, 1947.

18. Bushland, R. C., Wells, R. W., Radeleff, R. D. Effect on livestock of sprays and dips containing new chlorinated insecticides. *J. Econ. Entomol.* 41(4): 642–45, 1948.

19. Radeleff, R. D. Chlordane poisoning: symptomatology and pathology. *Vet. Med.* 43: 342–47, 1948.

20. Burin, G. J. Memorandum to Ellenberger, J., Chlorpyrifos registration standard. EPA., Office of Pesticides and Toxic Substances, May 25, 1984.

21. Bronowski, J. *Science and Human Values.* Harper and Row, New York, 1965.

22. OPP, EPA. *Analysis of the Risks and Benefits of Seven Chemicals Used for Subterranean Termite Control.* EPA-540/9-83-005, 1983.

23. OPP, EPA. *Guidance for the Re-registration of Pesticide Products Containing Heptachlor as the Active Ingredient.* Dec. 1986.

24. Ambrose, A. J., Christensen, H. E., Robbins, D. J., Lelland, J. R. Toxicological and pharmacological studies of chlordane. *Arch. Ind. Hyg. Occup. Med.* 7: 197–210, 1953.

25. Den Tonkelaar, E. M., Van Esch, G. J. No effect levels of organochlorine pesticides based on induction of microsomal liver enzymes in short-term toxicity experiments. *Toxicology* 2(4): 371–80, 1974.

26. Hodgson, E., Kulkarni, A. P. Fabacher, D. L. Robacher, K. M. Induction of hepatic drug metabolizing enzymes in mammals by pesticides: a review. *J. Environ. Sci. Health* 15(6): 723–54, 1980.

27. Fabacher, D. L., Kulkarni, A. P., Hodgson, E. Pesticides as inducers of hepatic drug-metabolizing enzymes. I. Mixed-function oxidase activity. *Gen. Pharm.* 11(5): 429–35, 1980.

28. Street, J. C. Environmental distribution, transformation and toxicological implications of pesticide residues. *Toxicol. Res. Proj. Dir.* 6: 8, 1981.

29. Mahon, D. C., Oloffs, P. C. Effects of sub-chronic low-level dietary intake of chlordane on rats with cirrhosis of the liver. *J. Environ. Sci. Health* 14(3): 227–45, 1979.

30. Epstein, S. S., Arnold, E., Andrea, J., Bass, W., Bishop, Y. Detection of chemical mutagens by the combinant lethal assay in the mouse. *Toxicol. Appl. Pharmacol.* 23: 288–325, 1972.

31. Shirasu, Y. Significance of mutagenicity testing on pesticides. *Environ. Qual. Saf.* 4: 226–31, 1975.

32. Ahmed, F. E. Environmental genetics: a model to investigate pollutants producing genetic damage. *Diss. Abstr. Int.* B-6(8): 3737, 1976.

33. DeSerres, F. J. Prospects for a revolution in the methods of toxicological evaluation. *Mut. Res.* 38: 165–76, 1976.

34. Bishun, N., Smith, N., Williams, D. Mutations, chromosome aberrations and cancer. *Clin. Oncol.* 4(3): 251–63, 1978.

35. Ames, B. N. The identification of chemicals in the environment causing mutations and cancer, in *Carcinogenesis*, Miller, E. C., Ed. Japan Sci. Soc. Press, Tokyo, 1979, pp. 345–58.

36. Burchfield, H. P., Storrs, E. E., Ziegler, J. A., Thomas, J. J., Thomas, B. J. Metabolism and carcinogenicity of organohalogens. *Toxicol. Res. Proj. Dir.* 4(5): 1979.

37. Ruttkay-Nedecka, J., Cerey, K., Rosival, L. Contribution to the evaluation of the chronic toxic effect of heptachlor. *Kongr. Chem. Pol'Nohospod. Pr.* 2: 9, 1972.

38. *Guidana for Re-registration* (see ref. 23).

39. Clegg, D. J. Animal reproduction and carcinogenicity studies in relation to human safety evaluation. *Toxicol. Occup. Med.* 45–49, 1979.

40. Yoshimura, M., Sugiyama, S., Noda, Y., Nakane, H., Doi, S. Organochlorine pesticide residues in the human body. *Med. J. Klin. Univ. Osaka* 2(3): 211–14, 1977.

41. Cranmer, J. S., Avery, D. L., Grady, R. R., Kitay, J. I. Postnatal endocrine dysfunction resulting from prenatal exposure to carbofuran diazinon or chlordane. *J. Environ. Pathol. Toxicol.* 2(2): 357–69, 1978.

42. Donovan, J. S., Schein, L. G., Thomas, J. A. Effects of pesticides on metabolism of steroid hormone by rodent liver enzymes. *J. Environ. Pathol. Toxicol.* 2(2): 447–54, 1978.

43. Kamenov, D. A. Analysis of the effect of pesticides on *Mus muscularus L.* and *Apodemus flavicollis Melch.* *Byull. Mosk. Ova. Ispyt. Prir. Otd. Biol.* 85(6): 24–30, 1980.

44. Cranmer, J., Cranmer, M., Morris, F., Goad, P. T. Prenatal chlordane exposure: effects on plasma corticosterone concentrations over the lifespan of mice. *Environ. Res.* 35: 204–10, 1984.

45. Bernstein, R. L., Serrato, K. M., Wininant, R. C. Chlorinated insecticides enhance the metabolic activation of chemical carcinogens as mutagens. *Amer. Soc. Microbiol.*, meeting abst. 79: 120, 1979.

46. Taira, M. C., San Martin de Viale, L. C. Porphyrinogen carboxylase from chick embryo liver in vivo effect of heptachlor and lindane. *Int. J. Biochem.* 12(5–6): 1033–38, 1980.

47. Del, C., Vila, M., San Martin de Viale, L. C. Effect of parathion, malathion, endosulfan and chlordane on porphyrin accumulation and alasynthetase in chick embryo liver. *Toxicology* 25: 323–32, 1982.

48. Denys, A., Guilbert, E., Levy, J. Characteristics of enzymatic induction provoked by chlordane. *Therapie (Fr.)* 30: 277–88, 1980.

49. Verman, S. R., Bansal, S. K., Gupta, A. K., Dalela, R. C. Pesticide induced hematological alterations in a freshwater fish, *Saccobranchus fossilis. Bull. Environ. Contam. Toxicol.* 22(4-5): 467-74, 1979.

50. Koller, L. D. Effects of environmental contaminants on the immune system. *Adv. Vet. Sci. Comp. Med.* 23: 267-95, 1979.

51. Spyker-Cranmer, J. M., Barnett, J. B., Avery, D. L., Cranmer, M. Immunoteratology of chlordane: cell-mediated and humoral immune responses in adult mice exposed in utero. *Toxicol. Appl. Pharmacol.* 62: 402-8, 1982.

52. Theby, J. E. A preliminary investigation of the effects of six pesticides on the living tissues of mice. *Trans. Kans. Acad. Sci.* 81(1): 83, 1978.

53. Christophers, A. J. Hematological effects of pesticides. *Ann. N.Y. Acad. Sci.* 160(1): 352-55, 1969.

54. Infante, P. F. Blood dyscrasias and childhood tumors and exposure to chlordane and heptachlor. *Scand. J. Work Environ.* 4(2): 137-50, 1978.

55. Slichter, S. J., Harker, L. A. Thrombocytopenia: mechanisms and management of defects in platelet production. *Clin. Hematol.* 7(3): 523-39, 1978.

56. Furie, G., Trubowitz, S. Insecticides and blood dyscrasias: chlordane exposure and self-limited refractory megaloblastic anemia. *J.A.M.A.* 235(16): 1720-22, 1976.

57. Scott, J. M., Weir, D. G. Drug-induced megaloblastic change. *Clin. Hematol.* 9(3): 587-606, 1980.

58. Cerey, K., Izakovic, V., Ruttkay-Nedecka, J. Effect of heptachlor on dominant lethality and bone marrow in rats. *Mut. Res.* 21: 26, 1973.

59. Aleksandrowicz, J. Leukemias and environment. Epidemiology of leukemias and investigations on the leukemogenic influence of the physiochemical environment. *Pol. Hyg. Lek.* 19(35): 1313-15, 1964.

60. Infante, P. F., Newton, W. A. Prenatal chlordane exposure and neuroblastoma. *New Engl. J. Med.* 293(6): 308, 1975.

61. Best, J. B., Morita, M. Planarians as a model system for in vitro tetratogenesis studies. *Tetratogen. Carcinogen. Mutagen.* 2:277-91, 1982.

62. Drummond, L., Chetty, N., Desaiah, D. Changes in brain ATP-ase in rats fed on chlordane mixed with iron-sufficient and deficient diet. *Drugs Chem. Toxicol.* 6(3): 259-66, 1983.

63. United Nations Labor Organization. *Environmental Health Criteria 38.* World Health Organization, Geneva, Switzerland, 1984, pp. 1-80.

64. Fishman, E., Gianutsos, G. Inhibition of 4-aminobutyric acid (GABA) turnover by chlordane. *Toxicol. Letters* 26: 219-23, 1985.

65. Ohno, Y., Kawanishi, T., Takahashi, A., Nakura, S., Kawashima, K. Comparisons of the toxicokinetic parameters in rats determined for low and high dose of chlordane. *J. Toxicol. Sci.* 11: 111-23, 1986.

66. Mrak, E. M., Chairman. *Report of the Secretary's Commission on Pesticides and Their Relationship to Health.* U.S. Department of Health, Education and Welfare, 1969, pp. 1-677.

67. *IARC Monographs on the Evaluation on the Carcinogenic Risk of Chemicals to Man: Some Organochlorine Pesticides.* International Agency for Research on Cancer, Lyon, 1974, 5: 1-235.

68. National Cancer Institute. *Preliminary Report on the Carcinogenesis Bioassay of Chlordane and Heptachlor.* U.S. Department of Health, Education and Welfare, Washington, D.C., 1975.

69. Epstein, S. S. Carcinogenicity of heptachlor and chlordane. *Sci. Total Environ.* 6(2): 103-54, 1976.

70. Epstein, S. S. Prevention—environmental exposure: an overview, including the role of pesticides. Third International Symposium on Detection and Prevention of Cancer, 1976.
71. Epstein, S. S. The carcinogenicity of organochlorine pesticides, in *Origins of Human Cancer, Book A*. Cold Spring Harbor Laboratory, New York, 1977, 4: 243-65.
72. Anon. Report on carcinogenesis bioassay of chlordane and heptachlor. *Amer. Ind. Hyg. Assoc. J.* 38(11): 10-44, 1977.
73. Infante, P. F., Epstein, S. S. Blood dyscrasias and childhood tumors and exposure to chlorinated hydrocarbon pesticides, in *Proc. Conference on Women and the Workplace*, Bingham, E., Ed. Society for Occupational and Environmental Health, Washington, D.C., 1977, pp. 51-74.
74. Reuber, M. D. Histopathology of carcinomas of the liver in mice ingesting heptachlor or heptachlor epoxide. *Exp. Cell. Biol.* 45(3-4): 147-57, 1977.
75. Reuber, M. D. Hepatic vein thrombosis in mice ingesting chlorinated hydrocarbons. *Arch. Toxicol.* 38(3): 163-68, 1977.
76. Reuber, M. D. Histopathology of liver carcinomas in (C57BL/6N × C3H/HeN) F1 mice ingesting chlordane. *J. Nat. Cancer Inst.* 63(1): 89-92, 1979.
77. Butler, W. H., Jones, G. Pathological and toxicological data on chlorinated pesticides and phenobarbital. *Ecotoxicol. Environ. Saf.* 1(4): 503-9, 1978.
78. Cohen, D. Guarding against cancer. *EPA J.* 4(3): 12-13, 1978.
79. Kraybill, H. F. Carcinogenesis induced by trace contaminants in potable water. *Bull. N.Y. Acad. Med.* 54(4): 413-27, 1978.
80. Robens, J. F. Tests for possible carcinogenicity of 20 pesticides in Osborne-Mendel rats and B6C3F1 mice. *Toxicol. Appl. Pharmacol.* 45(1): 236, 1978.
81. *IARC Monographs on the Evaluation of the Carcinogenic Risk of Chemicals to Humans.* International Agency for Research in Cancer, Lyon, 1979, 20: 45-65, 129-154.
82. Pendygraft, G. W., Schlegel, F. E., Huston, M. J. Organics in drinking water: maximum contaminant levels as an alternative to the GAC treatment requirements. *Amer. Water Works Assoc. J.* 71(4): 174-83, 1979.
83. Sternberg, S. S. The carcinogenesis, mutagenesis and teratogenesis of insecticides. Review of studies in animals and man. *Pharmacol. Ther.* 6(1): 147-66, 1979.
84. Anon. GAO, Velsicol differ on chlordane. *Chem. Eng. News* 58(34): 25, 1980.
85. Axelson, O. Chlorinated hydrocarbons and cancer: epidemiologic aspects. *J. Toxicol. Environ. Health* 6(5-6): 1245-51, 1980.
86. Blejer, H. P. Is your patient's job killing him? *Med. Times* 108(7): 91-95, 1980.
87. Saleh, M. A. Mutagenic and carcinogenic effects of pesticides. *J. Environ. Sci. Health* 15(6): 907-27, 1980.
88. Shabecoff, P. Pesticide still cancer risk despite caution, U.S. says. *New York Times*, p. A-12, Apr. 20, 1987.
89. Pub. L. 80-104, 61 Stat. 163.
90. Pub. L. 92-516, 86 Stat. 973.
91. EPA Office of Pesticides and Toxic Substances. *Guidance for the Re-registration of Pesticide Products Containing Chlordane as the Active Ingredient.* Environmental Protection Agency, Washington, D.C., 1986.
92. Baxter, D., Feldman, J., Epstein, S. Petition to EPA for emergency suspension and cancellation of registrations for the pesticides chlordane/heptachlor and aldrin/dieldrin. National Coalition against the Misuse of Pesticides, Washington, D.C., 1987.
93. Memorandum of understanding between Velsicol Chemical Corporation and EPA, Aug. 11, 1987.

94. Taylor, R. E. Velsicol agrees with EPA to halt sale of anti-termite chemical pending test, *Wall Street Journal*, Aug. 11, 1987.

95. Anon. Maker to stop sales of 2 cancer-tied pesticides, *New York Times*, Aug. 12, 1987.

96. Letter from U.S. Senator Spark Matsunaga to William Ruckelshaus, Administrator, EPA, Apr. 15, 1983.

97. Mattision, D. R. Heptachlor in breast milk—to feed or not? Dept. Obstetrics and Gynecology, University of Arkansas, Little Rock, Mar. 29, 1986.

98. Anon. Newsbreak: Heptachlor alert. *University of Arkansas for Medical Sciences*, Little Rock, 1(3): 1-3, 1986.

99. Kutz, F. W., Yobs, A. R., Strassman, S. C. Organochlorine pesticide residues in human adipose tissue. *Bull. Soc. Pharm. Environ. Pathol.* 4(1): 17-19, 1976.

100. Kutz, F. W., Yobs, A. R., Strassman, S. C. Racial stratification on organochlorine insecticide residues in human adipose tissue. *J. Occup. Med.* 19(9): 619-22, 1977.

101. Carson, R. *Silent Spring.* Houghton Mifflin Co., Boston, 1962, p. 297.

8

PESTICIDAL CHEMICALS — ORGANOPHOSPHATES

INTRODUCTION: POLLUTION ON A PERSONAL LEVEL

The problem of indoor air pollution arising from the intentional use of pesticides is increasing in severity and frequency. Contributing to the problem are aggressive advertising urging the use of pesticides to control relatively innocuous pests; lack of teaching and knowledge about alternative, less hazardous methods of pest control; use of persistent and toxic chemical products; and construction of buildings with poor or no outside ventilation.

Pesticide commercials prey on the fears and the ignorance of the public: with ploys such as a roach that fills the borders of a television screen. We are led to believe that if we don't "protect" our families with pesticides, we are remiss in our duties. Given the choice of the hazard from roaches or the hazards from many pesticides, those concerned with health and the environment, including myself, will choose the former. There are time-honored methods of dealing with insect pests, none requiring toxic chemicals. Cleanliness, closed containers for foods, prompt removal of garbage, and plugging of access routes are easily and safely achieved. Because of the promotion of "easy," "fast," "professional" services, we are losing our common sense abilities to care for ourselves. In our desire to have a weed-free lawn or an insect-free building, we lose the connection between our actions and the consequences for the rest of the world. Is a weed-free lawn intrinsically superior to one with clover, fescue and dandelions?

To allow a person to dispense a toxic agent in or around ones home or place of work defies common sense. Few people know the identity of products they pay to have applied in their homes or allow to be used in their work areas. Fewer still understand the hazards associated with these chemicals.

Insects are some of the oldest and most resistant life forms on earth. What makes anyone believe that a chemical that harms insects will not also harm other forms of life?

Pesticides are regulated by the U.S. Environmental Protection Agency, under the law called FIFRA. FIFRA stands for the Federal Insecticide, Fungicide and Rodenticide Act. All of the products registered under FIFRA are designed for one thing only, and this is to *kill*. Insecti*cides* kill insects, fungi*cides* kill fungi, rodenti*cides* kill rats and mice, but also other animals such as foxes, wolves, voles, etc. Missing from the acronym is herbi*cide*, designed to kill plants, and also regulated under FIFRA. But, as is the case with pesticides in general, there are overlapping effects, as we have learned from a number of the herbicides, such as 2,4-D and 2,4,5-T, the components of "Agent Orange" used in Vietnam, which killed their users as well.

ORGANOPHOSPHATE EFFECTS:

Exposure to organophosphate (OPO^4) products, whether used as war agents or as pesticides, results in predictable toxic effects due to neurotransmitter interference. Acetylcholine is the chemical nerve transmitter normally found at the connection of three main divisions of the nervous system. OPO^4 pesticides cause inhibition of cholinesterase, the enzyme that normally reverses the action of acetylcholine, thus allowing accumulation of acetylcholine and un-reversed stimulation of nerve tissue and effector organs. Effects may be produced by multiple small doses, resulting in signs and symptoms comparable to a single larger exposure.

The *muscarinic* nerve receptors, located in the postganglionic para-sympathetic nerve endings, are found primarily in smooth muscle, the heart, and endocrine glands, with stimulation resulting in the following symptoms: wheezing, nausea, vomiting, abdominal cramps, frequent and/or involuntary defecation and urination, and visual disturbance.

Stimulation of the *nicotinic* receptors, located in the autonomic ganglia and at the endings of skeletal muscle nerves, produces symptoms of weakness, fatigue, muscle cramps, and involuntary muscle twitching and fasciculation. Weakness of the muscles of respiration contributes to loss of respiratory efforts and dyspnea. Autonomic effects include rapid heartbeat, which may mask the muscarinic bradycardia, increased blood pressure, and pallor.

Delayed neurotoxic effects usually begin in the distal lower limbs with sensory disturbance and motor weakness.[1] Neurological damage may lead to ataxia, increasing weakness, and flaccid paralysis. Some recovery may occur with discontinuation of exposure, but repair is slow and often not complete.

Involvement of the nerves controlling urination and defecation, resulting in loss of bladder and bowel control, has been described in both animal models and humans. These findings are some of the most serious chronic alterations of body function reported by patients poisoned with chlorpyrifos,[2,-3] the most serious life-threatening effects obviously being respiratory and cardiac damage.

Central nervous system symptoms of acetylcholine accumulation secondary to OPO[4] exposure include headache, anxiety, confusion, slurred speech, tremors, incoordination, generalized weakness, restlessness, sleep disturbance, nightmares and excessive dreaming, emotional instability, neurosis, apathy, and seizures. OPO[4] poisoning may result in permanent neurobehavioral abnormality.[4]

The symptoms of enhanced sensitivity to environmental agents, including multiple chemical sensitivity (MCS) are often attributed to neurosis, despite a change in a person's otherwise normal function in *temporal* relation to pesticide exposure. The psychologist/psychiatrist "may be in the best position to detect and validate symptoms of intoxication from 'safe' chemicals that do not affect most people."[5] Research detecting immunological abnormalities in persons exposed to pesticides [6-7] lends credence to a physicochemical basis of MCS.

The potential for adverse effects is much greater for infants and children because of their relatively smaller size, their propensity to crawl about on the floor or ground and to put foreign objects in their mouths.[8] The problem of loss of mentation is especially troubling, in that the person becomes mentally, physically, and economically disadvantaged for life.[9]

Summary of Effects:

- *Muscarinic*
    Bronchial constriction   } { wheezing
    Increased bronchial secretions   } { chest tightness
    Increased salivation
    Increased lacrimation
    Increased sweating
    Increased peristalsis   } { nausea, vomiting, diarrhea,
    Increased gastrointestinal   } { abdominal cramps, tenesmus,
        tone   } { involuntary defecation
    Bradycardia — heart block
    Bladder muscle contraction — frequent and involuntary urination
    Pupillary constriction visual disturbances
- *Nicotinic*
    Weakness
    Fatigue
    Involuntary muscle twitching and fasciculations
    Muscle cramps. (The weakness of the muscles of respiration contributes
        to the dyspnea and loss of respiratory efforts.)
    Autonomic effects:
        Tachycardia (which may mask the muscarinic bradycardia)

           Increased blood pressure
           Pallor

- *Central nervous system*
      Slurred speech
      Tremors
      Loss of coordination
      Generalized weakness
      Convulsions
      Headache
      Sleep disturbance
      Restlessness
      Anxiety
      Emotional instability
      Neurosis
      Nightmares and excessive dreaming
      Confusion
      Apathy

Depression of the circulatory and respiratory centers is the main cause of coma, respiratory failure, and death.

Exposure to organophosphate pesticides can result in a *local* effect on a target organ system, bronchial constriction and increased bronchial secretions in the case of an inhaled pesticide, and eye symptoms from airborne contamination.

Cumulative inhibition of cholinesterase may be produced by multiple small doses, resulting in signs and symptoms comparable to a single larger exposure.

Exposed persons may display any of the constellation of symptoms, and thus present a problem of correct diagnosis for a physician or health official. Failure to make the connection between a patient's illness and pesticide exposure has serious consequences. Patients have received ineffective (and often harmful) treatment with such drugs as antihistamines, tranquilizers, and sedatives. If long-term sequelae are to be avoided, treatment of the acute phase with atropine must be undertaken. Equally important is cessation of exposure, with removal of the offending pesticide via clean-up, and often including evacuation of the contaminated premises.

CONFOUNDERS OF DIAGNOSIS: FORMULATIONS

Marketing of similar products under different trade names, and the promotion of odorless and slow-release encapsulated formulations, make diagnosis difficult.

Unrecognized migration and persistence of indoor pesticide products, even when used according to direction, results in contamination.[10–11]

Pesticides that are not water soluble are often formulated in an organic solvent base (such as xylene) and/or with an emulsifying agent, and thus add the toxicity of the carrier chemical(s), and increase the potential for uptake. Fat soluble pesticides are absorbable through the intact skin and have potential to bioaccumulate and biomagnify. Those products with a benzene-ring structure and/or a chlorinated benzene ring structure are retained within the fat stores of the body, and allow for prolonged biological action.

HISTORY OF DEVELOPMENT OF ORGANOPHOSPHATE CHEMICALS

The organophosphate pesticides (OPO[4]) were developed in Germany during World War II.[12] These products included TEPP (tetraethyl pyrophosphate, developed as a nicotine substitute), followed by Tabun (dimethyl phosphoroamidocyanidate) and Sarin (isopropyl methylphosphonofluoridate) — the chemical "nerve agents" that have been employed in warfare. The developer IG Farben, not content with tests on monkeys, confirmed lethality by testing Tabun on prisoners at Auschwitz.[13] Thus, the OPO[4] chemicals, such as chlorpyrifos and diazinon,[14–15] became direct descendants of the nerve agents.

CHLORPYRIFOS (Dursban):

Because of widespread use, special attention is paid to Dursban (chlorpyrifos — O,O-diethyl-0-3-5-6-trichloro-2-pyridyl phosphorothioate), a chemical with three chlorine atoms on the central nitrogen-containing benzene ring. Chlorpyrifos was developed by Dow Chemical Company, and is sold by Dow-Elanco, a division of the pharmaceutical corporation Eli Lilly. Chlorpyrifos is sold under a number of trade names, such as Dursban, Lorsban, Empire, Equity, Lentrek, Lock-On, Pagent, Killmaster II, Tricel, Pyriban, Contra-Insect, Pamol, Dorsan, and Pyrinex. Chlorpyrifos is one of the most widely used insecticides on the market, and is employed in homes, schools, nurseries, hospitals, restaurants, office buildings, etc. for general pest control of such insects as fleas, roaches, etc. Chlorpyrifos has been promoted for structural termite control, to replace chlordane and heptachlor, which were removed from sale due to health concerns including carcinogenicity, persistence, and bioaccumulation.

HISTORY OF CHLORPYRIFOS

The story of the development of chlorpyrifos is an interesting one. A 1963 report from the Biomedical Research Laboratory of Dow Chemical Company states that the product possessed "rather potent cholinergic properties." Testing concluded: "Judging by general appearance and weight gain, doses of 2.0 g/kg body weight produced no untoward physiological effects.... However, when the material was administered as a concentrated solution (50% Dowanol DPM), four out of six animals failed to survive 2.0 g/kg active."[16]

Knowing that the reaction in animals was increased with solvent usage, and knowing that it possessed cholinergic properties, one might properly question its safety. The animal observations were based on *appearances and weight gain*, not cholinesterase tests. The 2.0 g/kg dosage produced "no untoward physiological effects," but death of two thirds of the animals might have predicted a concern.

In 1964, the Dow Chemical Co. and Pitman-Moore Research Laboratories in Indianapolis tested Dursban on dogs by dipping the animals in solutions varying from 0.0125 to 0.1% solutions. Of the 40 dogs tested, 12 were pregnant, 10 whelping during the dipping test, and two afterward. The pups born to four of the pregnant females accounted for 22% of the deaths, either at birth or soon thereafter. Another 51% of the pups died of "pneumonia", with a total cumulative mortality of 62%. Despite the findings, the conclusion was that "based on the results of these tests, dip solutions prepared from an emulsified concentrate of Dursban are safe for use in adult male and female dogs in concentrations up to 0.1% regardless of size." The further conclusion was that "concentrations up to 0.1% of Dursban can be used to dip bitches at any state of pregnancy without interfering with gestations, parturition or puppy viability."[17]

In 1966, 1967, and 1968, Dow Chemical Co. conducted outdoor tests of spraymen exposed to an 0.5% suspension or emulsion of chlorpyrifos, employing spray pressures ranging from 250 to 125 per square inch. It was concluded that: "Inasmuch as a compound that produces a measurable decrease in plasma or red blood cell cholinesterase level in spraymen is not considered to be suitable material for use in public health work in the absence of an epidemic or the immediate threat of one, the conclusion is that Dursban emulsion or suspension formulations are not acceptable for use as a premises larvicide treatment."[18] Despite that finding, the use of chlorpyrifos is widely used on premises.

Exemplifying a cavalier approach to public health, the above trials took place in premises and vacant lots "normally encountered in a subtropical, lower socioeconomic neighborhood." An average of 29.1 gallons of 0.5% Dursban was applied per premise, at 250 pounds per square inch of pressure. "The high pressure resulted in micronization and considerable splashback of the finished spray," necessitating discontinuation of the program after two weeks because of cholinesterase depression in the spraymen. There is no indication that any inquiries were made into the medical status or cholinesterase levels of the unsuspecting residents whose premises had been sprayed. In light of the emerging scandal regarding experimentation on humans with radioactive materials, a lack of follow-up of these unwitting subjects is no less unethical. It may be argued that the use of these pesticidal products without specific informed consent is a violation of moral and ethical principles. In all fairness, it should be noted that Dow Chemical Corporation is not alone in testing chemicals on disadvantaged populations. In 1989, Eli Lilly, the manufacturer of terbuthiuron (sold under the trade name of Spike) sprayed 70 acres of coca fields in Peru, endangering the region's ecosystem.[19]

VOLATILITY

In 1965, Dow undertook a series of rat exposure tests "to simulate dwellings treated with *high* insecticidal residues and extended coverages similar to those used in malarial areas of Africa (100 milligrams per square foot with 75% of the inside area treated." The results showed that both plasma and red-cell cholinesterase levels were depressed in the animals, but recovered upon removal from exposure. Ominously, "housefly bioassays showed considerable vapor toxicity for at least 174 days, and lethal contact effects for 183 days, at the time of the termination of the test." Furthermore, after 81 days, air samples showed concentrations of as much as 0.04 micrograms per liter. A second test, with 40 milligrams per square foot of treatment, controlled house flies for 70 days or more "by contact with the combined residue and vapor."[20] The results noted that Dursban concentrations "decreased less than half during the span of the test." And, "Dursban has affinity for surfaces in the vapor phase, *condensing and recondensing₊ moving along certain channels*, thus this sampling device may never register the true concentration." The document states further: "The concentrations detected by our air sampling device varied between 4.2 and 2.3 per cent of the saturated vapor concentration."

A publication of the Dow Chemical Company shows a pesticide application being made next to a stove.[21] Given the volatilization of Dursban, concern is raised about the dynamics of translocation when the oven was heated, and the subsequent effects upon the family in that home.

The duration of biological activity was confirmed later, noting, "chlorpyrifos is volatile from surfaces such as glass, steel, and green foliage where it is not sorbed tightly and therefore, the half-life may be hours or days under such conditions. In the presence of sunlight and water, and in the absence of organic particulate matter chlorpyrifos is transient due to degradation. However, in dry soil, stored grain, plywood surfaces, and organic matter in soil and water, chlorpyrifos may remain active for weeks or months." It states further: "This is because the volatility, hydrolytic, and other dissipation forces are less effective and there is sequestering and strong binding of chlorpyrifos by absorptive forces to such surfaces."[22]

This work done in the 1960s is all the more interesting when contrasted with the 1980 creation and registration of Dursban L. O. (low odor), employing a "patented formulation technology with greatly reduced solvent levels".[23] Given that smell is one of the critical senses alerting an animal or person to the presence of danger, the production of a low-odor formula would block at least one of the mechanisms used to protect against exposure to toxic chemicals.

PERSISTENT CONTAMINATION

Tests to determine how much of the pesticide could be *removed* for various surfaces were done, with unsurprising results indicating that more chemical could be removed from a smooth surface such as vinyl than from a rough surface such as a rug. Despite the fact that chlorpyrifos is soluble in solvent, and most tests to recover non-water soluble chemicals from various surfaces employ a solvent to do wipe tests, this exercise employed "a piece of Curity Gauze Diaper." This study then went on to say: "The wipe test is not intended to indicate how much surface-applied chemical is present," and "No exposure thresholds have been established for surface concentrations of chlorpyrifos."

The question of how much chemical remains in and on various surfaces and how much could be absorbed through the skin of a child playing on a floor is of concern. Given the "condensing and recondensing, moving along certain channels," described two decades ago, there is indeed cause for concern.

Dow workers had already shown that "Dursban is very residual on porous and fairly inert surfaces such as wood and grain and only moderately residual on foliar surfaces....It is volatile enough to form insecticidal residues on surfaces of nearby untreated objects and is stable except under fairly rigorous conditions of alkalinity and acidity."[24] An application of 40 mg per square foot of the chlorpyrifos formulation, Dowco 179, produced 100% mortality for the control of cockroaches through ten weeks.[25] Dursban was claimed to give effective control of cockroaches for 19 weeks on unpainted metal, and 60 weeks on tile surfaces.[26]

In response to concerns from the U. S. Department of Agriculture concerning "safety of children exposed to lawns treated with Dursban 2E," the reply was that "the chemical was not readily absorbed through the skin and hence would not be a safety problem." Although it was stated "we do not have any data on the persistence of the Dursban on grass foliage," it was also said that "Dursban is not easily broken down and is quite persistent under certain circumstances where it is tightly absorbed."[27] Persistence was confirmed by a second Dow employee, who said it was "resistant to leaching or movement by soil water...,"since "Application of 6 inches of water with an artificial rain maker did not move the compound from the top 1 inch of loam soil where it was originally placed."[28] The 1987 Dow Technical Data Sheet indicates an average soil half-life of 68 ± 13 days, chlorpyrifos being "retained on paper, wood and painted surfaces."

In the absence of sunlight and water, as within a building, the half-life can be expected to be *at least* as long as out-of-doors. Additionally, where frequent applications are made, often on a monthly basis, the total amount of chlorpyrifos can be expected to accumulate as the rate of application exceeds the rate of breakdown.

As with all pesticides, chlorpyrifos is required to be used according to the label. Nowhere on the label is there any caution about the frequency of use.

ABSORPTION

A 1973 Dow-sponsored study of dogs, using Dursban as a flea-control product,[29] employed 5 test animals (3 male, 2 female), plus 2 control animals, with 4 animals exposed only to the solvent vehicle. Plasma cholinesterase levels were significantly depressed in test animals and remained so for 14 to 42 days after cessation of direct application. The final two paragraphs of the study state:

> Erythrocyte and plasma cholinesterase values obtained to indicate systemic absorption of Dursban did not reflect significant decreases in erythrocyte cholinesterase values. However, at periods of 1 and 14 days during the study and 14 and 42 days following their final treatment, *significant decreases occurred in plasma cholinesterase values.* (underlined for emphasis). The latter returned to pre-test values 74 days following the final treatment of 88 days following the initial exposure to Dursban. (sic)

> In the absence of clinical toxicity, the significance of depression of cholinesterase values as a measure of toxicosis is uncertain. However, this test appears to be a reliable indication of systemic absorption of Dursban.

The "absence of clinical toxicity" evidently did not consider weight loss by two dogs exposed to Dursban through 4 weeks of observation, and two exposed through the 8 week point.

Percutaneous absorption of Dursban was demonstrated in rabbits exposed to 316, 630, 1260, 2510, and 5000 mg/kg doses applied for 24 hours, after which the skin of the animals was washed with soap and water.[30] The formulation contained 43% chlorpyrifos in a formulation containing 28% dipropylene glycol monomethyl ether and other unknown components. There were 2 animals of each sex in each treatment group. The results showed:

Dose mg/kg	#Dead / #Treated	Time of Death
316	0/4	-0-
630	2/4	1 death at 3 days, 1 death 14 at days
1260	1/4	1 death at 4 days
2520	2/4	2 deaths at 3 days
5000	4/4	3 deaths at 3 days, 1 death at 4 days

The discussion of the results noted "slight diarrhea was observed in three animals that subsequently died, and incoordination of the hind limbs was observed in two animals from the highest group prior to death." These findings are characteristic of cholinesterase inhibition and neurological involvement.

The results of acute oral administration of chlorpyrifos to groups of 5 rats was noted in this same report:

Animals in cages 2-8 have body tremors at 1.5 hours past dosage. These animals are also listless and jump if an attempt is made to move them. At 5.5 hours past dosage all animals have severe body tremors. The next day, the technician noted: "Animals in cages 5-7 are very listless and urine soaked."[31]

Urinary incontinence and incoordination of the legs are evidence of peripheral neuropathy.

An oral test of chlorpyrifos, by Dow, done 3 months later noted:

"2.5 hrs — all have body tremors." The day following, the technician noted; "body tremors top 3 doses; lethargy top 2 doses; dark eye secretions top 2 doses; urine soaked top dose." One day later the notes state: "all levels body tremors and hyperreactive."[32]

Classic tests for delayed neurotoxicity were undertaken in 1978, administering chlorpyrifos to hens in a single oral dose, as well as both positive and negative controls.[33] Traditionally the test must be done with single doses at least as great as the LD-50 of the compound, with the birds to be protected from cholinergic effects with full doses of atropine.

The reported results leave a number of questions unanswered: The table listing acute oral toxicity of chlorpyrifos to adult white leghorn hens, showed a 66.7 percent mortality for dosages of both 50 and 100 mg/kg. With prior atropinization, the table of clinical conditions showed normal neurological function conditions by the second day in hens given 50 mg/kg and normal by the fourth day in those given 100 mg/kg. The text however, indicates "completely recovered" after 31 to 55 hours for the 50 mg/kg group, and 79 to 127 hours for the 100 mg/kg group. Additionally, one hen in the 50 mg/kg group was described with classical neurological symptoms: "it appeared depressed, was weak, and preferred recumbency," but recovered and "appeared normal by day 9 and remained normal until sacrificed on day 24." This disparity of results is not explained further.

In the positive control group given the organophosphate TOTP (tri-o-tolyl phosphate), all hens showed swollen, degenerate, fragmented axons and demyelination of the spinal cord with the dorsal and lateral spinal columns principally affected.

The report noted no histological evidence of delayed neurotoxic effects in the control or the hens given only atropine. Yet it stated "Some birds in *all test* groups had minimal to marked focal lymphocytic infiltrate in the spinal cord and/or sciatic nerve," and concluded: "The lesions probably antedated the experimental treatments and were considered evidence of neural lymphomatosis."

Either all of the animals were normal prior to the experiment, or they weren't. It leaves unanswered questions of adequate science to conclude no neurotoxicity when an experiment shows normal findings in the control groups, and abnormalities of the spinal cord and/or sciatic nerves in *all* of the test groups.

BIOLOGICAL PERSISTENCE

Residues of chlorpyrifos, and its metabolite 3,5,6-trichloro-2-pyridinol, were found in cattle treated with a single skin application of 2 ml/100 lb. body weight.[34] Chlorpyrifos residues in fatty tissue ranged from 0.28 ppm to 1.1 ppm at days one and seven. Levels in muscle, liver, and kidney were reported less than 0.18 ppm. The pyridinol moiety was found predominantly in the liver and kidneys, reaching a maximum of 1.0 and 0.86 ppm at 7 days, and less than 0.24 ppm in fat. Essentially nondetectable levels were reported for 35 days. Three animals were killed at day one and every 7 days thereafter, providing data on only three animals for each time period. Fat retention of Dursban or its degradation products in animals had been found "for approximately 40 days after treatment with 0.5% spray," although it was stated that "this concentration is considerably higher than that anticipated for practical field use." The 1987 recommended level for use on farm animals was 10 mg/kg, or 1.0%.

A petition to establish a tolerance on meat animals was proposed with an experimental use of Dursban 44 (containing 43% chlorpyrifos) applied as a single application to swine. A 70-day holding period following treatment and prior to marketing was proposed.

The continuing use of chlorpyrifos in the human habitat must be contrasted with that recommended for breeding swine. It was urged that chlorpyrifos be applied as a varnish to metal *covered* areas to control cockroaches to avoid the "risk of contaminating farm animals."[35]

Dursban became approved for use on cattle and sheep, in poultry houses,[36] and on a variety of food products, including corn, sorghum, soybeans, sugar beets, sunflowers, and peanuts.[37]

Considering that one of the main uses of corn, soybeans, sunflowers and peanuts is for the production of food oils, there is a question if these foods are contaminated with chlorpyrifos. The standard EPA/ FDA consumer "Market Basket" tests of food for pesticide contamination have not routinely included assays for chlorpyrifos.

Interior use of chlorpyrifos results in contamination and a cycle of exposure, with carpets furniture and bedding acting as absorbent reservoirs. A major source of exposure, especially for children, is skin contact with contaminated surfaces. When applied outdoors, pesticide residues are carried indoors on the feet of animals and humans.

Heat, whether from sunlight or a heating system, can volatilize chemicals, resulting in deposition on colder surfaces not usually thought of as absorbent, such as the outside of appliances, mirrors, and windows, especially at night. As heat increases, a new cycle of re-circulation takes place. For this reason, when monitoring is needed, wipe samples should always be obtained, in addition to air samples, to obtain a full measure of the reservoir of exposure.

THE QUESTION OF DIOXIN CONTAMINATION

In June of 1989, the EPA requested from Dow "specific analysis of chlorpyrifos for potential impurities structurally related to dibenzodioxins." The EPA cited the rationale: "The Agency is concerned that an impurity structurally related to TCDD (2,3,7,8-tetrachloro-p-dioxin) may form during the manufacture of chlorpyrifos, based on the starting materials used and other impurities known to occur."[38] Despite requests to the EPA, under the Freedom of Information Act, information concerning potential dioxin contamination remains unknown. An

EPA notice issued three years later requested data "in all discipline areas (product chemistry, residue chemistry, toxicology, environmental fate and ecological effects).

The subject of contamination is of more than theoretical interest on two accounts. Illnesses suffered by Gulf War veterans are similar to illnesses developed by civilians who have been exposed to pesticides.[39-40] The use of pyridostigmine, related to the carbamate group of pesticides, plus the use of chlorpyrifos and other organophosphate pesticides may be one of the keys to the illnesses suffered by these military personnel. As with Vietnam veterans, pesticide exposure again may be a significant factor in disease causation.

A second medical problem, involving what is called the "Post Polio Syndrome," became recognized but a decade ago, occurring in the young as well as the elderly after a stable course of 20 to 40 years. Nerve damage caused by the polio virus leaves victims with fewer functioning nerve cells than those unaffected. It follows that a person with a nervous system previously damaged by the polio virus will be less tolerant of a subsequent neurotoxic insult such as an organophosphate pesticide.

A thorough inquiry into organophosphate pesticide exposure and its possible causation in each of these situations should be undertaken.

OTHER USES

Testing had revealed that Dursban was not effective against mites, and gave unpredictable results against beetles and their larvae. By 1971, Dow Chemical Company recommended DOWCO 179, the active ingredient in Dursban for "widespread use in the control of mosquitos" and other insects in areas visited by birds, despite the warning of other workers who reported "relatively high degree of accumulative toxicity for an organophosphate."[41]

The methyl form of chlorpyrifos is marketed as Reldan for use on stored grain. Early in 1994, the substitution of the methyl form of chlorpyrifos with the ethyl form resulted in the contamination of 16.8 million bushels of oats owned by General Mills. Before the error was discovered, 160 million boxes of cereals such as Cheerios, Lucky Charms, Booberry, and Trix had been made, with over 2/3 on grocery shelves. While the substitution cost General Mills as much as $87.5 million, the question of potential harm spread throughout the food chain remains unanswered. Use of the remaining 15 million bushels of sprayed oats as animal feed has been proposed,[42] despite retention of chlorpyrifos, and possible ill-effects upon the animals consuming the contaminated feed.

DISCUSSION

The costs accruing from exposure to pesticides are great. These costs include those of direct medical care, loss of income, loss of mentation, loss of enjoyment of life, condemnation of some structures, and cost of litigation in many cases. The real number of persons who are symptomatic as a result of exposure to OPO[4] pesticides is unknown. Many of the illnesses mimic those with other causes. In the United States, no systematic pesticide application records are kept. Only case-by-case reports of persons residing and working in buildings that have been contaminated with pesticides are available, many of these only after litigation has been instituted. California is one of the few states to maintain records concerning pesticide injuries. There is no international repository for documented pesticide-caused illnesses.

The extensive body of information concerning the OPO[4] pesticides should alert the regulatory and public health fields to the hazards of use, especially where interior contamination can result. Larger problems of loss of non-target species, development of resistant pest strains, food, soil, and water contamination, and world-wide reports of endocrine disruption across multiple species[43] should force radical changes in pesticide production and use.

Alternative and effective methods of pest control have been available for decades, and should be promoted. The interior use of toxic chemicals, viewed in terms of personal and economic loss, increased medical burdens and costs, and loss of intellectual capacity, demonstrates that benefits may be far less than the hazards.

A reading of Rachel Carson's book, *Silent Spring*, written over three decades ago proves that history is of little help; the unnecessary, indiscriminate, and wide-spread promotion of toxic pesticidal chemicals continues. She wrote:

The question is whether any civilization can wage relentless was on life without destroying itself, and without losing the right to be called civilized.[44]

REFERENCES

1. Kaplan, J. G., Kessler, J., Rosenberg, N., Pack, D., Schaumburg, H. H. Sensory neuropathy associated with Dursban (chlorpyrifos) exposure. *Neurology*. 43: 2193–2196, 1993.
2. Sherman, J. D. Letter to William Reilly, Head EPA, re: Seven cases of neuropathy following Dursban exposure, including urinary and fecal incontinence. June 11, 1991.
3. Sherman, J. D. Invited testimony before the U. S. Senate Subcommittee on Toxic Substances,

Environmental Oversight, Research and Development of the Committee on Environment and Public Works. Senate bill S849, ISBN # 0-16-035481, pp. 27–28, 85–98. May 9, 1991.

4. Steenland, K., Jenkins, B., Ames, R. G., O'Malley, M., Chrislip, D., Russo, J. Chronic neurological sequelae to organophosphate pesticide poisoning. *Amer. J. Pub. Health.* 84(5): 731–736, 1994.

5. Rosenthal, N. E., Cameron, C. L. Exaggerated sensitivity to an organophosphate pesticide. *Amer. J. Psychiatry* 148: 2, 1991.

6. Thrasher, J. D., Madison, R., Broughton, A. Immunological abnormalities in humans exposed to chlorpyrifos: Preliminary observations. Arch. Environ. Health 48(2): 89–91, 1993.

7. Broughton, A., Thrasher, J. D., Madison, R. Chronic health effects and immunological alterations associated with exposure to pesticides. *Comments Toxicol.* 4(1): 59–71, 1990.
   Fenske, R. A., Black, K.G., Elkner, K. P., Lee, C. L., Methner, M. M., Soto, R. Potential exposure and health risks of infants following indoor residential pesticide applications. *Amer. J. Public Health.* 80: 689–693, 1990.

9. O'Brien, M. Are pesticides taking away the ability of our children to learn? J. Pest. Reform 10(4):4–8, 1990/91

10. Wright, C. G., Leidy, R. B., Dupree, H. E. Mission impossible? *Pest Control Technol.* 50–54, 1984

11 Fenske, R. A., Black, K. G., Elkner, K. P., Lee, C. L., Methner, M. M., Soto, R. Potential exposure and health risks of infants following indoor residential pesticide application. *Amer. J. Pub. Health.* 80: 689–93, 1990.

12 DuBois J. E., Jr. *The Devil's Chemists.* Beacon Press, Boston, 1952.

13. Sasuly, R. *I. G. Farben.* Boni & Gaer, New York, 1947,p. 125.

14 Meselson, M., Robinson J. P. Chemical warfare and chemical disarmament. *Sci. Amer.* 242(4): 38–47, 1990.

15 Rosenstock, L., Keifer, M., Daniell, W. E., McConnell, R., Claypoole, K. Chronic central nervous system effects of acute organo-phosphate pesticide intoxication. *Lancet* 338: 223–227, 1991.

16. Anon. Toxicological properties of O,O-diethyl-O-3,5,6-trichloro-2-pyridyl phosphorothioate. Biochemical Research Laboratory, Dow Chemical Company. Sept. 26, 1963.

17. The Dow Chemical Co. Clinical toxicity of Dursban insecticide in the dog after multiple dipping. Pittman-Moore Research Laboratories, Indianapolis, IN, Nov. 11, 1964.

18. Eliason, D. A., Cranmer, M. F., von Windeguth, D. L., Kilpatrick, J. W., Suggs, J. E., Schoof, H. F. Dursban premises applications and their effects on the cholinesterase level in spraymen. *Mosquito News.* 29(4): 591–595, 1969.

19 Isikoff, M. Peruvian coca fields sprayed in test of plan. *Washington Post,* p. A-16, Mar. 22, 1989.

20. Doty, A. E., Kenaga, E. E. Bioassay and chemical analyses of Dursban and other insecticidal vapors from simulated premise treatments. Bioproducts Department, Dow Chemical Co., Summary. July 2, 1965.

21. Harrison, R. P., Whitney, W. K. Laboratory and field performance of DURSBAN insecticide against cockroaches and other household pests. *Down to Earth* 23(3): 3–7, 1967.

22. Kenaga E. E., The environmental fate of chlorpyrifos as related to EPA PR Notice 70-15. Ag-Organics Department. Dow Chemical Co., March 20, 1972.

23. Naffziger, D. H., Sprenkel, R. J., Mattler, M. P. Indoor environmental monitoring of Dursban L. O. following broadcast application. *Down to Earth* 41(1): 1985.

24. Kanaga, E. E., Whitney, W. K., Hardy, J. L., Doty, A. E. Laboratory tests with Dursban insecticide. *J. Econ. Entomol.* 58(6): 1043–1050, 1965.

25. Schmolesky, G. E., Head Pesticide Department, WARF (Wisconsin Alumni Research Foundation). Letter to H. E. Gray, Dow Chemical Co Dec. 16, 1963.

26. Thomas, P. A. Dursban insecticide for the professional pest control operator. Dow Chemical Co.. p 1–7, 1969.

27. Lynn, G. E., Director Registration Section, Bioproducts Department, Dow Chemical Co. Letter to J. S. Leary, Jr., Pesticides Regulation Division, Agricultural Research Service, U. S. Dept. Agriculture. June 21, 1965.

28. Gray, H. E. Dursban A new organo-phosphorus insecticide. *Down to Earth* 21(3): 2, 26, 27. 1965.

29. Sharp, H., Warner, S., Robinson, V. The clinical toxicity of Dursban in the dog after multiple applications of an aerosol formulation. Dow Chemical Co.

30. Keeler, P. A., Jersey, G. C. Acute toxicological properties and industrial handling hazards of insecticide formulation M4237. Toxicology Research Laboratory, Dow Chemical Co. p. 19–20. 1976.

31. Ibid., p. 22

32. Ibid., p. 24.

33. Rowe, L. D., Warner, S. D., Johnston, R. V. Acute delayed neurotoxicological evaluation of chlorpyrifos in white leghorn hens. pp. 1–12 with appended tables. May 22, 1978.

34. McKellar, R. L., Dishburger, H. J. Determination of residues of chlorpyrifos and 3,5,6-trichloro-2-pyridinol in tissues of cattle receiving a single treatment of Dursban Spot-On. Agricultural Products Department, Dow Chemical Co., 1976.

35. Iglisch, I. Control of Blatta orientalis L. in a swine breeding unit by means of the "cover method." *Anz. Schaedlingskd Pflanzenschutz Umweltschutz* 64(6): 105–111, 1991.

36. Dursban Insecticides Technical Information. Technical Data Sheet. Dow Chemical Co. p. 1–12, 1987.

37 *Farm Chemicals Handbook '91*. Meister Pub. Co., Willoughby, OH., 1991, p. C-74

38 U. S. EPA, Registration Standard (Second Round Review) for the Reregistration of Pesticide Products Containing Chlorpyrifos. June 1989, p. 42,.

39 Sherman, J. D. Letter to Richard Christian, Deputy Director for Research and Technology Assessment, American Legion, Washington, D. C., 3 pages, dated August 15, 1992. (Copies were transmitted by the American Legion to the Veterans' Administration in August 1992, to L. Drash of the V. A. on August 20, 1993, and to the Senate Committee on Veterans' Affairs on April 11, 1994.)

40 Roberts, Lyman W., Colonel, Medical Service Corps, Office of the Surgeon General, Preventive and Military Medicine Consultation Division, Falls Church, VA. Facsimile transmission to Dr. Olsen, Veterans' Affairs, 2 page list of chemicals, April 6, 1994.

41 Kenaga, E. E. An evaluation of the safety of DOWCO 179 to birds in areas treated for insect control. Agricultural Department, Dow Chemical Co., p. 14. June 16, 1971

42 Walsh, S. A bowlful of lessons. *Washington Post*. p. 1, 10, August 13, 1994.

43 Colborn T., Clement, C., Eds. Chemically-induced alterations in sexual and functional development: The wildlife/ human connection. Princeton Scientific Publishing Company. Princeton. 1992, pp. 1–403

44. Carson, R., Silent Spring, Crest Book Edition, Houghton Mifflin, Co., New York. 1962, p. 95.

STRUCTURE-ACTIVITY RELATIONSHIPS OF CHEMICALS CAUSING ENDOCRINE, REPRODUCTIVE, NEUROTOXIC, AND ONCOGENIC EFFECTS — A PUBLIC HEALTH PROBLEM

*We shall require a substantially new manner
of thinking, if mankind is to survive.*
Albert Einstein

Thoughtful science, spanning international fields of biology has demonstrated that a number of biologically active chemicals have been wreaking havoc among wild animals, causing thinning of eggs, reproductive failure, reversal of female/male reproductive behavior, and malformations in off-spring, including both structural and developmental abnormalities.[1-2] This finding, involving not only science, but public policy, must involve the ramifications of similar effects in the human population.

Legitimate science lends compelling evidence to the concept that biologically active chemicals are connected to the increases in malignancy and other adverse effects in the human population. That environmental agents are related to cancer should come as no surprise. This connection has been documented since before the turn of the century,[3] and continues.[4-7]

Ominously, four recent papers point to the consequences of an environment increasingly laden with biologically active chemicals: an increase in cancer among the younger population,[8] including an increase in non-smoking related cancers,[9] an increase in breast cancer, despite improved methods of detection and treatment, in both pre- and post-menopausal women in most industrialized countries,[10] and endocrine-related developmental effects in both wildlife and humans.[11] An additional epidemic of disease involves both benign prostatic hypertrophy and prostatic cancer, tissues also sensitive to the effects of hormonal substances.

The interconnections between chemical exposure and adverse effects, among animals and humans, males and females, including reproductive abnormalities and failures, endocrine disruption, immunological dysfunction, neurological damage, and cancer is reaching a crescendo in the scientific and lay press.

Living systems function by being biologically active, with cells involved in reproduction, the immune, endocrine, and nervous systems among those most active. In most cases, adverse effects upon one system result in alterations in homeostasis in other body systems.

The final 1993 issue of *Science* featured "p53 Molecule of the Year". The accompanying editorial stated: "misshaped molecules caused by mutations in the gene may be brought back into the right shape by a drug that binds to the mutant p53 and pushed the mutant back into the shape of the normal molecule."[12] p53 is the tumor suppressor gene, mutations in which are found in about 50% of all cancers.[13] Rather than attempting to correct gene damage, preventing mutations in suppressor genes may be a more logical and cost-effect method to reduce the incidence of cancer. This may be accomplished in part by decreasing the load of xenobiotic carcinogen and endocrine disrupters in the environment.

Chemicals with similar structures frequently share similar functions. The general action of a chemical class will be similar, even with molecular substitution on the basic chemical. Chemical manufacturers themselves rely upon alterations in known chemical structures to create new, patentable products with specific biological actions. Looking at any class of chemicals, be it antihistamines, organophosphate pesticides or the non-steroidal anti-inflammatory (NSAID) agents will demonstrate that fact. It is critical that scientists, physicians, and policy makers understand the biology behind structure-activity relationships.

When one chemical is known to mimic an endocrine function, to cause cancer, or damage to an organ system, it must be assumed, until proven otherwise, that chemicals sharing similar structures may also cause these effects. This dictates that testing, monitoring and caution must be exercised in the use of such chemicals.

It is said that a picture is worth a thousand words. Translating chemical names into meaningful pictures provides a basis to understand structure-activity relationships.

Many of the chemicals addressed in this paper have been designed to kill something (pesti*cides*, herbi*cides*, fungi*cides*, and the like). Tamoxifen and Tace have been developed and sold because of demonstrated hormonal effects. All were *designed* to be biologically active, and with few exceptions, possess multiple actions in addition to those for which they were primarily marketed.

TABLE 9.1 Chemicals with Multiple Adverse Biological Effects.

Chemical Name (Figure Number)	Category and Footnote Reference
AMITROLE (7)	R^{14}, E^{15}, O^{16}
ATRAZINE (7)	R^{17}, C^{18}
BENOMYL (1)	R^{19}, R/T^{20}
BHC (3)	C^{21-22}
CHLORDANE (2)	$C/E/N/R/O^{23}$
CHLORPYRIFOS (Dursban) (2)	N^{24-25}, R^{26}
2,4-D, 2,4,5-T (7)	C^{27}, see also dioxin
DDD (Marketed as Rhothane, now discontinued in U. S. by manufacturer)	
DDT, DDE and DDD (4)	C^{28-30}, I^{31*-32}, R^{33-35}, R/C^{36-39}, E^{40-41}, N/C^{42}
DES (5)	R^{43-44}, C^{45-47}, E^{48-49}, I^{50}, T/C^{51}
DICOFOL (Kelthane) (4)	see DDT references
DIOXIN(s) (6)	R^{52-54}, I^{55}, C^{56-57}, E^{58-59}
HCB (1)	E^{60-61}
LINDANE (1) (gamma isomer of benzene hexachloride)	E^{62}, R^{63-64}, N^{65}, C^{66-67}
METHOXYCHLOR (4)	R^{68}, $R/C/E^{69}$, E^{70-71}
TAMOXIFEN (Nolvadex) (5)	C^{72-74}, R^{75}, E^{76}
NONYLPHENOL (1)	R/N^{77}, T^{78}
PBBs (6)	E^{79}, C^{80}, R^{81}, N^{82}
PCBs (6)	R^{83-85}, C^{86-88}, N^{89}, I^{90}, E^{91}
PENTACHLOROPHENOL (1)	R^{92}, I^{93}
PERTHANE (4)	See DDT
PHTHALATES (2)	R^{94}, T^{95-96}, C^{97}
STYRENE (1)	N^{98-99}, E^{100}, C^{101}, R^{102}
CHLOROTRIANISENE (Tace) (5)	$C^{103-104}$, R^{105}

C = Cancer; E = Endocrine; I = Immunological; N = Neurological;
R = Reproductive; T = Teratogenic; O = Other.
* Includes also PCBs and other pesticides.
All figures are located at the end of the text.
    While liver and kidney damage, cardiac, respiratory, immunologic, and other

adverse effects are important, in order to simplify and correlate adverse effects, this paper emphasizes structure-activity relationships of chemicals on the endocrine, reproductive, immunologic, and neurological systems, as well as the "end-result" of cancer. Several chemicals will be discussed in their historic context, and chemicals causing multiple adverse effects are indicated. Where available, the earlier literature will be emphasized.

Most of the above chemicals are fat soluble, with the potential for bioaccumulation, bio-magnification and persistence. Thus these chemicals, preferentially sequestered in the fatty portions of the body, have the potential for prolonged biological activity, persisting long after any "acute" exposure. The fat soluble chemicals may be absorbed, not only by ingestion and inhalation, but through the intact skin. With much discussion of adverse effects from a high fat diet, it is essential to understand that fats are reservoirs for- and carriers of- lipophilic xenobiotic chemicals. Thus pesticides, herbicides and industrial chemicals that pollute fish, meat animals and oil-containing plant crops can be expected to be carried in the food supply of humans. Lipophilic chemicals have the potential to bioaccumulate and biomagnify with each food transfer.

Where multiple contaminants are the rule, the effects may be additive and/or synergistic. Biological effects may occur not only in the animal/human ingesting them, but also in the offspring, frequently not manifest until adulthood.

A number of the chemicals have been either banned in the U. S., or their production and sale has been voluntarily withdrawn by the manufacturer. A number, notably chlordane and heptachlor, continue to be produced in the U. S. and sold overseas.[106]

Several chemicals, because of their class, persistence, and/or wide-spread use bear special discussion:

DES: The ultimate effects from the widespread contamination of the food-chain with DES will never be known. DES was used as growth agent to fatten cattle and poultry. It was prescribed post-natally to stop lactation,[107] and ultimately given to pregnant women in a misguided attempt to prevent miscarriage. The daughters of those women have borne the most tragic of outcomes, which have included genital malformations, infertility and genital cancers. The sons also did not escape, developing genital abnormalities, cancer and infertility.

Hearings before the U. S. Congress, brought about by mink farmers whose animals had lost hair and failed to reproduce as a result of having been feed DES-contaminated slaughter-house offal, provide a record of scandal[108], and yet prescription of this chemical to women continued for two decades.

Despite the knowledge of the adverse effects of DES, and the adverse effects of DES's chemical cousin, DDT, production and use of similar chemicals, such as kelthane (Dicofol) continues unabated, worldwide. Chlorotrianisene (Tace), a product, chemically similar to DES, was marketed as recently as 1987 to control postpartum breast engorgement. It is still marketed for atrophic vaginitis and post-menopausal symptoms.[109]

LINDANE and BENZENE HEXACHLORIDE. The chlorine atoms of Lindane are equatorially oriented, while those of BHC are arranged two equatorial and four axial. BHC (hexachlorocyclohexane) lacks pesticidal properties, while lindane has been marketed as a general pesticide, including as the pediculicide, Qwell. Interrelated effects are demonstrated in the literature.

NONYLPHENOL: Nonylphenol is a degradation product of nonionic surfactants[110], as well as an intentional additive in many commercial products, including pesticides, emulsifiable oils, and vaginal cremes. Birth defects have been associated with vaginal spermatocide use.[111] A nonylphenol-derived detergent, used in spermicide preparations was absorbed though the vaginal wall into the systemic circulation, and detectable in the milk of lactating rats and their pups within two hours of dosing.[112] The medical literature gives no information concerning possible absorption of nonylphenol by the male, during intercourse. Given the estrogenic properties of nonylphenol, a retrospective study of men with prostatic and/or breast cancer who were exposed to spermatocide may be in order.

In a pesticidal formulation, nonylphenol, the solvent classified as "inert" under FIFRA law, was found to be more toxic than the primary ingredient.[113] The emulsifying action of nonylphenol can be expected to increase the uptake of associated lipophilic chemicals and to enhance the action of other xenobiotic. Nonylphenol is lipophilic and translocates from water to sediment. The half-life of 2.5 days is markedly lengthened in the absence of microbiological degradation.[114]

DDT and congeners: Dichlorodiphenyltrichloroethane (DDT) is metabolized by progressive dechlorination to DDE and DDD. In addition to the endocrine, teratogenic, and carcinogenic effects, an additional effect may be wreaking harm in the population in terms of heart disease and stroke. Elevation in blood lipids is associated with elevated organochlorine residues.[115-117] While this finding has been noted for over four decades, emphasis has been on the alteration of a "life style" rather than controlling environmental contaminants. Despite the

ban on DDT in the United States and documentation of wide-spread harm to the biosphere, the World Bank has funded a DDT factory in Brazil.[118]

DDD: Rhothane, the commercial insecticidal product, has been discontinued in the United States. Less than a decade after DDT was banned in the United States, it was tested as a treatment for hyper-adrenalism in humans,[119] resulting in adrenal gland damage and atrophy in test animals,[120] as well as slowly developing hypercholesterolemia.[121]

DICOFOL: This product, marketed as Kelthane, and containing as much as 12% DDT was allowed to be sold via a loophole in the pesticide law, which allowed the DDT to be classified as an "inert ingredient." All products containing more than 0.1% DDT were cancelled in the United States as of 1986.[122]

Additional chemicals, widely used in commerce, have not been fully characterized as to toxicological findings; however their chemical structures are so similar to other chemicals with known adverse reproductive, endocrinologic, and carcinogenic effects, that they should be placed in a special category for study. Thorough animal testing should be available for assessment. Persons exposed to these chemicals, either during production or use, must be monitored for adverse effects. The chemicals under question should be restricted from dissemination until safety concerns have been satisfied.

No matter how minimal the dose of a carcinogen, no dose can be assumed to be safe.[123] We must assume that the same holds true for chemicals that mimic endocrine function, especially if exposure is chronic, or occurs during reproductive and growth periods. Exposure to multiple agents with endocrine, reproductive, immunologic, neurologic, and carcinogenic consequences, acting in an additive and/or synergistic fashion is more the reality.

Hundreds of studies and reports concerning the toxicology of major groups of chemicals are available. In general, there is sufficient information in the scientific literature to make informed decisions about the likely results from exposure to single chemicals and to members of a chemical family. With few exceptions, the effects resulting from exposure to combinations of chemicals have not been well studied.[124]

To ignore the existing biological, epidemiological and test data is folly. In human terms, failure to control chemical exposure will condemn the unprotected to disease and early death. Loss of mentation from neurotoxic chemicals removes creativity and productivity from the human milieu. Reduced solely to economics, the medical care and social support systems simply cannot finance the treatment, education, and care of an increasingly sick and impaired population. Preventive efforts make not only humane and scientific sense, but economic sense as well. Contamination of land and water impacts not only human health but that of domestic and wild species as well. It is well to remember that as *Homo sapiens*, we are members of the animal kingdom, and subject to the same rules as all other animals.

What are human beings without animals? If all animals ceased to exist, human beings would die of a great loneliness of the spirit. For whatever happens to the animals will happen also to human beings....All things connect.

Chief Seattle.

CHEMICAL DIAGRAMS

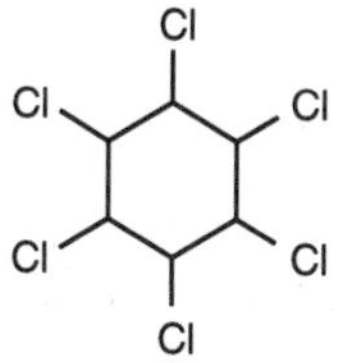

BHC
BENZENE HEXACHLORIDE

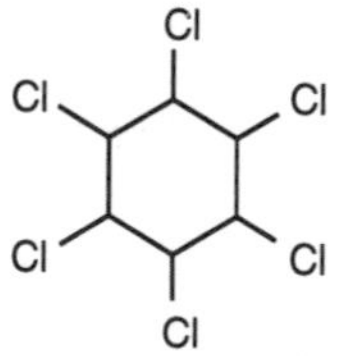

LINDANE
gamma isomer of BHC

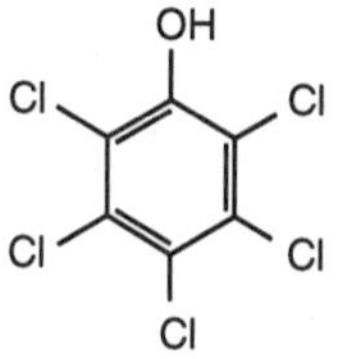

PENTACHLOROPHENOL

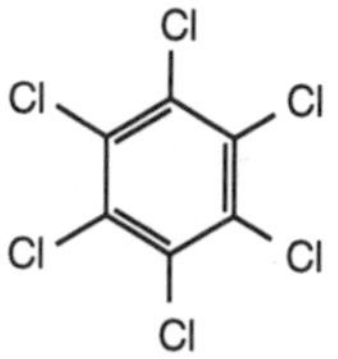

HCB
HEXACHLOROBENZENE

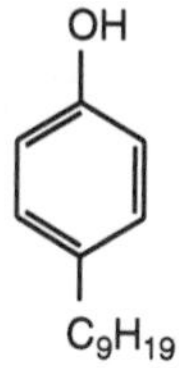

NONYL PHENOL

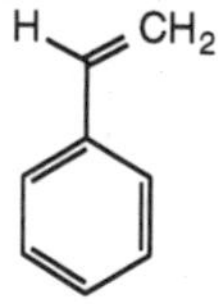

STYRENE
VINYL BENZENE

Figure 9.1. BHC, LINDANE, HCB, PENTACHLOROPHENOL, NONYL PHENOL, AND STYRENE.

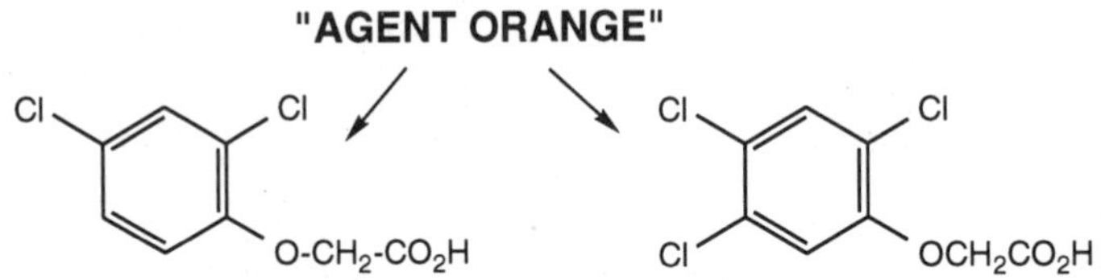

"AGENT ORANGE"

2,4-D
dichlorophenoxy-acetic acid

2,4,5-T
trichlorophenoxy-acetic acid

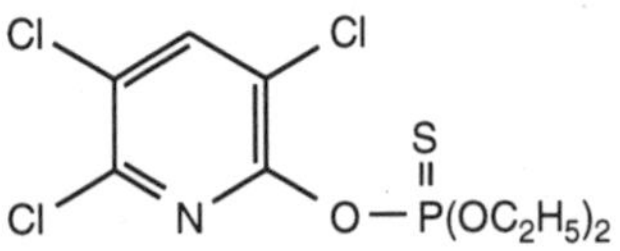

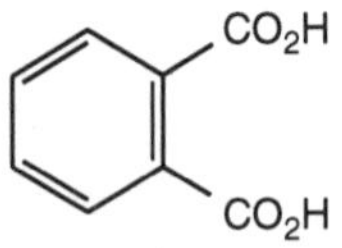

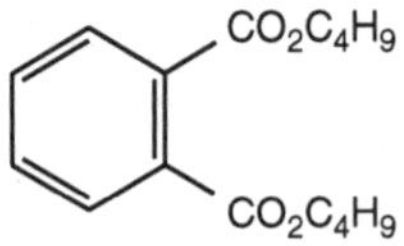

DURSBAN
chlorpyrifos

PHTHALATE

PHTHALATE, DIBUTYL

Figure 9.2. 2,4-D, 2,4,5-T, DURSBAN, PHTHALATE, and PHTHALATE DIBUTYL.

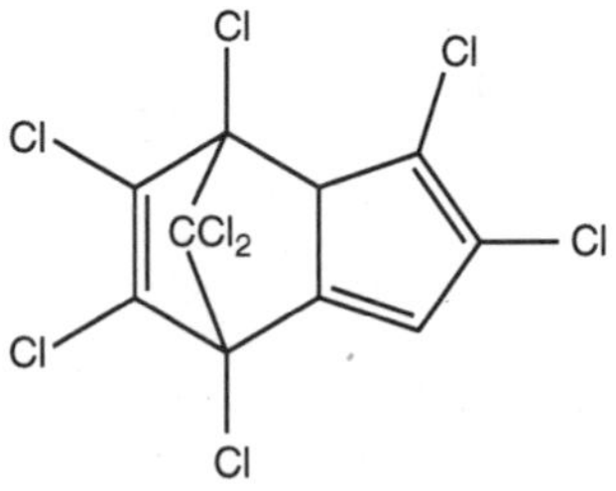

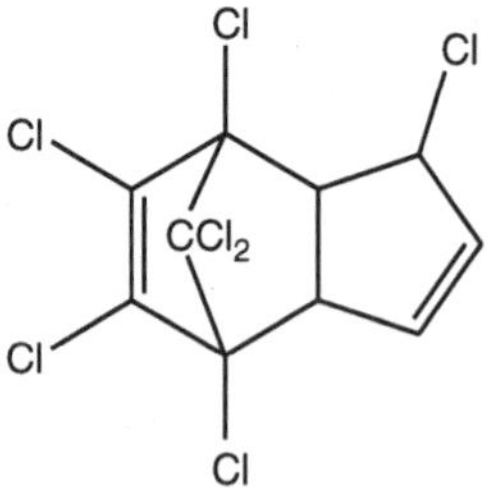

CHLORDANE

HEPTACHLOR

Figure 9.3. CHLORDANE and HEPTACHLOR.

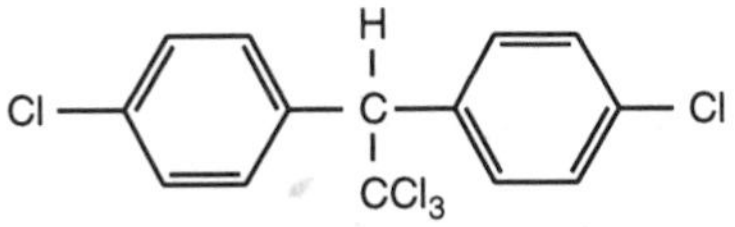

DDT
dichloro-diphenyltrichloroethane

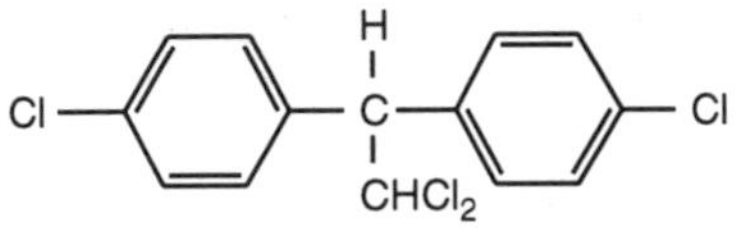

DDE
dichloro-diphenyldichloroethane

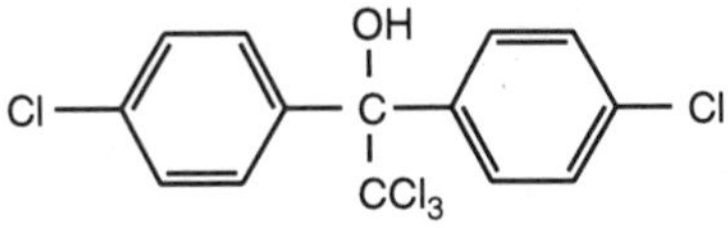

DICOFOL

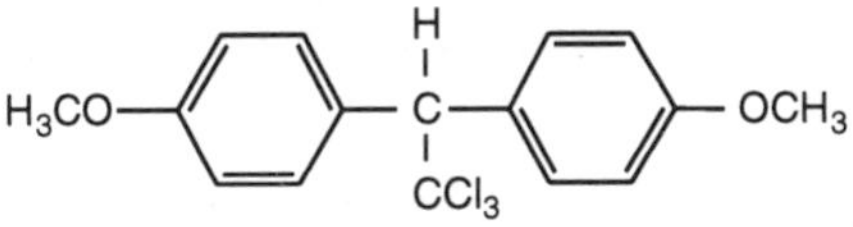

METHOXYCHLOR

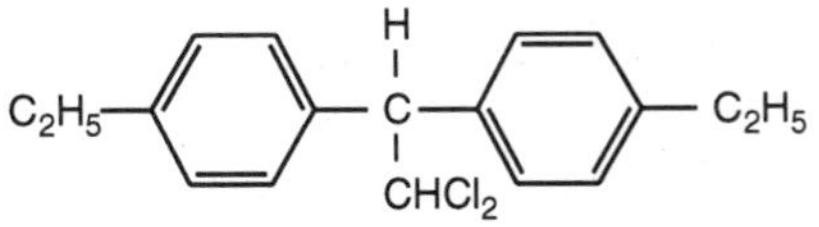

PERTHANE

Figure 9.4. DDT, DDE, DICOFOL, METHOXYCHLOR, and PERTHANE.

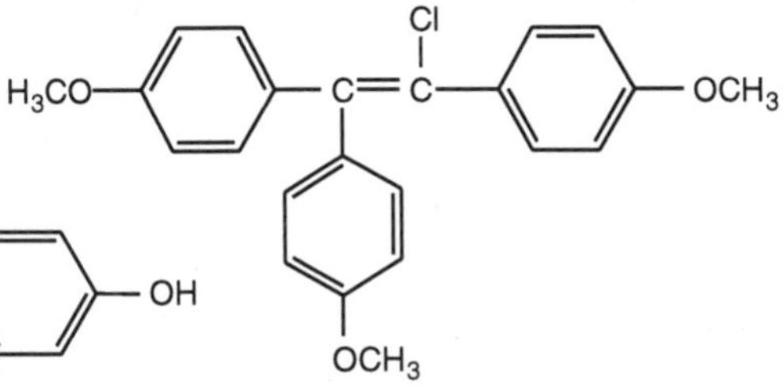

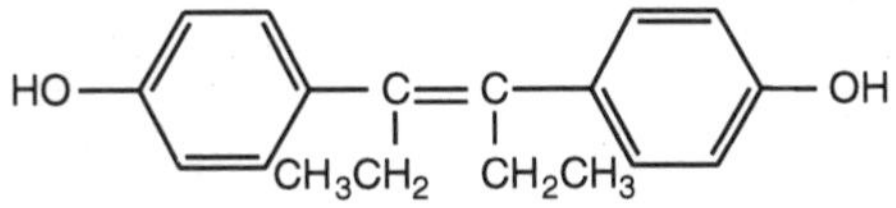

DES
diethylstilbestrol

TACE
chlorotrianisene

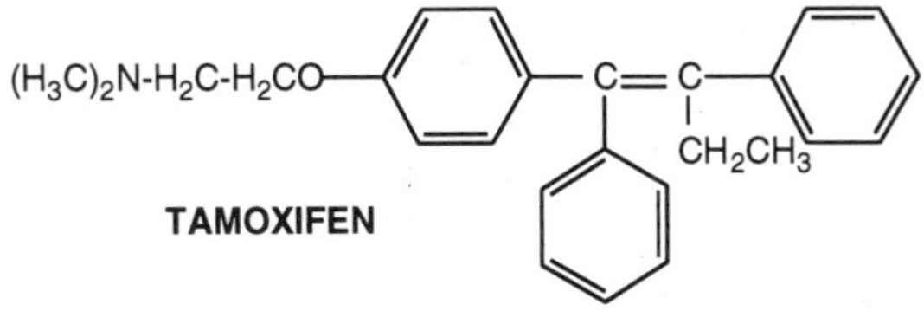

TAMOXIFEN

Figure 9.5. DES, TAMOXIFEN, and TACE..

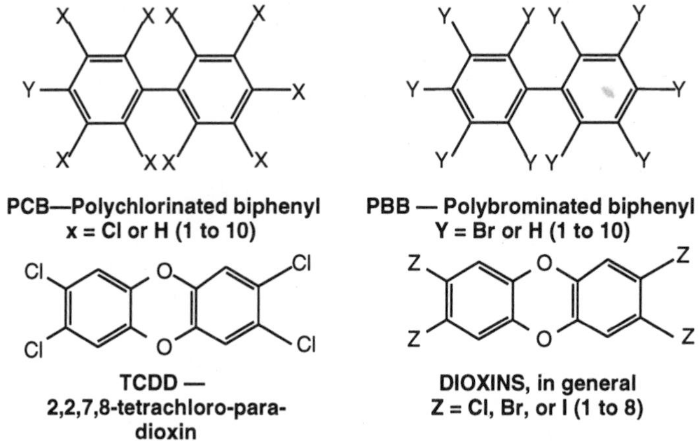

PCB—Polychlorinated biphenyl
x = Cl or H (1 to 10)

PBB — Polybrominated biphenyl
Y = Br or H (1 to 10)

TCDD —
2,2,7,8-tetrachloro-para-
dioxin

DIOXINS, in general
Z = Cl, Br, or I (1 to 8)

Figure 9.6. PCB, PBB, TCDD, AND DIOXINS in general..

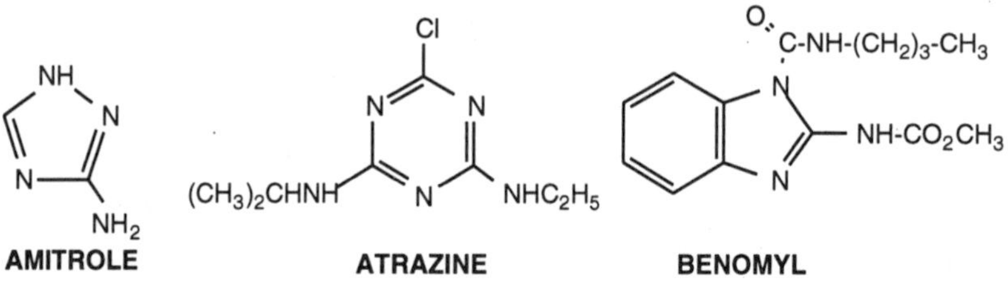

AMITROLE

ATRAZINE

BENOMYL

Figure 9.7. AMITROLE, ATRAZINE, and BENOMYL.

REFERENCES

1. Colborn, T., Clement, C., Eds. *Chemically-Induced Alterations in Sexual and Functional Development: The Wildlife/Human Connection.* Princeton Scientific Publishing Co., Princeton, N.J., 1992., 402 pp.
2. Nicolson, T. B., Nettelton, P. F., Spence, J. A., Calder, K. H., High incidence of abortions and congenital deformities of unknown aetiology in a beef herd. *Veterinary Record* 116: 281–284, 1985.
3. Rehn, L. Blasengeschwuelste bei Fuchsin-Arbeitern. *Arch. f. Klin. Chir.* 50: 588–600, 1895.
4. Anon. Editorial: Environmental cancer. *J. Amer. Med Assoc.* p. 836, Nov. 25, 1944.
5. Hueper, W. C. Environmental carcinogens and cancers. *Cancer Res.* 21: 842–857, 1961.
6. Sherman, J. D. Cancer—Our social disease. *CBE Environ. Rev.* p. 7–10, June 1978.
7. Eisenbud, M., Environmental causes of cancer. *Environment* 20(8): 4–16, 1978.
8. Adami, H-O, Bergstrom, R., Sparen, P., Baron, J. Increasing cancer risk in younger birth cohorts in Sweden. *Lancet.* 341(8848): 773–777, 1994.

9. Davis, D. L., Dinse, G. E., Hoel, D. G. Decreasing cardiovascular disease and increasing cancer among whites in the United States from 1973 though 1987. *J. Amer. Med. Assoc.* 271(6): 431–437, 1994.

10. Davis, D.L., Hoel, D. *Trends in Cancer Mortality in Inductrial Countries*, N.Y. Acad. Sci., 609:8, 1990.

11. Colborn, T., vom Saal, F. S., Soto, A. M. Developmental effects of endocrine-disrupting chemicals in wildlife and humans. *Environ. Health Perspect.* 101(5): 378–384, 1993.

12. Koshland, D. E., Molecule of the year. *Science.* 262:1953., 1993

13. Hollstein, M., et al. *Science* 253:49, 1991.

14. Tjalve, H. Fetal uptake and embryogenetic effects of aminotriazole in mice. *Arch. Toxicol.* 33: 41–28, 1974

15. Jukes, T. H., Shaffer, C. B., Antithyroid effects of aminotriazole. *Science.* 132: 296, 1960.

16. Ghiazza, G., Zavarise, G., Lanero, M., Farraro, G. Sister-chromatid exchange induced in chromosomes of human lymphocytes by trifluralin, atrazine and simazine. *Boll. Soc. Ital. Biol. Sper.* 60(11): 2149–2154, 1984.

17. Babic-Gojmerac, T., Kniewald, Z., Kniewald, J. Testosterone metabolism in neuroendocrine organs in male rats under atrazine and deethylatrazine influence. *J. Steroid. Biochem.* 33:141, 1989.

18. Pinter, A., Torok, G., Borzsonyi, M., Surjan, A., et al. Long-term carcinogenicity bioassay of the herbicide atrazine in F344 rats. *Neoplasma* 37(5): 533–544, 1990.

19. Hess, R. A., Moore, B. J., Forrer, J., Linder, R. E., Abuel-Atta, A. A. The fungicide benomyl (methyl 1-(butylcarbamoyl)-2-benzimidazole-carbamate) causes testicular dysfunction by inducing the sloughing of germ cells and occlusion of efferent ductules. *Fundam. Appl. Toxicol.* 17: 733–745, 1991.

20. U. S. EPA. Suspended, Cancelled and Restricted Pesticides. Feb. 1990.

21. Hoshizaki, H., Niki, Y., Tajima, H., Terada, Y., Kashahara, A., A case of leukemia following exposure to insecticide. *Acta Hematol. Jap.* 32(4): 672–677, 1969.

22. Hanada, M., Yutani, C., Miyaji, T. Induction of hepatoma in mice by benzene hexachloride. *Gann* 64(5): 511–513, 1973.

23. Sherman, J. D., *Chemical Exposure and Disease.*, Van Nostrand Reinhold., New York., pp. 138–156, 1988.

24. Kaplan, J. G., Kessler, J., Rosenberg, N., Pack, D., Schaumburg, H. H., Sensory neuropathy associated with Dursban (chlorpyrifos) exposure. *Neurology* 43: 2193–2196, 1993.

25. Rosenstock, L., Keifer, M., Daniell, W. E., McConnell, R., Claypoole, K. Chronic central nervous system effects of acute organophosphate pesticide intoxication. *Lancet.* 338: 223–227, 1991.

26. Deacon, M. M., Murray, J. S., Pilny, M. K., Rao, K. S., Dittenbar, D. A., Hanley, T. R., John, J. A., Embryotoxicity and fetotoxicity of orally administered chlorpyrifos in mice. *Toxicol. Appl. Pharmacol.* 54: 31–40, 1980.

27. Hayes, H. M., Tarone, R. E., Cantor, K. P., et al. A case-control study of canine malignant lymphoma: positive association with dog owner's use of 2,4-dichlorophenoxyacetic acid herbicides. *J. Nat. Cancer Inst.* 83: 1226–1231, 1991.

28. Wolff, M., et al. Blood levels of organochlorine residues and risk of breast cancer. *J. Nat. Cancer Inst.* 85: 648–652, 1993.

29. Reuber, M. D., Carcinomas of the liver in Osborne-Mendel rats ingesting DDT. *Tumori.* 64(6): 571–577, 1978.

30. Wassermann, M., Nogueira, D. P., Tomatis, L., Mirra, A. P., Shibata, H., Arie, G., Cucos, S., Wassermann, D. Organochlorine compounds in neoplastic and adjacent apparently normal

breast tissue. *Bull. Environ. Contam. Toxicol.* 15(4): 478–484, 1976.

31. Street, J. C., Sharma, R. P. Alteration of induced cellular and humoral immune responses by pesticides and chemicals of environmental concern: Quantitative studies of immunosuppression by DDT, Arochlor 1254, Carbaryl, Carbofuran, and methylparathion. *Toxicol Appl. Pharmacol.* 32: 587–602, 1975.

32. Wassermann, M., Wassermann, D., Kedar, E., Djavaherian, M. Immunological and detoxication interactions in p,p'-DDT-fed rabbits. *Bull. Environ. Contam. Toxicol.* 6: 426–435, 1971.

33. Carson, R., *Silent Spring*. Houghton Mifflin., New York. 1962.

34. Rappolt, R. T. Sr., Mengle, D., Hale, W., Hartman, B., Salmon, B., Kern County: annual generic pesticide input; blood dyscrasias; p,p'-DDE and p,p'DDT residues in human fat, placentas with related stillbirths and abnormalities. *Ind. Med. Surg.* 37(7): 513, 1968.

35. Burlington, H., Lindeman, V. F., Effect of DDT on testes and secondary sex characteristics of white leghorn cockerels. *Proc. Soc. Exper. Biol. Med.* 74: 48–51, 1950.

36. Shabad, L. M., Kolesnichenko, T. A., Nikonova, T. V., Transplacental and combined long-term effect of DDT in five generations of A-strain mice. *Int. J. Cancer* 11(3): 688–693, 1973.

37. Tomatis, L., Turusov, V., Day, N., Charles, R. T., The effect of long-term exposure to DDT on CF-1 mice. *Int. J. Cancer* 10(3): 489–506, 1972.

38. Turusov, V. S., Day, N. E., Tomatis, L., Gati, E., Charles, R. T., Tumors in CF-a mice exposed for six generations to DDT., *J. Nat. Cancer Inst.* 51(3): 983–997, 1973.

39. Clegg, D. J., Animal reproduction and carcinogenicity studies in relation to human safety evaluation. *Dev. Toxicol. Environ. Sci.* 4: 45–59, 1979.

40. Bitman, J., Cecil, H. C., Harris, S. J., Fries, G. F., Estrogenic activity of o,p-DDT in the mammalian uterus and avian oviduct. *Science* 162: 371–372, 1968.

41. Heinrichs, W. L., Geller, R. J., Bakke, J. L., Lawrence, N. L., DDT administered to neonatal rats induces persistent estrus syndrome. *Science* 173(997): 642–643, 1971.

42. Kashyap, S. K., Nigam, S. K., Karnick, A. B., Gupta, R. C., Chatterjee, S. K., Carcinogenicity of DDT (dichlorodiphenyl trichloroethane) in pure inbred Swiss mice. *Int. J. Cancer* 19(5): 725–729, 1977.

43. Greene, R. R., Burrill, M. W., Ivy, A. C. Experimental intersexuality: The production of feminized male rats by antenatal treatment with estrogens. Science 68: 30–131, 1938.

44. Whitehead, E. D., Leiter, E. Genital abnormalities and abnormal semen in male patients exposed to diethylstilbestrol in utero. *J. Urology* 125(1): 47–50, 1981.

45. Gardner, W. U., Allen, E., Malignant and non-malignant uterine and vaginal lesions in mice receiving estrogens and androgens simultaneously. *Yale J. Biol. Med.* 12: 213–234, 1940.

46. Shimkin, M. B., Grady, H. G., Carcinogenic potency of stilbestrol and estrone in strain C3H mice. *J. Nat. Cancer Inst.* 1: 119–128, 1941.

47. Herbst, A. L., Scully, R. E. Adenocarcinoma of the vagina in adolescence. *Cancer* 25: 745–757, 1970.

48. Herrick, E. H., Some influence of stilbestrol, estrone and testosterone propionate on the genital tract of young female fowls. *Poultry Sci.* 23(1): 65–66, 1944.

49. Wiberg, G. S., Stephenson, N. R. The detection of estrogenic activity in tissue of steers which have been fed diethylstilbestrol. *Canad. J. Biochem. Physiol.* 35: 1107–112, 1957.

50. Luster, M. I., Faith, R. E., McLachlan, J. A., Clark, G. C. Effect of in utero exposure to diethylstilbestrol on the immune response in mice. *Toxicol. Appl. Pharmacol.* 47(2): 279–285, 1979.

51. Vorheer, H., Messer, R. H., Vorheer, U. F., Jordan, S. W., Kornfeld, M. Teratogenesis and carcinogenesis in rat offspring after transplacental and transmammary exposure to

diethylstilbestrol. *Biochem. Pharmacol.* 28(12); 1865–1877, 1979.

52. Cless, D. J. Embryotoxicity of chemical contaminants in foods. *Food Cosmet. Toxicol.* 9(2): 195–205, 1971.

53. Neubert, D., Dillmann, I. Embryotoxic effects in mice treated with 2,4,5-trichlorophenoxyacetic acid and 2,3,7,8-tetrachloro-dibenzodioxin. *Arch. Pharmacol.* 272(3): 243–264, 1972.

54. Mably, T. A., et al. In utero and lactational exposure of male rats to 2,3,7,8-tetrachlorordibenzo-p-dioxin. 3. Effects on spermatogenesis and reproductive capacity. *Toxicol. Appl. Pharmacol.* 114: 118–126, 1992.

55. Vos, J. G., Moore, J. A., Zinkl, J. G. Effects of 2,3,7,8-tetrachlorodibenzo-p-dioxin on the immune system of laboratory animals. *Environ. Health Perspect.* 5: 149–162, 1973.

56. Van Miller, J. P., Lalich, J. J., Allen, J. R. Increased incidence of neoplasms in rats exposed to low levels of 2,3,7,8-tetrachlorodibenzo-p-dioxin. *Chemosphere* 6(10): 625–632, 1977.

57. Manz, A., Berger, J., Dwyer, J. H., Flesch-Janys, D., Nagel, S., Waltsgott, H., Cancer mortality among workers in chemical plant contaminated with dioxin. *Lancet* 338(8773): 959–964, 1991.

58. Rier, S. E., Martin, D. C., Bowman, R. E., Dmowski, W. P., Becker, J. L., Endometriosis in Rhesus monkeys (Macaca mulatta) following chronic exposure to 2,3,7,8-tetrachlorodibenzo-p-dioxin. *Fund. Appl. Toxicol.* 21: 433–441, 1993.

59. Romkes, M., Piskorska-Pliszczynska, J., Safe, S., Effects of 2,3,7,8-tetrachlorodibenzo-p-dioxin on hepatic and estrogen receptor levels in rats. *Toxicol. Appl. Pharmacol.* 87: 306–314, 1987.

60. Smith, A., Dinsdale, D., Cabral, J., Wright, A. Goitre and wasting induced in hamsters by hexachlorobenzene. *Arch. Toxicol.* 60: 343–349, 1987.

61. Loeber, J. G., van Velsen, F. L. Uterotropic effect of beta-HCH, a food-chain contaminant. *Food Addit. Contam.* 1(1): 63–66, 1984.

62. Raizada, R. B., Misra, P., Saxena, I., Datta, K. K., Dikshith, T. S. S. Weak estrogenic activity of lindane in rats. J. Toxicol. Environ. Health. 6: 483–492, 1980.

63. Chowdhurry, A., Venkatakrishna-Bhatt, H., Gautum, A. Testicular changes in rats under lindane treatment. *Bull. Environ. Contam. Toxicol.* 38: 154–156, 1987.

64. Tezak, Z., Simic, B., Kniewald, J. Effect of pesticides on oestradiol-receptor complex formation in rat uterus cytosol. *Fd. Chem. Toxic.* 30(10): 879–885, 1992.

65. Joy, R. M. Mode of action of lindane, dieldrin, and related insecticides in the central nervous system. *Neurobehav. Toxicol. Teratol.* 4: 813–823, 1982.

66. Jedlicka, V. Hermanska, Z., Smida, I., Kouba, A. Paramyeloblastic leukemia appearing simultaneously in two blood cousins after simultaneous contact with Gammexane (hexachlorocyclohexane). *Acta Med. Scand.* 56: 448–451, 1958.

67. Hoshizaki, H., Niki, Y., Tajima, H., Terada, Y., Kasahara, A. A case of leukemia following exposure to insecticide. *Acta Haematol. Jap.* 32(4): 178–183, 1969.

68. Cummings, A. M., Gray, L. E. Methoxychlor affects the decidual cell response of the uterus but not other progestational parameters in female rats. Toxicol. *Appl. Pharmacol.* 90: 330–336, 1987.

69. Reuber, M. D., Interstitial cell carcinomas of the testis in Balb/C male mice ingesting methoxychlor. *J. Cancer Res. Clin. Oncol.* 93(2): 173–179, 1979.

70. Fry, D. M., Toone, C. K., Speich, S. M., Peard, R. J., Sex ratio skew and breeding patterns of gulls: demographic and toxicological considerations. Studies *Avian Biol.* 10: 26–43, 1987.

71. Mitchell, R. H., West, P. R. Contamination of pesticide methoxychlor with the estrogen chlorotrianisene. *Chem. Ind.* 15: 581–582, 1978.

72. 61. Hardell, L. Tamoxifen as a risk factor for carcinoma of corpus uteri. *Lancet* 2: 563, 1988.

73. Fendl, K. C., Zimniski, S. J. Role of tamoxifen in induction of hormone-independent rat mammary tumors. *Cancer Res.* 52: 235–237, 1992.

74. van Leeuwen, F. E., Benraadt, J., Coebergh, J. W. W., et al. Risk of endometrial cancer after tamoxifen treatment of breast cancer. *Lancet* 343: 448–452, 1994.

75. Cunha, G. R., Taguchi, O., Namikawa, R. Teratogenic effects of clomiphene, tamoxifen, and diethylstilbestrol on the developing human female genital tract. *Human Pathol.* 18: 1132–1143, 1987.

76. Ferrazzi, E., Cartei, G., Mattarazzo, R., Fiorentino, M., Oestrogen-like effect of tamoxifen on vaginal epithelium. *Brit. Med. J.*, i:1351–1352, 1977.

77. Weinberger, P., Rea, M. *Nonylphenol: A perturbant additive.* Canad. Tech. Report of the Fisheries and Aquatic Sciences 990: 369–381, 1981.

78. U. S. EPA. *Chemical Hazard Information Profile: Nonylphenol.* p. 20. 1985.

79. Allen-Rowlands, C. F., Castracane, V. D., Hamilton, M. G., Seifter, J. Effect of polybrominated biphenyls (PBB) on the pituitary-thyroid axis of the rat. *Proc. Soc. Exper. Biol. Med.* 166: 506–514, 1981.

80. Sherman, J. D. Polybrominated biphenyl exposure and human cancer: report of a case and public health implications. *Toxicol. Indust. Health.* 7(3): 197–205, 1991.

81. Corbett, T. H., Beaudoin, A. R., Cornell, R. G., Anver, M. R., Schumacher, R., Endress, J., Szwabowska, M. Toxicity of polybrominated biphenyls (Firemaster BP-6) in rodents. *Environ. Res.* 10: 390–396, 1975.

82. Tilson, H.A., Cabe, P. A., Mitchell, D. Some neurotoxic effects of polybrominated biphenyl (PBB) compounds in rodents. *Health Perspect.* 20: 144, 1977.

83. Sager, D. B., Shih-Scaroeder, W., Girand, D. Effect of early postnatal exposure to polychlorinated biphenyls (PCB's) on fertility in male rats. *Bull. Environ. Contam. Toxicol.* 38: 946–953, 1987.

84. Green, S., et al Cytogenic effects of the polychlorinated biphenyls (Arochlor 1242) on rat bone marrow and spermatogonial cells. *Toxicol. Appl. Pharmacol.* 25: 482, 1973.

85. DeLong, R. L., Gilmartin, W. G., Simpson, J. G. Premature births in California sea lions, associated with high organic pollutant residue levels. *Science* 181(4105): 1168–1170, 1973.

86. Kimura, N., Baba, T. Neoplastic changes in the rat liver induced by polychlorinated biphenyl. *Gann.* 64: 105–108, 1973.

87. Bahn, A. K., Rosenwaike, I., Herrmann, N., Grover, P., Stellman, J., O'Leary, K. Melanoma after exposure to PCB's. *New Engl. J. Med.* 295(8): 450, 1976.

88. Falck, F., Jr., Ricci, A., Jr. Wolff, M., Godbold, J., Deckers, P. Pesticides and polychlorinated biphenyl residues in human breast lipids and their relation to breast cancer. *Arch. Environ. Health* 47(2): 143–146.

89. Murai, Y., Kuroiwa, Y. Peripheral neuropathy in chlorobiphenyls poisoning. *Neurology* 21: 1173–1176, 1971.

90. Vos, J. G., Van Driel-Grootenhuis, L., PCB-induced suppression of the humoral and cell-mediated immunity in guinea pigs. *Sci. Total Environ.* 1: 289–302, 1972.

91. Nelson, J. A., Effects of dichlorodiphenyltrichloroethane (DDT) analogs and polychlorinated biphenyl (PCB) mixtures on 17B-[H3] estradiol binding to rat uterine receptor. *Biochem. Pharmacol.* 23: 447–451, 1974.

92. Larson, R. V., Born, G. S., Kessler, W. V., Shaw, S. M., van Sickle, D. C. Placental transfer and teratology of pentachlorophenol in rats. *Environ. Lett.* 10(2): 121–128, 1975.

93. McConnachie, P. R., Zhalsky, A. C., Immunological consequences of exposure to pentachlorophenol. *Arch. Environ. Health.* 46(4): 249–253, 1991.

94. Gray, T. J. B., Gangolli, S. D., Aspects of the testicular toxicity of phthalate esters. *Environ. Health Perspect.* 65: 229–235, 1986.

95. Dillingham, E. O., Autian, J. Teratogenicity, mutagenicity, and cellular toxicity of phthalate esters. *Environ. Health Perspect.* 3: 81, 1973.

96. Ritter, E. J., Scott, W. J., Randall, J. L., Ritter, J. M., Teratogenicity of dimethoxyethyl phthalate and its metabolites methoxyethanol and methoxyacetic acid in the rat. *Teratology* 32(1): 25–31, 1985.

97. Kluwe, W. M., Carcinogenic potential of phthalic acid esters and related compounds: structure-activity relationships. *Environ. Health Perspect.* 65: 271–278, 1986.

98. Pratt-Johnson, J. A. Case report: retrobulbar neuritis following exposure to vinyl benzene (styrene). *Canad. Med. Assoc. J.* 90: 975–977, 1964.

99. Lillis, R., Lorimer, W. V., Diamond, S., Selikoff, I. J. Neurotoxicity in styrene production and polymerization workers. *Environ. Res.* 15: 133–138, 1978.

100. Mutti, A., Vescovi, P. P., Falzoi, M., Arfini, G., Valenti, G., Franchini, I. Neuroendocrine effects of styrene on occupationally exposed workers. *Scand. J. Work Environ. Health* 10: 225–228, 1984.

101. Maltoni, C., Failla, G., Kassapidis, C. First experimental demonstration of the carcinogenic effects of styrene oxide. *Med. Lavoro.* 10(5): 358–362, 1979.

102. Norppa, H., Vainio, H. Genetic toxicity of styrene and some of its derivatives. *Scand. J. Work Environ. Health* 9(2): 108–114, 1983.

103. Sherman, J. D. *Chemical Exposure and Disease.* Van Nostrand Reinhold, New York, 1988, pp. 128–129.

104. Hardell, L., Tamoxifen as a risk factor for carcinoma of corpus uteri. *Lancet* 2: 563, 1988.

105. Cunha, G. R., Taguchi, O., Namikawa, R. Teratogenic effects of clomiphene, tamoxifen and diethylstilbestrol on developing human female genital tract. *Human Pathol.* 18: 1132–1143, 1987.

106. Marquardt, S. *Exporting Banned Pesticides, Fueling the Circle of Poison.* Greenpeace, U. S. A., Washington, D.C., 1989, 115 pp.

107. Diethylstilbestrol. *Physicians Desk Reference*, 1973, p. 876.

108. U.S. House of Representatives. Committee of the Judiciary. Hearings on HR846, HR1568, HR2591, HR2592-HR2776, and HR2777 for the relief of various mink ranchers. March 2, 5, 9, 1951.

109. *Physician's Desk Reference.* Medical Economics Co, Inc., Montvale, N. J., 1993,, pp. 1401–1403.

110. Schaffner, C., Brunner, P. H., Giger, W. 4-Nonylphenol, a highly concentrated degradation product of nonionic surfactants in sewage sludge. *Commission Europ. Commun.* 9192:168–171, 1984

111. Jick, H., Waller, A. M., Rothman, K. J. Vaginal spermicide and congenital disorders. *J. Amer. Med. Assoc.* 234: 1329–1332, 1981.

112. Chvapil, M., Eskelson, C. D., Stiffel, V., Owen, J. A., Droegemueller, W., Studies on nonoxynol-9. II Intravaginal absorption, distribution, metabolism and excretion in rats and rabbits. *Contraception* 22(3): 325–329, 1980.

113. Weinberger, P., Rea, M. Nonylphenol: A perturbant additive. *Canad. Tech. Report Fish. Aquat. Sci.* 990: 369–381, 1981.

114. Sundaram, K. M. S., Szeto, S. The dynamics of nonylphenol in an aquatic model ecosystem. *Proc. Annual Wrkshp. Pestic. Residue. W. Canad.* pp. 31–42, 1981.

115. Laug, E. P., Nelson, A. A., Fitzhugh, O. G., Kunze, F. M., Liver cell alteration and DDT storage in the fat of the rat by dietary levels of 1 to 50 ppm DDT. 268–273, 1949.

116. Rashad, M. N., Klemmer, H. W., Association between serum cholesterol and serum

organochlorine residues. *Bull. Environ. Contam. Toxicol.* 15(4): 475–477, 1976.

117. Kreiss, K., Zack, M. M., Kimbrough, R. D., Needham. L. L., Smrek, A. L., Jones, B. T., Cross-sectional study of a community with exceptional exposure to DDT. *J. Amer. Med. Assoc.* 245(19): 1926–1930, 1981.

118. Anon., The World Bank's pesticide of choice. *Multinational Monitor* 7–9, Oct. 1989.

119. Luton, J. P., Mahoudeau, J. A., Bouchard, P. H., Thieblot, P. H., Hautecouverture, M., Simon, D., Laudat, M. H., Touitou, Y., Bricaire, H., Treatment of Cushing's disease by O,p-DDD. *New Engl. J. Med.* 300(9): 459–464, 1979.

120. Tullner, W., Adrenal atrophy effects in animals in *Proc. Chemotherapy Conf. on O,p-DDD*, Broder, L. E., Carter, S. K., Eds. Nat. Cancer Inst. 1970.

121. Luton, J. P., Mathieudefossey B., Bricaire, H., Cushing's syndrome: medical or surgical treatment? *Sem. Hosp. Paris Therap.* 53(1): 35–38, 1977.

122. 51 FR 19508 May 29, 1986.

123. Nisbet, I. T. C. Measuring cancer hazards: it takes a mouse to catch a rat. *Technol. Rev.* 78(2); 8–9, 1975.

124. Mehlman, M. A., Lutkenhoff, S. D. *Management of Chemical Mixtures.* Princeton Scientific Publishing Co., Princeton, N.J. 320 pp., 1990.

CHAPTER

10

SMOKING, A CONFOUNDING FACTOR IN DISEASE CAUSATION

There is no one, with the exception of decision makers in the tobacco and advertising industries, who believes that smoking is good for people, despite the admonition to join the Marlboro man, or believe, as the Virginia Slims promoters do, that "You've come a long way, baby." Smoking's adverse effects run the gamut from bad breath to untidy clothes and surroundings, burn damage to furnishings, increased fire hazard, reproductive dysfunction, and increases in heart disease, bronchitis, emphysema, and cancer. But to attribute the preponderance of cardiorespiratory disease and cancer to smoking alone is—like the tobacco industry itself—overkill.

The single-minded emphasis on blaming disease and cancer on cigarettes and the victims who smoke them delays and diverts attention from other well-documented and preventable disease-causing and carcinogenic factors.

It has been estimated that the two-pack-a-day smoker has a life expectancy about eight years shorter than that of the average nonsmoker.[1] While regrettable for the smoker, cigarette smoking should not bear the onus of being considered the only cause of pulmonary disease and malignancy.[2]

In assessing adult human cancer cases, it is uncommon for the physician to see a "pure" exposure to a single carcinogen. In addition to exosure to a multiplicity of chemicals that may initiate or promote cancer, there may be the confounding factor of "life-style" events, including one of the most talked of—smoking. In nearly every occupational lung cancer case, the confounding factor of smoking is raised. Although smoking is ubiquitous across most geographical areas, neither lung disease nor lung cancer is randomly distributed. Nor, it should be emphasized, is any disease randomly distributed in any population. This nonrandomness of disease, including cancer, has been demonstrated in studies of populations worldwide.[3-9]

Although smoking factors must be considered, one must analyze them along with other occupational and/or environmental factors, bearing in mind that:

- Until recently, smoking was more common in men than women, and the advertising appeal to the young woman smoker is relatively new. Men's work in general is heavier than women's and requires more physical effort, resulting in an increased rate and depth of respiration and thus a greater opportunity for the uptake of pollutants into the lungs.
- A synergistic, or at least an additive, effect between smoking and exposure to environmental and occupational pollutants is most likely.[9] It may be that some of the components of cigarette tars, which are similar to common industrial pollutants, may act as promoters and initiators of neogenesis, rather than as complete carcinogens.[10, 11] At least one cigarette filter, the Kent Micronite, contained asbestos (reference 61, p. 13)! An important mechanism of learning to smoke depends upon paralysis of the cilia of the tracheo-bronchial tree, suppressing the cough reflex and thus rendering other pollutants less likely to be removed from the lungs and more likely to be retained. One theory proposes that a worker who is exposed to pulmonary irritants may actually smoke more in an effort to achieve the ciliary calming effect of nicotine *because* he is exposed to the pollutants.[12] Persons who stop smoking report the phenomenon of increased cough as the sympathomimetic properties of nicotine diminish. Both this and the central nervous system stimulant effect of nicotine have been utilized in the marketing of nicotine-containing chewing gum.
- There was a real and progressive increase in lung cancer in industrialized countries before cigarette smoking became a widespread habit.[13, 14] Additionally, there has been a parallel association between industrial production and development and cancer in general.[15, 16]
- The more recent increase in lung cancer in woman may reflect the entry of women into industry, vis-à-vis "Rosie the riveter" of World War II fame, and, more recently, the acceptance of women in traditionally "male" jobs with concomitant increased exposure to industrial pollutants.[17]
- Smoking is more common in the blue collar trades than among professionals,[18] and cancer, with the exception of that of the breast, like disease in general, is more common among the poor.[19]
- Persons who enter and *stay* in occupations containing pulmonary irritants and pollutants probably have better than average lung function at the onset of employment (the "healthy worker" effect).[20–22]
- The corollary to that is, those who take up smoking probably have better lung function and better capacity to tolerate pollutants than those who never begin to smoke.[23, 24]
- Lung cancer diagnosed at autopsy has showed approximately the same incidence in both genders, contrary to the usual 3 : 1 male : female clinical ratio, and may reflect the spread of industrial pollutants into the general milieu.[25]

- Smoking is not associated with fibrotic lung disease, which is found in workers in the dusty trades and is independent of smoking.[26]
- Few smoking surveys have included occupational and/or environmental data. Additionally, most smoking surveys have been cross-sectional rather than longitudinal. Taking into consideration variation in human responsiveness, more attempts at longitudinal studies, determining pulmonary function and disease incidence over time, would be useful to assess the respective contributions and interactions of smoking and occupation.[27]
- The cancer risk of a person who has stopped smoking shows a rapid decline in incidence, approaching in 15 years an incidence of about 15% of that of the male who had continued to smoke.[28]
- Studies of smokers have rarely taken into consideration the fact that the adverse effects of many occupational and environmental pollutants continue after the exposure has ceased. In many cases studied, damage to the lungs was permanent and irreversible, and/or hazardous materials were retained for many years, the most notable cases being the pneumoconioses associated with exposure to coal dust, silica, and asbestos. The asbestos exposure of some 3 million World War II shipyard workers cannot be ignored by attributing cancers and disease in these workers, and in the men who served in the asbestos-laden ships, solely to smoking.[29] In 1976, Selikoff and Hammond, commenting upon the 200,000 to 300,000 tons of asbestos used yearly in the decade from 1930 to 1940, stated: "At present, approximately 850,000 tons (of asbestos) per annum were used. To the extent that this use is inadequately controlled, we shall see deaths of asbestosis and cancer in the year 2000 and beyond."[30] As with the tons and tons of asbestos that are in place and potentially friable, we cannot ignore the chemicals vented from industrial and municipal sources and attribute adverse health effects only to smoking.
- Malignancies in infants and children cannot be directly related to smoking.
- Malignancies that are not associated with lung function, by way of absorption and/or interference with normal pulmonary function, are for the most part independent of smoking. These include cancers of the breast, cervix, pancreas, ovary, testis, liver, and skin and probably the leukemias and lymphomas. Some malignancies, such as those of the kidney, may receive a dose of carcinogenic agent via transport from the lungs; so some chemicals common to occupational, environmental, and tobacco products may contribute to the cause of renal cancers. One must question, however, the contribution of the cigarette product itself compared to the dose of chemical transported by contaminated hands from benchtops to the cigarette and thence to the person.
- Although there have been skin cancers caused by the application of petroleum tars, and other components, that are common to tobacco products,

there is not a single record of cancer of the fingers attributed to cigarette smoke.[31,32]

- Some studies have proposed that the adverse effects of tobacco products stem from contaminants such as heavy metals and pesticides that[33,34,18] are common to industrial, agricultural, and environmental exposure.

- Production of lung cancer in animal models exposed to cigarette products is exceedingly rare. The study reported by Hammond et al.[35] was largely discounted by the National Academy of Sciences.[36,37] Rabbits exposed for $5\frac{1}{2}$ years to cigarette smoke in an inhalation chamber failed to show tumors related to smoking.[38] Probably the most significant lung effect produced in animals was a 40 to 50% incidence in hyperplasia after exposure to cigarette smoke for five days a week for a year; however, none of the animals developed malignancies.[39] Malignant changes have been produced on the skin of animals after painting with smoke condensates, but the U.S. Department of Health, Education and Welfare's *Bibliography of Smoking and Health*, which contains approximately 1200 literature citations, did not cite a single study concerning the induction of cancer in experimental animals.[40] No experimental data are available concerning the effects of side-stream smoke, main-stream smoke, or environmental smoke, or the comparisons between the particulate and gaseous phases of cigarette smoke.[41] This paucity of data is in stark contrast to that found on every other human carcinogen!

Cigarette smoking habits have been emphasized in practically every workers' compensation case that has gone to deposition and/or trial. In many cases it has been used as a diversion to dismiss compelling data linking chemicals with disease,[42] including lung cancer.[43]

Consider, for instance, a smoker who has been industrially exposed to asbestos, carbon black, iron dust, chromium, and benzene for a period of from 20 to 30 years, and develops a disease, perhaps cancer. How can the scientist separate the smoking effects from those of the other pollutants? If the person has a pulmonary disease or malignancy, smoking along with the other exposure factors must be considered contributory in one way or another. If the person developed leukemia (due to his benzene exposure) or skin cancer, his malignancy most probably would be considered as independent of his smoking history.

No review of the literature will indicate that the chemicals cited in the case above are *absent* an effect. Proportionally, more weight may be given to a particular exposure if the exposure *time* period and/or the *intensity* of that exposure was markedly different from one or more of the other exposures. In any assessment, all data are reviewed, and none may be discarded. In assessing human cancer cases, a composite of the exposure history (occupational, environmental, and smoking) must be coupled with information gained from case reports, epidemiological data, and animal and other laboratory studies. (See Chapter 9 for further information on the decision-making processes.)

All arguments aside concerning the relative contributions of environmental pollutants and smoking to cardiac, reproductive, pulmonary, and other health ills, the smoking habit should be, and is being, discouraged by public health authorities, by the establishment of nonsmoking areas in public facilities and work places and by various educational campaigns. Use of the addicting aspect of cigarettes has been proposed as a basis for legal action against tobacco companies, with the argument that cigarette advertising fails to warn of the problem of addiction.[44] Other legal attempts to recover damages from cigarette-associated conditions (bronchitis, emphysema, cancer, heart disease, allergies, burns, etc.) have been largely unsuccessful.[45-47]

The U.S. Congress, in an attempt to estimate the cost of smoking to Medicare and Medicaid, commissioned a study by the Office of Technical Assessment. In their critique of the OTA draft of smoking-related deaths and financial cos s, reviewers in general were critical of the U.S. government's overemphasis on epidemiological data based upon populations that do not represent industrial workers, and that in general fail to consider other causes of cardiopulmonary disease and death.[48-51]

The economic factors favoring the tobacco industry are difficult to overcome, with over 2 billion pounds of tobacco grown in the United States, representing about 20% of the world supply and about 5% of all U.S. exports.[52] Ominously, much of the Third World is experiencing economic problems and inadequate medical care, while receiving ''dirty industries'' exported from countries with more strict environmental and work regulations. Their added burden of cigarette promotion and consumption may be intolerable. There is evidence that U.S. governmental and private interests are increasing the exportation of cigarettes to these areas.[53,54] This policy, like the exportation of chemical and drug products banned from sale in the United States, must be addressed.

The complex economics of cancer, smoking, and environmental factors was expounded by Cairns in 1975, when he wrote: ''It could even be argued that few Western societies could afford to abolish a habit that creates a large secondary industry, generates considerable revenue and kills mostly older members of the population, who otherwise would draw on governmental welfare and social security benefits.''[55] If this cynical, although realistic, view is indeed correct, our public health sector will need to greatly increase its efforts to educate the public and obstruct the flow of tobacco products, as well as control environmental factors.

Efforts to limit tobacco advertising may be one of the few ways to remove the ''message'' from the public's eye, particularly from those most impressionable, the young. This has been done by some publications and other entities that have declined to accept the revenues from tobacco advertising.[56,57] One suggestion has been to market all cigarettes in identical brown paper wrappers with a uniform black ink label, which might help dispel the advertised ''differences'' between brands.[58] A proposal made in 1973 by the newly created

Consumer Product Safety Commission to ban cigarettes on the basis of provisions of the Hazardous Substances Act drew ire from the tobacco industry and got nowhere.[59] Nearly a quarter century after the release of the *Surgeon General's Report on Smoking and Health*, the tobacco industry spends roughly $2 billion a year on advertising, while the federal government continues a subsidy of $3.5 million a year to tobacco growers. By contrast, the federal contribution to the National Cancer Institute stands at approximately $1 million per year.[60] Unfortunately, in the face of these enormous expenditures, the problems of occupational and environmental pollution and smoking continue.[61]

It would do well for all to heed the words of W. C. Heuper, who over 30 years ago, in 1955, was head of the Environmental Cancer Section of the National Cancer Institute, chairman of the Cancer Prevention Committee of the International Union Against Cancer, and later recipient of the United Nations Cancer Award. He said:

> Finally, it may be noted that the evidence on hand justifies the viewpoint that, in arriving at a judgement in any medico-legal dispute requiring the assessment of liability for the development of a respiratory cancer, any evidence incriminating special occupational factors should be given preference over that possibly provided by a cigarette smoking history.[62]

REFERENCES

1. Gail, M., Measuring the benefit of reduced exposure to environmental carcinogens. *J. Chron. Dis.* 28: 135–47, 1974.
2. Breslow, L. Industrial aspects of bronchogenic neoplasms. *Dis. Chest* 28: 421–30, 1955.
3. Mason, T. J., McKay, F. W., Hoover, R., Blot, W. J., Fraumeni, J. F. *Atlas of Cancer Mortality for U.S. Countries: 1950–1969.* DHEW Pub. #(NIH) 75-780, U.S. Department of Health, Education and Welfare, Washington, D.C., 1979.
4. Mason, T. J., McKay, F. W., Hoover, R., Blot, W. J., Fraumeni, J. F. *Atlas of Cancer Mortality among U.S. Nonwhites: 1950–1969.* DHEW Pub. #(NIH) 76-1204, U.S. Department of Health, Education and Welfare, Washington, D.C., 1976.
5. Mason, T. J., Fraumeni, J. F., Hoover, R., Blot, W. J. *An Atlas of Mortality from Selected Diseases.* NIH Pub. #81-2397, U.S. Department of Health and Human Services, Washington, D.C., 1981.
6. Riggan, W. B., Van Bruggen, J., Acquavella, J. F., Beaubier, J., Mason, T. J. *U.S. Cancer Mortality Rates and Trends: 1950–1979*, Vols. I, II, III. National Cancer Institute/Environmental Protection Agency, U.S. Government Printing Office, Washington, D.C., 1983.
7. Segi, M. *Graphic Presentation of Cancer Incidence by Site and by Area and Population.* Segi Institute of Cancer Epidemiology, Nagoya, Japan, 1977, pp. 1–42.
8. Pickle, L. W., Mason, T. J., Howard, N., Hoover, R., Fraumeni, J. F., Jr. *Atlas of Cancer Mortality among Whites 1950–1980.* DHHS Publication No. (NIH) 87-2900, Washington, D.C., 1987.

9. Selikoff, I. J., Hammond, E. C. Relation of cigarette smoking to risk of death of asbestos associated disease among insulation workers in the United States. Presented at the meeting of the Working Group to Assess Biological Effects of Asbestos, Oct. 1972.

10. Hoffman, D., Wynder, E. L. A study of tobacco carcinogenesis. *Cancer* 27: 848–64, 1971.

11. Gori, G. *Status Report: Smoking and Health Program.* National Cancer Institute, Bethesda, Md., 1977, pp. 107–10.

12. Sterling, T. D. A critical reassessment of the evidence bearing on smoking as the cause of lung cancer. *Amer. J. Pub. Health* 65(9): 939–52, 1975.

13. Anon. Cigarettes: medical aspects. *Consumer Reports* 20(2): 67–73, 1955.

14. Heuper, W. C. *A Quest into the Environmental Causes of Cancer of the Lung.* U.S. Public Health Service Publication #452, 1955, pp. 1–54.

15. Hammond, E. C. Lung cancer and common inhalants. *Cancer* 7: 1100–109, 1954.

16. Epstein, S. S. *The Politics of Cancer.* Sierra Club Books. San Francisco, 1978, pp. 23–35.

17. Center for Disease Control. Update: lung cancer and breast cancer trends amoung women— Texas. *Morbidity and Mortality Weekly Report* 33: 2226–28, 1984.

18. Sterling, T. D., Weinkam, J. J. Smoking characteristics by type of employment. *J. Occup. Med.* 18(11): 743–54, 1976.

19. Cairns, J. *Cancer: Science and Society.* W. H. Freeman, San Francisco, 1978, pp. 46–47.

20. Kauffmann, F., Drouet, D., Lellouch, J., Brille, D. Occupational Exposure and 12-year spiro-metric changes among Paris area workers. *Brit. J. Ind. Med.* 39: 221–32, 1982.

21. Graham, B. L., Dosman, J. A., Cotton, D. J., Weisstock, S. R., Lappi, V. G., Froh, F. Pulmonary function and respiratory symptoms in potash workers. *J. Occup. Med.* 26:209–14, 1984.

22. Sterling, T. D., Weinkam, J. J. The "healthy worker" effect on morbidity rates. *J. Occup. Med.* 27(7): 477–82, 1985.

23. Leech, J. A. Ghezzo, H., Stevens, D., Becklake, M. R. Respiratory pressures and functions in young adults. *Amer. Rev. Respir. Dis.* 128: 17–23, 1983.

24. Tashkin, D. P., Clark, V. A., Coulson, A. H., Bourque, L. B., Simmons, M., Reems, C. Comparison of lung function in young non-smokers and smokers before and after initiation of the smoking habit: a prospective study. *Amer. Rev. Respir. Dis.* 128: 12–16, 1983.

25. McFarlane, M. J., Feinstein, A. R., Wells, C. K., Chan, C. K. The "epidemiologic necropsy": unexpected detections, demographyic selections, and changing rates of lung cancer. *J.A.M.A.* 258(3): 331–38, 1987.

26. Castellan, R. M., Sanderson, W. T., Petersen, M. R. Prevalence of radiographic appearance of pheumoconiosis in an unexposed blue collar population. *Amer. Rev. Respir. Dis* 131: 684–86, 1985.

27. Becklake, M. R. Chronic airflow limitation: its relationship to work in dusty trades. *Chest* 88: 608–17, 1985.

28. Doll, R. Practical steps towards the prevention of bronchial carcinoma. *Scot. Med. J.* 15: 433–47, 1970.

29. Anon. *Business Week,* p. 31, Sept. 29, 1975.

30. Selikoff, I. J., Hammond, E. C. Environmental cancer in the year 2000. *7th National Cancer Conf. Proc.*, Los Angeles, 1973, pp. 687–96.

31. Heuper, W. C. Environmental lung cancer. *Ind. Med. Surg.* 20(2): 49–62, 1951.

32. Heuper, W. C. *A Quest into the Environmental Causes of Cancer of the Lung.* Public Health Monograph #36, U.S. Public Health Reports, Vol. 71(1), 1955.

33. Smadkowski, D., Schultze, H., Schaller, K. H., Lennert, G. Ecological significance of heavy metal content of cigarettes. Lead, cadmium and nickel analysis of tobacco and the gas and particulate phase. *Arch. Hyg. Bacteriol.* 153(1): 1–8, 1969.

34. Cralley, L. J., Leinhart, W. S. Are trace metals associated with asbestos fibers responsible for the biologic effects attributed to abestos? *J. Occup. Med.* 15: 262–66, 1973.
35. Hammond, E. C., Auerbach, O., Kirman, D., Garfinkel, L. Effects of cigarette smoking on dogs. *Arch. Environ. Health* 21: 740–53, 1970.
36. Committee on Biologic Effects of Atmospheric Pollutants. National Academy of Sciences, Washington, D.C., 1972, pp. 178–79.
37. Horing, D., Chairman. *Environmental Tobacco Smoke: Measuring Exposure and Assessing Health Effects.* National Research Council, National Academy Press, Washington, D.C., 1986, pp. 1–337.
38. Holland, R. H., Kozlowski, E. J., Booker, L. The effect of cigarette smoke on the respiratory system of the rabbit. *Cancer* 16(5): 612–15, 1963.
39. Ketkar, M. B., Reznik, G., Mohr, U. Pathological alterations in Syrian golden hamster lungs after passive exposure to cigarette smoke. *Toxicol.* 7: 265–73, 1977.
40. National Clearinghouse for Smoking and Health. *Bibliography on Smoking and Health.* U.S. Department of Health, Education and Welfare, Washington, D.C., 1971.
41. U.S. Surgeon General. *The Health Consequences of Involuntary Smoking.* U.S. Department of Health and Human Services, Washington, D.C., 1986, p. 249.
42. Sterling, T. D. Does smoking kill workers or working kill smokers? OR, the mutual relationship between smoking, occupation and respiratory disease. *Int. J. Health Serv.* 8(3): 437–52, 1978.
43. Williams, R. R., Stegens, N. L., Goldsmith, J. R. Associations of cancer site and type with occupations and industry from the Third National Cancer Survey Interview. *J. Nat. Cancer Inst.* 59(4): 1147–81, 1877.
44. Garner, D. W. *Cigarette dependency and civil liability of cigarette manufacturers: a modest proposal.* 58 S. Cal. L. Rev. 1423, 1425 (1980).
45. Anon. Tobacco product liability. *J.A.M.A.* 255(8): 1034–37, 1986.
46. Rose, P., Ream, D. Tobacco litigation. *For the Defense.* 1987, pp. 7–13.
47. *Detroit News*, pp. 1-A, 7-A, Aug. 26, 1987, citing case of Joseph Palmer, dec. v. Liggett and Meyers Tobacco Co., decision by 1st Circuit Court of Appeals in Boston.
48. Kronebusch, K. *Preliminary draft: Smoking-related deaths and financial costs.* Office of Technical Assessment, U.S. Congress, May 10, 1985.
49. Sterling, T., Arundel, A., Weinkam, J. A review of: Smoking-related deaths and financial costs. Simon Fraser University, Burnaby, B.C., June 1985.
50. Sherman, J. D. Critique of the OTA and Sterling, et al. Assessment of smoking-related deaths and financial costs to O.T.A. August 1985.
51. Lave, L. Comments on the OTA and Sterling, et al. reports. Carnegie-Mellon University, June 28, 1985.
52. Fielding, J. E. Smoking: health effects and control. *New Engl. J. Med.* 313(9): 555–61, 1985.
53. Wolfe, S. M., Ed. Exporting death: U.S. forcing cigarettes on Asia. 3(7): 12, 1987.
54. Jones, H. A. plague from the west threatens the third world. *Amer. Lung. Assoc. Bulletin*, pp. 7–10, Apr. 1980.
55. Cairns, J. The cancer problem. *Scientf. Amer.* pp. 64–78, Nov. 1985.
56. Foote, E. Advertising and tobacco. *J.A.M.A.* 245(16): 1667–68, 1981.
57. Fisher, L. Ed. 37 magazines where greed is not an issue. *Amer. Lung Assoc. Bull.*, pp. 3–4, Apr. 1982.
58. Personal communication from Charles Nichols, J.D.
59. Anon. *Wall Street Journal*, p. 9, Aug. 24, 1973.
60. Anon. *Sierra* 72(4): 13, 1987.
61. Warner, K. E. *Selling Smoke: Cigarette Advertising and Public Health.* American Public Health Association, 1986, pp,. 1–112.
62. *A Quest* (see ref. 14), p. 45.

11

CAUSE-AND-EFFECT DECISIONS BASED UPON SCIENTIFIC OPINION

If human society cannot be carried on without lawsuits, it cannot be carried on without penalties.

—Aristotle

It is implicit in United States law, and English common law, upon which it is based, that one must refrain from harming the person or property of another, or from infringing upon another's rights. One who fails to perform this duty— to do no harm—is legally responsible for want of care and/or skill. When damage occurs because of the breach of such obligation, compensation is in order. As we live in a complex society that is based upon law rather than revenge, and that accepts monetary compensation in lieu of an "eye for an eye," there is the need for legal processes, to resolve questions of whether rights have been infringed upon or harm has been done.

The failure to exercise reasonable and prudent conduct commensurate with the circumstances may be termed *negligence*. This does not imply any malice or intent to do harm, but suggests that harm has resulted from lack of care. Questions of cause and effect and of assessing responsibility are implicit in this legal process.

Strict liability, in turn, does not imply negligence or fault when harm has occurred. It places responsibility on the party whose product or activity resulted in harm. This might include such activities as crop dusting that results in contamination of an adjacent property or population, engaging in industry in the midst of a populated area, or chemical contamination of air or water,[1] as well as sale and use of products and processes not fit for the intended or foreseeable uses, and which result in harm.

In the context of chemically caused disease, this failure, negligence, can happen in a variety of situations, such as the following:

- Failure to uphold professional standards, as, for example, when a physician prescribes a drug or combination of drugs that harms a patient, when it can be demonstrated that other professionals in similar circumstances would have refrained from doing so, because of known or predictable harm.
- The action of a professional pest control operator who, because of improper choice of chemicals and/or improper application, causes harm to a home and/or its occupants.
- Use of a product in agriculture, construction, and/or industry that causes harm to those handling it.
- Marketing and/or promotion of a product that is defective, as to composition, intended use, labeling, or failure to inform of side effects.
- Discharge of chemicals into the air, water, or soil in such a manner that harm occurs to others.
- Failure to warn of adverse effects in a timely manner.
- Failure to take or recommend remedial action in a timely manner.
- Failure to provide for informed consent. It is recognized that a person (worker, consumer, patient) has a right to be master of his or her own body, and that if there is likelihood of harm, the person must be informed of it and give consent before exposure.

The concept of informed consent underlies the workplace Right-To-Know Laws, passed in many states, and the General Duty Clause of the OSHA Regulations,[2,3] in addition to being the basis for the FDA Regulation requiring that drug advertising and labeling warn of possible side effects.[4] The informed consent requirement was specifically extended to the consumer in the case of birth control pills.[5]

For many people, the law and legal liability are best understood through extreme examples. In the Film Recovery Systems case,[6] in Illinois, company officials were sentenced to 25 years in prison and fined $10,000 each for failure to perform their duty.[7] The company was in the business of recovering silver from photographic negatives, using a cyanide system that released poisonous cyanide gas from the vats. Although labels contained warnings (created years earlier as a result of legislation and prior litigation), Film Recovery Systems, Inc. hired employees who could not read or speak English. One employee died and several more were injured as a result of the company's failure to warn and protect the employees.

Many other far more subtle situations exist that demand expert testimony and scientific explanation. In all of these, there are the questions of whether harm has been done, and, if so, whether it can be traced to a particular chemical, or process and, lastly, whether the party involved caused the damage by either commission or omission.

NEED FOR OPINIONS

In the context of cause-and-effect decisions concerning chemicals, information exists on nearly every common industrial chemical and drug and may be used to assess probable risk of harm. These data, coupled with knowledge of chemistry, physiology and toxicology, intended use, and dispersal, plus exposure data, enable one to make an educated analysis, to form an opinion, about probable cause-and-effect relationships. As data are accumulated over time and become concordant, a point is reached where the opinion may be confirmed by scientific fact.

It is necessary to make information and opinions available despite imperfections and gaps in knowledge. In the scientific arena, a lack of complete and final data does not prevent the scientist, the engineer, the professional, or the business person from making a decision based upon the body of information available at the time, and then taking the appropriate action. Vaccinations are given, and planes are flown, despite lack of "proof" of either total effectiveness or complete safety. Physicians prescribe medications to sick patients before "conclusive" results of medical tests are available, and engineers construct bridges and tunnels before every factor has been tested and known. An opinion, based upon available knowledge, analysis, and logic, is necessary to bridge the gaps where data are either imperfect or absent.

From a purely scientific standpoint, we can wait for "just one more piece of data" and thus delay a decision or an action. Legally, however, decisions must be made and judgments rendered for one side or the other. It is in this context that scientists must formulate opinions "within a reasonable degree of scientific certainty" to aid the legal process.

Cause-and-effect opinions concerning harm need not be restricted to ex post facto decisions. In the context of chemically caused disease, they are also properly applicable to *prevention* of harm, by restraint in promotion and use of chemicals and/or by regulatory efforts.[8-11]

The need to perform an assessment to make a decision now does not mitigate the need for continued collection of data for future decisions. The converse, to wait until "all the data are in," not only delays a decision-making process but is unacceptable in the context of prevention of harm, and in some contexts may be a violation of law.[12] In the legal arena, it is nowhere more true that justice delayed is justice denied.

CONCEPT OF PROOF

Proof must satisfy different criteria, depending upon the circumstances in which it is applied. The difference between scientific proof and legal proof of causation is not a simple distinction. Scientifically, we establish a hypothesis and through testing determine whether the hypothesis is valid. The more often we test, and the larger the sample size, the closer we get to an absolute value or global law.

The test required by most courts when considering workers' compensation and tort actions is that it must be *more likely than not* that a cause-and-effect relationship exists. The probability must be greater than 50%; that is, the odds are tipped to the side of effect being related to cause. This concept becomes even more complicated when the question presented is not one of whether an action/agent/chemical caused a particular result, but whether it made a *contribution* to the end result. A contribution may have the same legal effect as a single cause even though it was not the sole cause of a particular event.

The "beyond a shadow of doubt" test, applied to criminal cases in the legal arena, is a more strict interpretation than is required for proof in the context addressed in this book, and it is important to understand the different applications of these two concepts of proof. Scientists, accustomed to the laboratory, are often hesitant to attempt an opinion concerning a medical–legal matter, erroneously believing that 100% certainty is required.

HAZARD + EXPOSURE = RISK

Risk with Harmful Effect = Causation Opinion

Where there are positive animal studies, case reports, and/or epidemiological data, as well as the opportunity for *exposure* to a known *hazardous* agent and the passage of time, it is possible to make a determination of *risk*. The next step—in the context of one who has suffered harm from the combined factors of hazard, exposure, and risk—is a *causation* determination. In humans, this connotes the precipitation, hastening, or aggravation of a disease state, or of death. Environmental harm decisions follow in a similar manner.

Hazard

Hazard is the potential for an adverse effect, and includes whether the effect is reversible or not, and whether it is life-threatening or not. A detailed discussion of factors considered in determining the hazardousness of a chemical is presented in Chapter 2.

Body of Data

Chemical hazard determination is based upon a body of information, including consideration of some or all of the following properties:

- Chemical structure and physical properties
- Structural activity comparisons to other known chemicals
- Reactivity and water/oil partitioning
- Site of entry, route through the body, and modes of excretion
- Anatomical and/or histochemical effects
- Irritative and/or sensitization effects
- Immunological effects
- Biochemical and metabolic effects
- Reproductive and/or teratogenic effects
- Mutagenic and/or genotoxic effects
- Behavioral effects (either direct or secondary to other effects)
- Carcinogenic effects
- Animal data
- Case reports
- Epidemiological studies

Exposure

Exposure is difficult to assess, especially in human populations. Strict dose-response is mainly applicable to animal studies. Very few, if any, measurements are made, or have been made, of industrial populations or of environmentally exposed populations, particularly on a longitudinal basis. (Longitudinal studies follow the same population over a period of time, whereas cross-sectional studies measure various aspects of a population at a given point in time. The latter, cross-sectional, are easier, less costly, and obviously faster to accomplish than longitudinal studies.) Unfortunately, most environmental measurements that have been made represent isolated "points in time" and do not necessarily characterize the day-in, day-out exposure of the population. An expanded discusion of the concept of exposure is presented in Chapter 4.

Although the determination of exposure is difficult, it can be evaluated by the collection of information, where known, in several areas:

- Intensity and duration of the exposure
- Mode of exposure (dermal, respiratory, gastrointestinal, parenteral)
- Physical activity at the time of exposure (increased physical activity may increase the rate and amount of chemicals taken into the lungs, by virtue of deeper and more frequent respiratory efforts)
- Age at the time of exposure (children, the elderly, the infirm, and people of child-bearing potential require special attention)
- Type of exposure: single or chronic
- Length of exposure: intermittent or continuous

- Time from initial exposure to expression of results
- Sequence of exposure to various chemicals

Risk Assessment and Determination of Causality

Risk is based upon information concerning hazard and exposure, and it in turn determines whether there is sufficient information to establish causality. This logical process includes questions such as the following:

- Is there a disease or an abnormality present?
- Has there been opportunity for exposure?
- Has either the duration or degree of exposure been sufficient to cause the condition?
- Is the outcome biologically plausible?
- Are there dose–response or time–response relationships?
- Is there an overall consistency of the various data?
- Is there concordance of the evidence in animals and humans?
- What is the strength of the data; that is, are there similar results in multiple studies?
- Are there any confounding causative and/or interfering factors?
- Is there any other explanation(s) to better explain the outcome?

HUMAN VARIATION IN RESPONSE TO RISK

In the human population, with its heterogeneous makeup, there are variations in the response to risks. We do not yet understand the mechanisms whereby some members of a population may be ''immune'' or ''resistant'' to the effects of various chemicals, nor do we understand why or how others succumb to chemically caused disease. We do know that, as in test animal populations, not all of the individuals in a human population will succumb to the effects of a chemical at the same time or at the same dose. In general, however, with either increasing dose or prolongation of time, increasing numbers of a population will express a reaction to a chemical.[13]

Based upon known data, we can predict that certain chemicals present a risk for exposed populations. What we cannot do is determine prospectively just *when*, and exactly *which*, members of the population will be effected.

Retrospectively, when we consider a person who has been harmed, we can— using animal test data, human case reports, and epidemiological data—reach an opinion as to most likely causation. In the special context of occupationally caused disease, the above factors may serve as guidance as to whether certain chemicals precipitated, aggravated, or accelerated a disease process. Ideally, these opinions of risk should then trigger action: prevention of future harm and/ or compensation for past exposures that resulted in harm.

Evaluation of the questions noted above provides a common sense basis for formulating an opinion in the decision-making process. Some additional thought is necessary in considering the "special circumstances" given in the NIOSH publication *A Guide to the Work-Relatedness of Disease*, which states that "special circumstances must be considered. Were there any unusual events at work that reduced the effectiveness of protective equipment? Of ventilation? Of safe work practices? If the employee is a woman, are there special risks to women from exposure to the agent? If so, this factor must be evaluated."[14]

Actually, there does not need to be a failure of protective equipment or of ventilation for exposure to occur. In these cases, the onset of disease is only altered. Moreover, the special mention of problems affecting women in the workplace is meritorious, but it must not be assumed that the male worker, or his progeny, are without risk for damage.

Some of the concepts about industrial exposure have been misunderstood. It has been asserted that for a disease to be considered occupationally related it must, for instance, occur in a particular part of the body. This is not true. Although some chemicals are more likely to cause illness in a particular organ (carbon tetrachloride damaging the liver, methyl alcohol causing blindness, mercury poisoning the kidneys), many industrial chemicals have multiple effects on the body. Well-recognized examples are: asbestos causing fibrotic lung disease as well as cancers of the pulmonary and gastrointestinal systems; lead causing poisoning of the bone marrow, gut, and nerves; and chlordane affecting the liver, immunological system, and bone marrow. It is true that the recognition of disease, especially malignancy, in a certain part of the body has heralded the recognition of a causal relationship, but it must not be assumed that a chemical causes an adverse effect in one part of the body only. Most biologically active chemicals cause reactions not only in various organs, but in various species as well.

Another misconception is that for a disease to be considered related to occupation, it must arise in a category of workers more frequently than it does in the general population of comparable age and sex; but this is to deny cases of any disease detected early in the exposure history, and to demand after the fact data when there may be none. The NIOSH guide states: "the work-relatedness of a disease must be based on an evaluation of the *available information*" (italics added for emphasis). Gaps in the information base do not negate information that is available, and lack of information does not prove lack of association. This concept is particularly applicable in considering an individual who may have suffered harm from exposure to a new drug or a new product introduced into the workplace or the home; here, the early cases may be the sentinel ones. In general, workers have greater and more prolonged exposures than the general population; but if contamination has spread from the work milieu to the general environment, the ''background'' level of illness may approach that of the workplace. This background of contamination is applicable to the illness associated with living near an asbestos factory, a uranium mine, a smelter, or a chemical factory.

The common claim that a given person's level of, say, DDT, dioxin, chlordane, or PCB is only slightly above ''background'' is a spurious argument. These particular chemicals are all man-made. There is no ''normal'' background level. All levels are a matter of contamination. The question then is, at what level of contamination and for what duration of time can harm occur? Chemicals that occur in the natural environment, such as arsenic, asbestos, coal, and so forth, may also spread, contaminating the general environment and contributing to ''background'' levels that by themselves or added to the workplace exposures may be unbearable. This was certainly the case in an asbestos mining area in Quebec where the entire town was contaminated.[15]

In practical matters, one seeks answers to such questions as:

- Can the dust reach the lungs, the solvent the bone marrow, the dye the kidneys, and the irritant the skin?
- If the effect is neurological, is the chemical soluble in lipids? Is the product fibrogenic, anesthetic, toxic to the kidneys or liver, and so on?
- Have similar associations been found in animals by experimentation? This means, for instance, the production of a malignancy in another species by the same chemical. Given the variable response of different species, but the basic DNA of all, the critical finding is the production of malignancy of any organ, not necessarily the same organ, across species differences.
- Has a dose–response or a time–response relationship been demonstrated, either in test animals or in human populations?
- Have similar reports of a cause-and-effect association been reported in the scientific literature, and if so, how many times, and for how long?
- And, finally, could the disease of the person under consideration be due to any other factors? If yes, do the factors outweigh the exposure under consideration?

SOURCE OF INFORMATION: EPIDEMIOLOGY

When epidemiological data are available, one should consider them along with other sources of information, keeping in mind the constraints of the epidemiology process, which considers the incidence, prevalence, and distribution of an abnormality in a given population.

One must be careful to avoid over-reliance on epidemiological studies to reach opinions and decisions concerning human health and disease matters. Epidemiological studies of disease are always after the fact, and harm must accrue before data are available. If the disease is newly recognized, and/or no studies have been conducted, there will be no epidemiological data. One cannot equate lack of epidemiological data with failure to support a possible cause-and-effect relationship. One can only say that there are no data, or the data are insufficient. Positive studies may support a relationship, but a negative study does not negate a possible relationship. Epidemiological studies rely on the characteristics of the population under consideration. The reactions of an individual may be an exception to that population.

The need for caution in employing epidemiology has been well stated by Dr. Eula Bingham, the director of the Occupational Safety and Health Administration, who urged that when "public policy mandates preventive health care, waiting for epidemiological data is unacceptable, because it means waiting to count dead bodies."[16]

SOURCE OF INFORMATION: INDUSTRY AS A WHOLE

Frequently, we must consider an industry, *as a whole*. For example, the foundry, coking, rubber, petrochemical, and welding industries have consistently shown increased rates of disease among their workers in case-control and epidemiological studies. The character of the industries themselves, containing multiple sources of exposure, makes it difficult to isolate a single "guilty" chemical in the work milieu, as well as difficult to impossible to determine the contribution of the various components.

The overlap between chemicals produced by the main petrochemical industry and those in secondary industry is illustrated below and in Figures 11.1-11.4.

We may utilize data from toxicological studies of individual chemicals that are used (when they are known) or generated within an industry. These data may come from animal studies or from effects on humans following exposure to the individual components. Several industries, and a number of occupations within specific industries, have been determined to be inherently dangerous, although it is not yet possible to pinpoint a specific causative agent or agents in some.

Illustrative are the high rates of cancers of workers in the coking part of the steel industry, with those employed at the tops of the coke ovens having the highest rates.[17] An apparent difference in respiratory cancer rates between white and nonwhite workers was revealed to be due not to increased susceptibility linked to race, but to job classification.[18] The coke workers were at increased risk not only for respiratory cancer, but for cancers of the kidney,[19] colon,

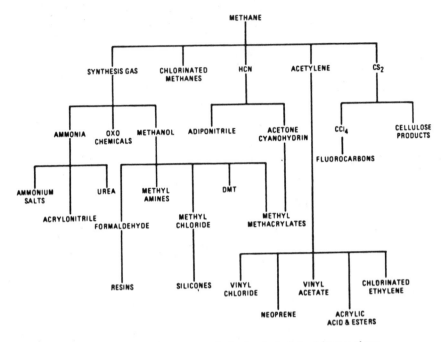

Figure 11.1. Some important synthetic chemicals derived from methane.

pancreas, buccal cavity, and pharynx as well.[20] Parallel findings of increased rates of disease have been reported in the aluminum,[21,22] chemical,[23,24] foundry,[25,26] welding,[27,28] and oil field[29] industries, and indeed in most industries that have mixed exposures to toxic chemicals.

Occupational diseases are not confined to workers in the obviously "dirty" occupations, or to those in the "smoke stack industries." Hospital operating room personnel, exposed to anesthetic gases, were found to suffer a higher rate of pregnancy loss, malformations in their infants, and cancer than nonexposed coworkers.[30-32] Similarly, hospital personnel handling equipment sterilized with ethylene oxide and nurses working with kidney dialysis equipment sterilized with formaldehyde have developed a variety of adverse effects, including skin reactions, asthma, reproductive disorders, and cancer.[33-35] This toxicity is in addition to that experienced by the patients whose blood courses through these machines, contaminated by various plastics components and sterilizing chemicals. Hemolysis and other reactions have been documented in these patients, already sick with kidney disease.[36] Although both formaldehyde and ethylene

Figures 11.1-11.4 reproduced with permission from Riegel's *Handbook of Industrial Chemistry*, 8th Ed., edited by James A. Kent. Van Nostrand Reinhold, 1983.

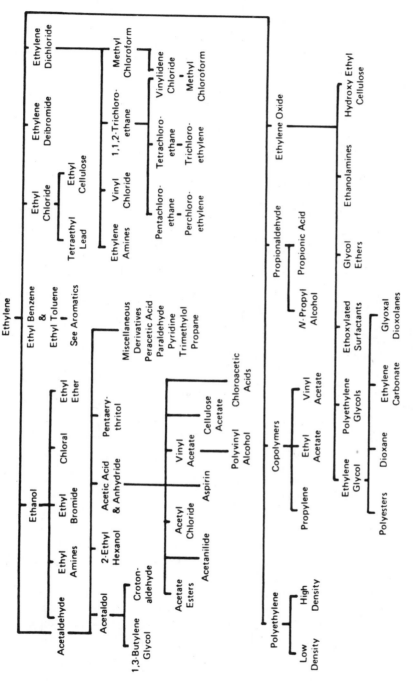

Figure 11.2. Major ethylene derivatives.

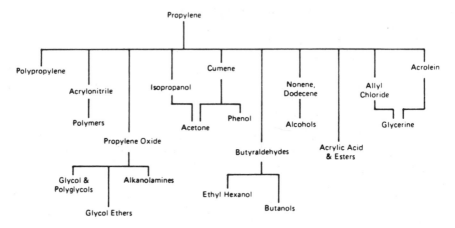

Figure 11.3. Some chemicals derived from propylene.

oxide are effective as sterilants, it is clear than an alternative method of sterilization is needed to protect both the patient and the personnel. On the oncology services, some of the chemotherapeutic agents administered to cancer patients do not represent an unmixed blessing, as a number of the anticancer drugs can result in malignancy in those handling them.[37]

These differences in rates of disease expression between industries, and between different job classifications within a single industry, underscore the emphasis made in Chapter 2 that illnesses are not random events.

STATE OF THE ART

A collection of information about any subject that has been assembled over time provides a historical overview on when and how the information has been developed, thus showing the state of the art of the subject area for the time period under study. With such a compilation, the knowledge that was available at a specific time concerning a specific chemical or process can be assembled and reviewed. Such a chronological retrieval of information provides a fascinating perspective on the breadth and depth of scientific thought, frequently across continents and languages. State-of-the-art information is considered as a cumulative, whole body of data encompassing what is known from all sources.

The concept of state of the art has been utilized to establish fault in situations where knowledge of adverse effects was available, but there was a failure to

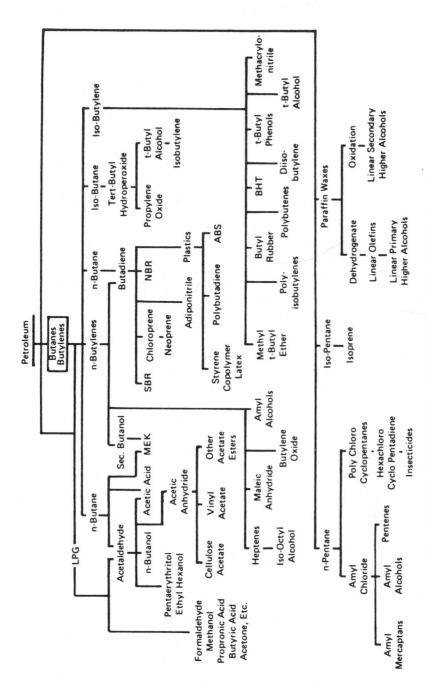

Figure 11.4. Chemicals from butanes, butylenes, LPG, and higher aliphatic hydrocarbons.

209

warn against exposure to a product or its use, or failure to remove a hazardous product from the market. Prospectively, review of data on a chronological, state-of-the-art basis can be used not only to determine information gaps, but to develop "worst-case" or "most likely" outcomes, and thus serve to warn of harm when the collective data indicate the likelihood of harm.

One of the best documented and most illustrative instances of such state-of-the-art knowledge has been the accumulation of information about asbestos. By at least 1942, there was sufficient information, in the form of well over 100 articles in the English-language literature alone, to indicate the hazardous nature of asbestos, and to demonstrate to any prudent person that exposures to it needed to be kept to a minimum and its uses curtailed. That year, Hueper characterized asbestos as a suspect human carcinogen and called for the use of substitute products (as well as the control of other chemicals with carcinogenic properties).[38] Unfortunately asbestos was cheap, and it found its way into such products as children's modeling clay, textured paint, fake snow for movies, and other uses where its fire-retardant properties were immaterial.

We can only speculate on the sparing of life and freedom from illness that would have occurred, had such asbestos companies as Johns Manville curtailed production and marketing. And we may ask, would this curtailment have been any more destructive to the company than the millions spent for legal claims brought by users and workers, and for attorneys' fees? Was the adverse publicity or the stigma of declaring bankruptcy to avoid claims worth it? Are there other disasters such as this waiting to be documented and litigated, based on state-of-the-art knowledge? I fear that the answer is yes.

In assessing literature reports, one gives weight to data collected over time and from multiple sources that have reached the same or similar conclusions. We do not live in a vacuum, and science and medicine are not conducted as single events. The body of data concerning formaldehyde, hormonal agents, and various pesticides and drugs, as well as asbestos, has been generated over the decades, by a multitude of independent scientists, working in a number of countries. This body of data grows with each addition and with the passage of time.

Evaluators should be cognizant of the source or financial sponsorship of research data. Questions of objectivity have arisen—regarding the conduct of research, generation of data, and even withholding of data—when honest disclosure was not in the best interest (usually financial) of the sponsoring agent.[39, 40]

The conclusions that asbestos and formaldehyde are a hazard to living matter and are carcinogenic cannot be denied. The threat of strokes to women taking birth control pills likewise was described in numerous reports in the scientific literature, backed up with physiological and pharmacological explanations. Less well known, however, are the data generated by researchers concerning the possible teratogenic potential of a drug called Bendectin[R], which contained dicyclomine and doxylamine, drugs of the antihistamine class. Although it was demonstrated that these antihistamines acted as "weak" teratogens, and certainly produced fewer dramatic deformities than thalidomide produced, it has been difficult to cite a single animal test or a single human epidemiological study to make a final determination of its teratogenicity. Taking into consideration the *body* of data generated concerning the drug, however, a physician would be hard-pressed to recommend the product to a pregnant woman. One can only conclude that the drug's manufacturer, under pressure of the accumulated scientific data, as well as the pressure of many lawsuits, came to the same decision when it was removed from sale in the United States in June 1983.

Opinions will vary on both sides of the litigation table about which data are of greater importance, and which will decide differences under contention. Not one of any of the cases cited in this book was decided without a range of differing opinions. The use of opinions aside, the important principle is that the court, not the street, is the appropriate place to resolve disputes, and that the opinions offered by experts on either side are an integral part of the judicial process.

DEGREES OF IMPAIRMENT

Emphysema, bronchitis, obstructive and restrictive lung disease, hypertension, and liver and kidney impairment usually progress imperceptibly. Thus, the degree of impairment at a particular point in time is difficult to determine. A "little" or a "lot" of emphysema, kidney disease, and so forth, is nearly impossible to measure until it is accompanied by changes in functional ability. The development of these types of conditions is frequently slow and little noticed by the affected person until a critical point is reached, with an interference in the customary ability to lead one's life. Frequently the patient attributes a slowing or decrease in his or her usual routine to "getting a little older," or a bout of the "flu," and dismisses the early symptoms as unimportant. Most of us enjoy good health, frequently denying the appearance of poor health. It is the interference with normal function that is critical. When this interference becomes persistent, irreversible, and/or progressive, an individual can be deemed to be disabled.

Attempts have been made to quantify impairment, and disability, particularly for legislative and compensatory uses, with varying degrees of utility.[41] Impairment implies alteration in health status, whereas disability implies that the impairment has resulted in a change in one's capacity to meet personal, social, and/or occupational demands. There having been a change from the proof of negligence required under common law to such remedies as legislation and statutory workers' compensation laws, the two are sometimes considered separately, although experience and common sense connect them. Although a 1972 congressionally sponsored National Commission on State Workman's Compensation Laws issued a statement that all work-related injuries or diseases should be covered by the compensation system, this has not come to pass. And so in the field of hazardous chemicals, the litigation continues, requiring opinions as to causation of illness and determinations of disability when a person is unable to carry on his or her life.

If a person's disability is found to be due to a chemical exposure, then decisions must be made about protection from additional exposure, treating the patient (if the disease processes are not irreversible), and, with foresight, taking steps to prevent harm to other exposed persons. Additionally, where harm has occurred, decisions about compensation may have to be made.

This concept of care, from known risk to prevention, a worthy principle of public health, is at variance with the following statement in the AMA guide: "While a physician properly may make an inference about degree of risk based on medical information, it is not proper for the physician to determine the acceptability of that risk. Such a decision is of a personal, social, legal, economic, business, or insurance nature and is properly made by non-physicians based on non-medical considerations."[42] This narrow view is in opposition not only to the ethical admonition of the Hippocratic oath "to do no harm," but is at odds with much experience, legislation, litigation, and public policy that has aimed to mitigate against risk.[43-48] Physicians are encouraged to warn against the risks of smoking, obesity, alcohol, and drug abuse; so why should they hesitate to warn of the risks of chemical exposures when these risks can be reasonably ascertained and further harm prevented?

SPECIAL PROBLEMS IN CAUSATION

The concept of using malignancy as a marker disease is useful; for, as with pregnancy and venereal disease, one never has "a small amount" of it. The end point of cancer, which is demonstrable as abnormal cells with the potential for autonomous growth and metastasis, is very real and definable. For this discussion, cancer will be used as an example of a disease in which decisions about causation clearly can be made, not only from a scientific, but from an ethical point of view.

Sequence of Neoplasia

Histological and biochemical studies show that neoplasia proceeds in a sequential fashion:

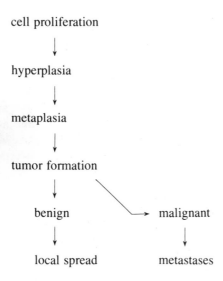

A cell does not react to a chemical insult in an infinite number of ways. Cellular reaction, like other biological processes, is orderly, with the following stages: *cell proliferation*, an increased numbers of cells; *hyperplasia*, an abnormal enlargement due to an abnormal numerical increase in cells; *metaplasia*, a change in structure or the inclusion of abnormal components; and the *malignant* stage, meaning disposed to cause harm or death. Malignancy is used interchangeably with the term cancer, the first term derived from the Latin word root, the latter from the Greek. *Metastases* are the transference of cancerous cells from one part of the body to another by way of the bloodstream or lymphatics, or along other tissue planes.

A benign (noncancerous) tumor may cause death if it blocks or interferes with a vital life process. Thus a benign tumor of the brain, contained within the skull which has a fixed volume, can kill by virtue of it space-occupying and pressure-causing characteristics, just as a malignant tumor can kill. Like a malignant growth, a benign tumor of any of the internal organs can obstruct or erode into vessels and cause hemorrhage, resulting in death. Malignancy carries the added peril of abnormal cells, with their characteristic autonomous growth and ability to spread.

Many carcinogenic chemicals can produce both benign and malignant tumors. This is an important consideration in evaluating results in test animals. A chemical that stimulates the production of benign tumors also has the potential to stimulate growth, and thus cause malignant tumors, either in other test animals and/or humans. The stimulation of growth, in and of itself, should trigger further investigation.

Through the mechanism of cell proliferation, more single-strand DNA becomes available for adduct formation, with less time available for repair of the adducts. Thus there are increased opportunities for "mistakes" in the otherwise orderly reproductive process of cell formation. These mistakes become expressed as DNA breaks, cross stranding, inhibition of repair, sister-chromatid exchanges, chromosome abberations, and gene deletions, as well as changes in activating, repressing, and/or competing enzyme systems. The results of these multiple expressions are a deranged function and, in some cases, neoplasia, either benign or malignant.

Chronic irritation of cells, from formaldehyde exposure, with increased need for repair and concomitant interference with repair, helps explain not only the hyperplasia that results, but the additional developments of metaplasia and carcinogenesis as well.[49]

Asbestos "corns" were described on the hands of workers handling this material. While these are "benign" lesions, and are examples of nonmalignant effects, they are in keeping with the metaplastic effects that were described as early as 1927,[50] and of course were predictive of the carcinogenic effects that became so well documented.

CARCINOGENIC RISK IN HUMANS

In assessing the carcinogenic risk of various chemicals for humans, it is well to keep in mind that in many countries, for both scientific and regulatory purposes, a chemical or an agent is considered a carcinogen if it produces: (1) the induction of neoplasms that are not usually observed, (2) the earlier induction of neoplasms that are commonly observed, and/or (3) the induction of more neoplasms than are usually found.

A chemical may be considered a *presumptive* human carcinogen if it shows:

1. Production of malignant tumors (a) in multiple species or strains of animals, and/or (b) by multiple experiments (routes and/or doses), and/or (c) to an usual degree (with regard to incidence, site, type, and/or precocity of onset).[51]
2. Additional evidence of reactivity, including dose–response, mutagenicity, teratogenicity, and/or structural–activity effects.

The International Agency for Research in Cancer, in trying to coordinate information and provide a basis for regulatory efforts and decision making, has stated: "In the absence of adequate data on humans, it is reasonable, for practical purposes, to regard chemicals for which there is sufficient evidence of carcinogenicity in animals as if they presented a carcinogenic risk to humans."[52] This concept of presumption of risk may be used by regulatory agencies charged with prevention of disease, by decision makers to avoid harm, and by the legal system in assessing responsibility when harm has occurred.

Proof of a human carcinogen is based upon case reports, which may be reinforced by epidemiological studies. With few exceptions, most of the human carcinogens were established first by way of human case reports, with supportive animal and epidemiological data. This has been true of asbestos workers, arsenic miners and users, chromium miners, vinyl chloride production workers, benzidine dye workers, bis-chloromethyl ether handlers, and others. After the sentinel cases were reported, some epidemiological studies were undertaken in a few selected exposed populations. But, an epidemiological exercise, waiting for "body counts" before protective procedures are put in place, obviously is not helpful for those whose exposure continues. To wait for the results of epidemiological studies in order to take precautions, establish regulations, and/or assess legal causation is clearly unacceptable in terms of disease prevention, or assessing legal responsibility for injured or dead victims.

DECISION MAKING IN CANCER CASES

It is generally accepted among scientists that there is no threshold for a carcinogen, and that all carcinogens must be considered as able to cause cancer, given a sufficient lifetime.[53] Practical considerations in humans concern the level of exposure at which cancer occurs, ways to consider variable exposures, and exposure to a variety of agents. Theoretically, it takes but one interference with the DNA of a cell to initiate the carcinogenic process. If many cells are affected, then the latency period for expression of cancer may be shortened, and the progress of the malignancy accelerated because of its multicentric origins.

In considering cancer causation in a person, one looks at the presence and exposure potential of carcinogenic agents that have caused malignancy in other humans and/or test animals. It is important to understand that carcinogens may cause neoplasia at one site in a test animal, and at a different site in a human. Likewise, carcinogens are not cell specific or site-specific.

OPINION AND DECISION MAKING IN GENERAL

In assessing causality of illness due to a toxic exposure, the first task is to determine whether the person is sick or not; and then, if sick, whether there is an identifiable exposure agent or situation that can explain the illness. One then considers medical records, exposure history, animal data, case reports, and epidemiological studies as to causality of a disease or condition, asking the following questions:

- Do the various data make logical sense?
- Has the condition been found in other persons similarly exposed?
- Have animals studies been positive?
- Is the biochemistry of the agent compatible with the effect seen in the human?
- Are there alternative, more likely factors that would explain the given effect?
- Are the data concordant?

From the viewpoint of pure science, cataloging the effects of chemicals involves the accumulation of sufficient information to determine whether a chemical is hazardous or not, and whether persons are at risk. From a social standpoint, greater urgency develops. Some effects are irreversible, some fatal. If a chemical is found to be hazardous, then exposure implies potential for harm. The ideal is to avoid known risks and prevent harm. Where risk–benefit or cost–benefit analysis is used, and where these factors accrue to different parties, as they usually do, the institution that benefits or profits from the use of a chemical product at the risk of another must compensate that person or group if damage is done. This provision of compensation for harm done is the basis of the legal system in the United States.

It could take years and untold funds and resources to accumulate sufficient data to satisfy a scientific standard of absolute proof of causation. In the intervening time, the courts must render decisions. For these to be informed decisions, they must rely upon scientific information and opinions for guidance. The legal decision is ultimately a social one (with economic aspects where loss of functional ability or fault is a factor), and this decision is made by a jury and a judge with the aid of information supplied by the informed opinions of various scientists.

The concept of "within a reasonable degree of scientific and/or medical certainty" is based upon inductive logic. For a prudent scientist, it provides the basis for formulating an opinion even though further investigation may be necessary. Learning to consider data in this context is the responsibility of not only the forensic scientist, but other scientists and physicians whose opinion performs a service to the court and the adjudicative process.

For all the claimed "objectivity" of "pure" science, a scientific fact is merely a collection of data existing in time. As information and opinions concerning these data are accepted and/or rejected, a body of knowledge accumulates revealing patterns of consistency and what we call fact—the basis, as we have said—for informed legal decisions. Every just and orderly society must make assessment of fault and allocate compensation for damage, else it degenerate into a society of outlaws, where only the strong prevail. When scientists lend their information, knowledge, and expertise to the courts, the ultimate result is a more enlightened decision-making process.

REFERENCES

1. Prosser, W. L. *Law of Torts*. West Publishing Co., St. Paul, Minn., 1971, Chap. 13, Sec. 78.
2. 29 USCS sec. 654.
3. 29 CF4 1910.120
4. Pub. L. 87-781, 76 Stat. 780.
5. 21 CFR 310.501(a) and 35 FR 9001.
6. Cited in 427 Mich. 1, 398 NW 2nd 882 (1986), and in Leviton, Policy considerations in corporate criminal prosecutions after People v. Film Recovery Systems, Inc., *ABA National Institute on Worker's Compensation; A Review of Costs, Emerging Developments and Remedies*, 1986, p. 186.
7. Anon. Company officials get 25-year terms in cyanide murder, *Chemical Worker*, p. 5, Aug. 1985.
8. Nelson, N., Chairman. *Principles for Evaluating Chemicals in the Environment*. National Academy of Sciences, Washington, D.C., 1975, pp. 1-451.
9. Johnson, H. W., Chairman. *Decision Making for Regulating Chemicals in the Environment*. National Academy of Sciences, Washington, D.C., 1975, pp. 1-232.
10. Costle, D. M. Pollution's "invisible" victims: why environmental regulations cannot wait for scientific certainty. Remarks by EPA Administrator Costle before the National Coalition on Disease Prevention and Environmental Health, Washington, D.C., Apr. 28, 1980.
11. Albert, R. E., Train, R. E., Anderson, E. Rationale developed by the Environmental Protection Agency for the assessment of carcinogenic risks. *J. Nat. Cancer Inst.* 58(5): 1537-41, 1977.
12. Peters, G. A. Warnings and safety instructions. *Hazard Prevention* 21-22, May/Jun. 1985.
13. Littlefield, N. A., Farmer, J. H., Gaylor, D. W. Effects of dose and time in a long-term, low-dose carcinogenic study. *J. Environ. Pathol. Toxicol.* 3: 17-34, 1979.
14. Kusnetz, S., Hutchinson, M. K. *A Guide to the Work-Relatedness of Disease*. NIOSH, U.S. Department of Health, Education and Welfare, 1979, p. 25.
15. LeDoux, B. *Asbestosis: East Broughton, Province of Quebec, Canada*. Jan. 1949, pp. 1-52.
16. Ruttenberg, R., Bingham, E. A comprehensive occupational carcinogen policy as a framework for regulatory activity, in *Management of Assessed Risk for Carcinogens*, Nicholson, W. J., Ed. N.Y. Acad. Sci. 363: 13-20, 1981.
17. Lloyd, J. W. Long-term mortality study of steelworkers: V. Respiratory cancer in coke plant workers. *J. Occup. Med.* 13(2): 53-68, 1971.
18. Lloyd, J. W., Lundin, F. E., Jr., Redmond, C. K., Geiser, P. B. Long-term mortality study of steelworkers. IV. Mortality by work area. *J. Occup. Med.* 12(5): 151-57, 1970.
19. Redmond, C. K., Ciocco, A., Lloyd, J. W., Rush, H. W. Long-term mortality study of steelworkers. VI. Mortality from malignant neoplasms among coke oven workers. *J. Occup. Med.* 14(8): 621-29, 1972.
20. Redmond, C. K., Strobino, B. R., Cypess, R. H. Cancer experience among coke by-product workers. *Ann. N.Y. Acad. Sci.* 271: 102-15, 1976.

21. Litvinov, N. N., Goldberg, M. S., Kimina, S. N. Morbidity and mortality in man caused by pulmonary cancer and its relation to the pollution of the atmosphere in the areas of aluminum plants. *Acta Union. Internat. Contra Cancrum* 19: 742–45, 1963.

22. Lloyd, J. W., Lemen, R. A., Miller, C., Brown, D. P. Cancer experience of workers employed in the steel and aluminum industries. Presented at Internat. Metalworkers Federation World Conf. on Health and Safety in the Metal Industry, Oslo, Norway, 1976.

23. Lemen, R. A., Johnson, W. M., Wagoner, J. K., Archer, V. E., Saccomanno, C. Cytologic observations and cancer incidence following exposure to BCME. *Ann. N.Y. Acad. Sci.* 271: 71–89, 1976.

24. Selikoff, I. J., Hammond, E. C., Eds. Brain tumors in the chemical industry. *N.Y. Acad. Sci.* 381: 1–364, 1982.

25. Ritta-Sisko, K., Hernberg, S., Karava, R., Jarvinen, E., Nurminen, M. A mortality study of foundry workers. *Scand. J. Work Environ. Health* 1: 73–89, 1976.

26. Egan-Baum, E., Miller, B. A., Waxweiler, R. J. Lung cancer and other mortality patterns among foundrymen. *Scand. J. Work Environ. Health* 7(4): 147–55, 1981.

27. Charr, R., Pulmonary changes in welders: a report of three cases. *Ann. Intern. Med.* 4: 806–12, 1956.

28. Friede, E., Rachow, D. Symptomatic pulmonary disease in arc welders. *Ann. Intern. Med.* 54: 121–27, 1961.

29. Gottlieb, M. S. Lung cancer and the petroleum industry in Louisiana. *J. Occup. Med.* 22(6): 384–88, 1980.

30. Corbett, T. H. Cancer and congenital abnormalities associated with anesthetics. *Ann. N.Y. Acad. Sci.* 271: 58–62, 1976.

31. Corbett, T. H. *Cancer and Chemicals.* Nelson Hall, Chicago, 1977, pp. 210.

32. Brodeur, P. Annals of chemistry: A compelling intuition, *The New Yorker*, pp. 122–49, Nov. 24, 1975.

33. Hogstedt, C., Malmqvist, N., Wadman, B. Leukemia in workers exposed to ethylene oxide. *J.A.M.A.* 241(11): 1132–33, 1979.

34. Hemminki, K., Mutanen, P., Saloniemi, I., Niemi, M. L., Vainio, H. Spontaneous abortions in hospital staff engaged in sterilizing instruments. *Brit. Med. J.* 285: 1461–63, 1982.

35. Yager, J. W., Hines, C. J., Spear, R. D. Exposure to ethylene oxide at work increases sister-chromatid exchanges in human peripheral lymphocytes. *Science* 219: 1221–23, 1983.

36. Goh, K., Cestero, R. V. M. Health hazards of formaldehyde. *J.A.M.A.* 247(20): 2778, 1982.

37. Vianio, H. Inhalation anesthetics, anticancer drugs and sterilants as chemical hazards in hospitals. *Scand. J. Work Environ. Health* 8(2): 94–107, 1982.

38. Hueper, W. C. *Occupational Tumors and Allied Diseases.* Charles C. Thomas, Springfield, Ill., 1942, pp. 399–405, 826–50.

39. Epstein, S. S. *The Politics of Cancer.* Sierra Club Books. San Francisco, 1978.

40. Brodeur, P. *Outrageous Misconduct.* Pantheon Books, New York, 1985.

41. A.M.A. *Guides to the Evaluation of Permanent Impairment*, 2nd ed. American Medical Association, Chicago, 1984, pp. 1–245.

42. Ibid., p. x.

43. Nagle, R., Selikoff, I. J., Spatz, D., Bale, A. Tort litigation for asbestos-associated disease, in *Disability Compensation for Asbestos-Associated Disease in the United States*, Selikoff, I. J., Ed. Environmental Sciences Laboratory, New York, 1982, pp. 330–33.

44. Wakefield, J. Education of the public, in *Persons at High Risk of Cancer*, Fraumeni, J. F., Jr., Ed. Academic Press, New York, 1975, pp. 415–34.

45. Johnson, H. W., Chairman. *Decision Making for Regulating Chemicals in the Environment.* National Academy of Sciences, Washington, D.C., 1975, pp. 1–232.

46. Ashford, N. A. *Crisis in the Workplace: Occupational Disease and Injury.* M.I.T. Press, Cambridge, Mass., 1976, pp. 39–41, 44–45, 372–73, 417, 508–10.

47. Berman, D. M. *Death on the Job: Occupational Health and Safety Struggles in the United States.* Monthly Review Press, New York, 1978, pp. 74–116.

48. Michigan Public Health Code—Act 368, Sect. 5611, 5613, 1978.

49. Graftstrom, R. C., Fornace, A. J., Jr., Autrup H., Lechner, J. F., Harris, C. Formaldehyde damage to DNA and inhibition of DNA repair in human bronchial cells. *Science* 220(4593): 216–18, 1983.

50. McDonald, S. Histology of pulmonary asbestosis. *Brit. Med. J.* 2(2): 1025–26, 1927.

51. *IARC Monographs on the Evaluation of the Carcinogenic Risk of Chemicals to Humans: Preamble.* International Agency for Research on Cancer, Lyon, France, 1982, 29: 19–20.

52. Ibid., p. 20.

53. Gaylor, D. W. The ED^01 study: summary and conclusions. *J. Environ. Path. Toxicol.* 3: 179–83, 1979.

CHAPTER

12

WHAT CAN I DO NOW?

It goes somehow against the grain to learn that cost-benefit analyses can be done neatly on lakes, meadows, nesting gannets, even whole oceans.

—Lewis Thomas[1]

A slow awakening of the ecological ethic is occurring worldwide, as more and more of the earth becomes contaminated, threatening life itself. In the name of "progress" and "convenience," damage has occurred to the world's resources and to the plants, animals, and humans dependent upon clean air, water, and soil for sustenance. The enormous scope of the problem, and the difficulty of making changes in or stopping the headlong race toward ecological disaster, have led to discouragement and feelings of impotency in many individuals who would like to lessen the chemical burden on all people. What follows are some simple, commonsense suggestions, within the reach of nearly everyone. What may be needed most is an awareness, a sense, of the ecological ethic, such as this book attempts to impart.

There are specific things that individuals can do to guard their own health and that of humankind while trying to prevent further ecological damage and preserve an intact world. There is an understandable overlap between persons in the following categories, but they have been separated here for emphasis.

CONSUMERS

Think before you buy. Know what you are buying, where it came from, and where it will ultimately come to rest! Is its usefulness worth the risk? Know what you are using, and understand the source, the effects, and the fate of chemicals that you may use. It is only through knowledge that intelligent decisions can be made.

The householder may bring into the home consumer products containing toxic chemicals that may be hazardous in and of themselves, or may combine with other products in the home to form harmful by-products. Products must be examined not only for the "active ingredients," but for other components as well, such as carriers, dyes, solvents, spray propellants, and the like.[2]

Among the consumer articles that may contain hazardous chemicals are the following products:

- Solvents used to clean carpets, furniture, and clothing.
- Cleaning compounds used in bathrooms and kitchens (caustics, acids, and abrasive compounds).
- Pesticides applied by the homeowner or by a pest control company (see Chapter 7 for more information on this topic).
- Spray cans emitting not only the intended product, but the pressurizing chemical as well. These chemicals are used in beauty, home, and pharmaceutical preparations, and include such propellants as the chlorofluorocarbons (Freons), nitrous oxide, and methylene chloride. Although now discontinued, the carcinogen vinyl chloride was used in the past as a propellant in products ranging from hair sprays to paint sprays. The continued use of the chlorofluorocarbons has been criticized on the basis of depleting the ozone layer of the atmosphere. On a more local basis, these substances can induce cardiac arrhythmias when inhaled in an industrial setting or intentionally with abuse, even with the use of aerosolized medications.[3-5]
- "Beauty" preparations, including hair sprays, dyes, shampoos, and nail preparations, which may contain formaldehyde, acrylics, solvents, fungicides, and so on.
- "Soaps" that are actually detergents, which may be identified as "beauty bars" or "bath bars" (as distinguished from soaps) and may be a source of sensitivity reactions. Some of these bath products contain dyes, perfumes, and antifungal and antibacterial agents that may cause reactions.
- Packaging materials, such as plastic containers used for foods and pharmaceuticals, that may leach chemicals intrinsic to the containers into the foods or pharmaceutical agents themselves. This is a particular problem where fat-containing and alcohol-containing products come into contact with the container.[6-8]

In general, before you buy, ask yourself these types of questions:

- Do I want dyes added to food, for cosmetic purposes, that may be harmful in a variety of ways, including causing cancer?[9]
- Do I know the effect of the various chemicals used in animal feed? Do they harm the animal handlers? Do the chemicals persist in our meat supply?[10]

- How many calories am I "saving" by using chemical sweetening agents? Is the substitution of a chemical for the approximately 18 calories contained in a teaspoon of sugar a savings at all? What are the biological effects of saccharine, cyclamates, and aspartame? Am I one of those people unable to metabolize phenylalanine, and could my or my child's symptoms be due to use of this chemical?

Other products may not be particularly harmful in the form sold to the consumer, but it is worth considering the manufacture and disposal of such products, and asking ourselves:

- Do we need dyed and/or printed toilet paper? What of the worker who made the dye, and the paper worker who applied it, and what of the water supply that the dye will be dissolved in? Is this "enhanced" product a false and unneeded vanity?
- Do we need fancy paper towels? What of the worker who manufactured and applied the dyes and the resins, and where will this product end up? In a dump? A municipal incinerator?
- Could we reuse glass containers rather than wrap things in plastic? What of the manufacture of these products? And, once again, what of their disposal, releasing toxic decomposition by-products if they are burned?[11]
- If we must have incinerators and landfills, can we recycle paper, glass, plastics, and metal so that we can decrease the bulk of materials disposed of, while conserving resources?[12]
- Do we need to carry our groceries home in a new paper or plastic bag every time? Could reusable carriers be employed, sparing the manufacture and disposal of some of these containers?
- Can we follow the lead of some states, and require the use of returnable beverage bottles and cans, ridding the landscape of litter and conserving both power and resources?

In essence, can we do better with less "convenience" and less "decoration," which provides little in the way of usefulness and value while adding much to the waste of resources and contributing to pollution? And who bears the costs of this convenience and vanity?

It is useful for the consumer, like the industrial worker, to keep a record of chemicals used in his or her milieu as a way to determine any adverse effects. Table 10-1 is a suggested form. The manufacturer of the product, the frequency of use, and the time relationship between use and symptoms should be noted.

Table 12-1. Consumer Product Use Record.

Product	Use, from to	Any Problems
Dog flea powder	Jan.–Aug. 1979	None noted
Oven Cleaner	Monthly	Coughing in user, ?-able wheezing in child
Ureathane pillows	1980	Asthma in user
Vinyl paint	1984	Skin and eye redness
Dog flea powder	Jan.–Aug. 1984	Wheezing in user
Pest control spraying	Sept. 1985	Blood dyscrasia
Hair spray	1975–83	Scaling of ears
Carbon paper	1977–81	Itching of face and hands
Room spray	1985	Wheezing and headache
Insect spray	1984	Diarrhea and wheezing

WORKERS

Compared to some groups, the milieu of the worker is relatively finite, in that processes and the variety of chemicals is fairly restricted (while the concentration may be high). Therefore it is not so difficult for the worker to develop a chemical data base. Whether the worker is employed in heavy industry or an office, it is recommended that one:

- Know the identity of every chemical, and the makeup of every product that you handled in the course of your work. Know the manufacturer's/supplier's name.
- Insist on seeing the Material Safety Data Sheets (MSDSs). If safety measures are suggested on the MSDS's, such as using respirators, protective clothing, glasses, etc., insist on having them.
- Do not rely solely on the information supplied with MSDSs. This is minimal information. Use your library for more extensive and specific information, and contact appropriate governmental agencies, such as the Department of Labor, OSHA, NIOSH, EPA, CPSC, FIFRA, etc., for governmental documents concerning occupational health hazards.
- Maintain records of exposures to all chemicals, complete with dates and any reactions or illnesses. (Understand that the latency for some conditions, such as those associated with asbestos or silica, may extend for decades, these records should be maintained for the lifetime of the worker.)
- Maintain records of illnesses.
- Compare chemical use records and illness records with other workers in your job category or plant, or local or international union.

- When you believe that there are unsafe practices, document the practices under question, report the situation to federal or state OSHA, and if it is of imminent hazard, refuse to do the job until an investigation is complete.
- Discuss workplace exposures with your physician and insist that your concerns are included in your medical record. If blood, urine, or other tests are needed, the workers should request that they be done. If similar tests are done by the employer, the workers should insist on a copy of the results.

AGRICULTURAL WORKERS

The general suggestions for workers, just listed, are applicable to persons working in the field of agriculture irrespective of whether trees, vegetable crops, grains, or animals are being grown. It must be emphasized that cautions apply to the home gardener as well. Home gardening accounts for a significant portion of the domestic food supply, and is a prime leisure activity, affording considerable opportunity for exposure to toxic chemicals. The special problems inherent in commercial agriculture require some additional discussion, however. These comments apply to field workers, farm families, and decision makers in the general field of agriculture, whether they be on the family farm or on a large agri-business plantation.

There are between 2.5 and 5 million farmworkers in the United States. Their rates for occupational illnesses are among the highest among all sections of the work force. California reported a disability rate among farmworkers of 11.9/1000 workers, twice the rate for any other employed group. The life expectancy for migrant workers is 20 years less than the national average. Migrant workers have a rate of maternal and infant mortality 125% higher than the national average. While poor living conditions and lack of sanitary facilities are factors in these appalling statistics, the widespread use of pesticidal chemicals adds to this burden.[13,14] Not only do agricultural workers suffer acute illness (and death) from pesticides such as the organophosphates, but they suffer higher rates of cancer as well.[15-17]

Chemicals used on forests, plants, and grain, as well as those used around animals and incorporated into animal feed, affect not only the intended targets of insects, weeds, fungi, and rodents, but also nontarget species including farmers, field hands, and the consumer of these contaminated foods.[18] These concerns have been expressed in the scientific literature and the lay press as well, giving some indication of the need for public concern and action.[19-25] The fact that pesticides and agricultural drugs are "approved for sale" does not mean that they are risk-free, not even if they have been adequately tested prior to use,[26-29] and not even when they are used according to directions.

In the field of agriculture it is especially incumbent upon the decision maker to obtain and understand information about the products that are proposed for use. Unfortunately, many farmworkers are poorly educated and may not be able to read, and many do not speak English. A warning given in a language that is not comprehended is no warning at all! Language difficulties should not stop attempts to supply adequate and understandable information about the use and toxicity of these products. It follows, therefore, that those in agriculture, like workers in general, should:

- Know the makeup and identity of every chemical product handled.
- Insist on seeing the labels for pesticidal products.
- Know the correct reentry time when areas have been sprayed, and insist on observing these times. (These times may not guarantee 100% safety, but they provide at least a minimum standard.
- Do not rely solely on the information supplied with the label. Seek information from the library and from various governmental agencies, such as EPA, FIFRA, and the Department of Labor, OSHA.
- Maintain records of pesticide use and exposures, with dates and amounts used.
- Maintain records of illnesses. (See Table 10-1, above.)
- Compare use of chemicals and illness records of other workers in your area.
- When you believe that there are unsafe uses of chemicals and/or unsafe practices (for instance, reentry too soon), report the situation to federal or state EPA, and if there is an imminent hazard, refuse to do the job until an investigation is complete.
- If you become sick, discuss the use of agricultural chemicals with your physician, being sure that this information is entered into their medical record. If pesticidal or other agricultural chemical poisoning is suspected, be certain that samples are obtained and tested.
- Use protective clothing, gloves, and respirators when working in treated areas or when handling contaminated products.
- Where herbicides are used along rights-of-way, highways, etc., determine safer ways for control of unwanted growth, such as cutting or planting alternative species.
- Search for alternatives to pesticides, as well as alternatives to growth-promoting hormones and antibiotics that are used in animal feeds.

DWELLING STRUCTURES

The chemical milieu of a home or an apartment may be varied and may contain a number of toxic chemicals, but exposure, with some exceptions, is rarely of the intensity of the workplace. Few of us live in homes with day-to-day atmospheres comparable to those of a foundry, a chemical factory, or a welding shop. Chemically caused diseases are more commonly associated with workplace exposure than with home exposure, although some exceptions to this finding have been described in Chapter 6. With the burgeoning use of chemicals in all phases of our lives, and the goal of minimizing exposures, it is incumbent upon the consumer to understand as much as possible about sources of chemical exposure in the home.

Considering the structure of a dwelling only, we must realize that some products, if included in a home, can become sources of toxic exposure for prolonged periods of time. These sources can be traced in a dwelling from the ground up, starting with the house itself and including its contents.

The following are sources of chemical exposure in a home:
- Wood-treating chemicals such as arsenicals, pentachlorophenol, the organo-tins, and other agents that are applied to control rot, termites, and fungal growth. These chemicals possess toxicities involving the liver, skin, and nervous system, as well as carcinogenicity.
- Plastic and spray coatings applied for soundproofing, waterproofing, or decorative purposes. These may offgas components during and after application.
- Foam rubber in carpet underlay and in furniture and mattresses, which may emit urethane or toluene diisocyanate (TDI) fumes, and be a source for the growth of molds.
- Formaldehyde-containing glues in particle board and plywoods, which offgas formaldehyde.
- Urea formaldehyde foam insulation (UFFI), which offgases formaldehyde. The use of these products has been stopped in most areas, but there are a number of homes from which the UFFI has not been removed.
- Asbestos wrapping on pipes, around furnaces and heaters, and as a component of asbestos–vinyl floor tile.

One would not ordinarily consider floor tile as a source of asbestos exposure because the fibers are embedded in a plastic matrix; but when sanded or otherwise disturbed, as during repair or floor resurfacing, these fibers may be released.[30] Janitorial workers who use heavy buffers on this type of floor may represent a group at risk. It can be expected that the wear and tear of use will release asbestos fibers, albeit slowly.

- Dyes, fireproofing chemicals, and plastics in carpets, upholstery, and wall coverings.
- Paints, grouts, and coatings containing vinyls, acrylics, urethanes, solvents, heavy metals, etc.
- Pesticides, such as chlordane, heptachlor, and aldrin, which may be applied to and under a house foundation for control of termites. This family of

chemicals is carcinogenic, as well as toxic to the liver and the immune and nervous systems. (See Chapters 7–9.)

These are but a few of the obvious chemical-containing products used in the construction and furnishings of homes and apartments.

PESTICIDE AND LAWN CHEMICALS

A relatively new industry that is dependent upon the application of toxic chemicals is the widely promoted pesticide and "lawn care" industry. The success of this $2.2 billion dollar industry, which grew 20% annually in a recent decade,[31] can be laid to extensive advertising, as well as to irrational concerns about insects and weeds.

There are few insects that cannot be controlled with barrier methods such as screens, garbage cans, and closed food containers. By using these methods, plus cleaning up sources where insects feed and breed, we should be able to control most flies, mosquitos, roaches, and similar insects.

Concerning the passion for uniform and weed-free lawns, it is well to bear in mind that few humans or animals have become sick or died as a result of weeds. The chemicals used on residential and commercial turf are a source of contamination that extends from those who manufacture and apply the chemicals to those who come in contact with the turf and, finally, to the lakes, streams, and water supplies that receive the runoff. Serious damage is occurring to fish and plants that depend upon pollution-free water for life because some 8 million homeowners use these products in the United States and Canada.[32] Additionally, the combined effects from pesticide drift and industrial pollution are turning some areas of the world into unlivable wastelands.[33]

It is incumbent upon every person making decisions about the use of these products, for either home or commercial use, to consider the risks and benefits of these products and "services." With pesticides, as with all products, consumers have certain responsibilities:

- Know the name, manufacturer, and *generic* identification of the pesticidal product.
- Read the entire label before using it yourself, or allowing anyone else to use it.
- Know the identity of pesticidal products used in your neighborhood, as well as the identity of products proposed for use by federal, state, or local agencies.
- Utilize your library, and request data on the health effects of pesticides from EPA–FIFRA.

- Urge passage of legislation requiring registration of pesticide applicators, notification of proposed use of pesticides, and posting of informational signs in areas of outdoor and public indoor pesticidal use.[34,35]

Are a weed-free lawn and an insect-free building worth the use of these toxic chemicals? Consider these parameters of their use:

- They are derived from nonrenewable petrochemical sources.
- There is exposure of workers and of the environment during the manufacture of pesticides.
- There is exposure of applicators.
- There is exposure of residents, pets, and workers in buildings and grounds that have been treated.
- There are increasing problems with contamination of the water table by runoff after rain, and dispersal by wind of these products.

Herbicides may be supplied singly or in combination, along with other products such as fertilizers, antifungal agents, and insecticides.

Common chemicals promoted for home and commercial "lawn care" include the following herbicides:

- 2,4-D: 2,4-dichloro phenoxy acetic acid, dimethylamine salt
- MCPP: 2-(2-methyl 4-chlorophenoxy) propionic acid, dimethylamine salt
- Dicamba®: 3,6-dichloro-o-anisic acid, dimethylamine salt

The first, 2,4-D is recognized as half of the preparation called Agent Orange, which was sprayed in Vietnam. (The other half, in addition to 2,4-D, was 2,4,5-T.) These three chemicals (2,4-D, MCPP, and Dicamba) are variations on the basic phenoxy herbicides (see Figure 2-3), and have been linked with the following ill effects:

- In humans: peripheral neuropathies (numbness, tingling, loss of muscle function); convulsions and central nervous system damage; cancer; birth defects; skin and eye irritation.
- In plants: premature flowering, ripening, and setting of fruit on nontarget species when exposed to *low* levels, thus resulting in loss of a crops; killing of nontarget species due to drift; damage from vapors carried on the wind, to such species as shrubs, trees, broad-leaf plants, etc.
- In animals: birth defects and/or death of fish, birds, and small animals including pets; neurological damage in dogs, cats, and other small animals including pets; mutagenic changes (damage to the cellular DNA).

The organophosphate compounds, which were developed in Germany during World War II, act as nerve poisons. Hundreds of deaths and countless sick have resulted from exposure to this family of insecticides. Most people survive an acute exposure *if* the correct diagnosis is made, and *if* therapy is instituted. Animals (birds, cats, dogs), the "nontarget" species, receive neither. These chemicals kill insects and other animals by interfering with nerve transmission; so a variety of symptoms can occur, from those listed on the Material Safety Data Sheet to respiratory failure, coma, and death. Of greater concern are the permanent and delayed neurotoxic effects that have been reported in the scientific literature, including permanent EEG (electroencephalographic) changes, decrease in mentation, tremors, and weakness.[36-38]

We speak of risk–benefit analysis and hazard assessment in connection with business and scientific efforts. There is no question that these pesticides, indeed all pesticides, carry hazard ("-cide" means "to kill"). Before we decide to use these pesticides, we need to determine any benefit that can accrue versus the risks, not only to the user but to other humans and to other species. It is difficult to believe that crab grass and other weeds impact any of us adversely. Indeed, if the aesthetics of weeds in the grass is too much to bear, they can be *dug* up, or mowed over. As far as insects are concerned, if we encourage a bird population and leave them unmolested by chemicals, we can control these unwanted pests. If we poison the environment and kill the birds, we will be in a vicious cycle of chemical dependency.

Two more groups that are exposed to pesticides need to be addressed: persons who produce and market them, as well as those who apply them. Is it worth the risk to the health and welfare of these others, as well as ourselves, to have weed-free lawns and "no bugs" when other methods are available?

In conclusion, we need to address an oft-quoted remark: "If it weren't safe to use, EPA wouldn't allow it to be sold." Most testing of pesticides, with few exceptions, is undertaken by the manufacturer—a situation, in some instances, akin to having Dracula guard the blood bank. These industry-generated data are largely unpublished and difficult to obtain. Once a pesticide has been registered for sale, it can take years, and be very difficult, to get it removed from commerce.

Despite this difficulty and delay, pesticides *have* been removed from the market after it was discovered that they caused illness, death, and ecological damage. The list of pesticides restricted or removed from sale includes: DDT, heptachlor, endrin, aldrin, tabun, sarin, and parathion, the last bearing the distinction of causing the greatest number of fatal poisonings. Only recently, chlordane was voluntarily withdrawn from the market by the manufacturer. Removal from sale does not mean removal from the environment because many pesticidal chemicals are persistent and biomagnify in the environment.

FOOD

The problem of chemical contamination of the food supply is extremely complex.[39] Adequate coverage of this subject would require a separate volume. Chemicals in food are discussed only in general, mainly to give the reader questions to be addressed. More detailed sources of information are available in references given in Chapter 13.

SUMMARY

A bland acceptance of advertising and failure to appreciate the ramifications of chemical manufacture, use, and disposal allow unecological and unhealthy practices to be promoted and to flourish without critical assessment.

An ecological ethic must be developed to avoid the pitfalls of depletion and contamination of the world's natural resources. The spread of contamination from the manufacturing, use, and disposal of chemical products promoted as essential for enjoyment of the ''good life,'' is contributing to a worsening of living conditions for countless species.[40]

Going a step further, we must ask whether it is worth the depletion of resources, the exposure of those who handle and manufacture the items, and the contamination of our air and water to have such things as:

- Weed-free lawns
- Disposable bottles and cups
- Candy dyed bright colors
- Fancy wheel covers
- Blemish-free vegetables
- Uniformly sized and uniformly dyed fruits

Can we really afford these cosmetics, or are there alternative ways to satisfy the needs of all of the world's citizens without depleting our resources and without harming our air, water, soil, other species, and ourselves?

REFERENCES

1. Thomas, L. *The Lives of a Cell.* Bantam Books, New York, 1974, p. 121.
2. Johnson, A. Unnecessary chemicals. *Environment* 20(2): 6–11, Mar. 1978.
3. Baselt, R. C., Cravey, J. H. A fatal case involving trichlorofluoromethane and dichlorodifluoromethane. *J. Forensic Sci.* 13: 407, 1968.
4. Azar, A., Zapp, J., Reinhardt, C. F., Stopps, G. J. Cardiac toxicity of aerosol propellants. *J.A.M.A.* 215: 1501, 1971.
5. Garriott, J., Petty, S. Death from inhalant abuse: toxicological and pathological evaluation of 34 cases. *Clin. Toxicol.* 16(3): 305–15, 1980.
6. Jaeger, R. J., Rubin, R. J. Plasticizers from plastic devices: extraction, metabolism, and accumulation by biological systems. *Science* 170: 460–62, 1970.

7. Autian, J. Plastics, in *Casarett and Doull's Toxicology*, Doull, J., Klassen, C. D., Amdur, M. O., Eds. Macmillan, New York, 1980, pp. 531–56.

8. U.S. Food and Drug Administration, P. L. 85-929, 1958.

9. Brooks, J., Chairman, Committee on Government Operations. The regulation by the Department of Health and Human Services of carcinogenic color additives. U.S. House of Representatives, Oct. 5, 1984.

10. Brooks, J., Chairman, Committee on Government Operations. The regulation of animal drugs by the Food and Drug Administration. U.S. House of Representatives, July 24–25, 1985.

11. Karasek, F. W., Dickson, L. C. Model studies of polychlorinated dibenzo-*p*-dioxin formation during municipal refuse incineration. *Science* 237: 754–56, 1987.

12. Peterson, C. Garbage: the new alchemy of high-tech trash, *Washington Post*, p. B-3, June 7, 1987.

13. McCarthy, C. The migrant poor: working without the basics, *Washington Post*, p. H-9, June 2, 1985.

14. Riley, J. Pesticides: no haven from harm in the fields, *Nat. Law J.*, pp. 1, 26, 28, June 9, 1986.

15. Hardell, L., Sandstrom, A. Case-control study: soft tissue sarcomas and exposure to phenoxyacetic acids or chlorophenols. *Brit. J. Cancer* 39: 711–17, 1979.

16. Hoar, S. K., Blair, A., Holmes, F. F. Boysen, C. D., Robel, R. J., Hoover, R., Fraumeni, J. F., Jr. Agricultural herbicide use and risk of lymphoma and soft-tissue sarcoma. *J.A.M.A.* 256(9): 1141–47, 1986.

17. Colton, T. Herbicide exposure and cancer. *J.A.M.A.* 256(9): 1176–78, 1986.

18. The regulation of animal drugs (see ref. 10).

19. Shabecoff, P. Pesticide plight, *New York Times*, May 22, 1987.

20. United Farm Workers of America, La Paz, Calif. 93570. See various issues of the publication *Food and Justice*.

21. Meyerhoff, A. H. Want the pesticide industry in your milk?, *The New York Times*, May 26, 1987.

22. Shabecoff, P. Herbicide 2,4-D tied to cancer in workers. *New York Times* News Service, published in the *Maui News*, p. A-12, Aug. 31, 1986.

23. Boraiko, A. A., Ward, F. The pesticide dilemma, *National Geographic*, pp. 145–83, Feb. 1980.

24. Schneider, K. Congress looks to the American table amid questions on food safety, *New York Times*, p. 10, June 22, 1987.

25. National Coalition Against the Misuse of Pesticides (NCAMP) 530 7th Street, S.E., Washington, D.C. 20003.

26. National Academy of Science. *Regulating Pesticides in Food: The Delaney Paradox*. Washington, D.C., 1987.

27. Bercz, J. P. *Temporal Variability of Toxic Contaminants in Animal Diets.* Environmental Protection Agency, Health Effects Research Laboratory, Cincinnati, Ohio 81-205 973, 1981.

28. Tattersall, A. Is EPA registration a guarantee of pesticide safety? *J. Pesticide Reform* 40–42, Spring 1986.

29. Curtis, J. Pesticide exposure and the role of the physician. Northwest Coalition for Alternatives to Pesticides, P.O. Box 1393, Eugene, Oreg., 1986, pp. 1–79.

30. Murphy, R. L., Levine, B. W., Al Bazzaz, F. J. Lynch, J. J., Burgess, W. A. Case reports: floor tile installation as a source of asbestos exposure. *Amer. Rev. Respir. Dis.* 104: 576–80, 1971.

31. Rollins, Inc. (Orkin). Annual report, 1985.

32. Ragsdale, R. *USA Today*, p. 2A, Apr. 10, 1985.

33. Bovard, J. A silent spring in Eastern Europe, *New York Times*, p. C-1, Apr. 26, 1987.

34. Ordinance No. 1984-0-31, Wauconda, Ill., July 3, 1984.

35. Christoffel, T. Grassroots environmentalism under legal attack: dandelions, pesticides and a neighbor's right-to-know. *Amer. J. Pub. Health* 75(5): 565-67, 1985.
36. Senanayake, N., Karalliedde, L. Neurotoxic effects of organophosphorus insecticides. *New Engl. J. Med.* 316(13): 761-63, 1987.
37. Gadoth, N., Fisher, A. Late onset of neuromuscular block in organophosphorus poisoning. *Ann. Intern. Med.* 88: 654-55, 1978.
38. Desi, I., Dura, G., Gonczi, L., Kneffel, Z., Strohmayer, A., Szabo, Z. Toxicity of malathion to mammals, aquatic organisms and tissue culture cells. *Arch. Environ. Contam. Toxicol.* 3(4): 410-25, 1976.
39. Macfadyen, J. T. A chemical feast: are pesticides in our food making us sick? *Sci. Illustr.* 1(2): 16-23, Apr./May 1987.
40. Gove, N., Spiegel, T., Bond, W. H. Air; An atmosphere of uncertainty, *National Geographic*, pp. 502-36, Apr. 1987.

CHAPTER

13

SOURCES OF INFORMATION AND DATA COLLECTION

The medical, industrial, and engineering literature contains numerous articles on occupational and environmental hazards in general (foundry, rubber workers, welding, smelting, mining, etc.), as well as on specific chemicals.

Information is available for most chemicals in general industrial use. Data and regulatory statutes are available from federal and state governmental sources such as EPA, OSHA, CPSC, NIOSH, FDA, NIEHS, and so forth. Additionally, valuable information can be had from patent searches.

Data are also available from standard computer searches such as: CHEMLINE, TOXLINE, MEDLINE, CANCERLINE, CHEMTOX, etc.[1-3] It must be emphasized that the computerized data bases rarely extend back to the time of the first information on a chemical. This early information is obtained by the time-honored process of obtaining each item on the bibliographic "tree," back to the first mention of a subject. Quite often, we find that information concerning a subject has been known for decades. Although more recent scientific techniques may be more sophisticated, the early toxicology reports remain valid in most cases. There exists enough information on nearly every chemical used in commerce to determine its potential for risk.

The references listed below are suggested as principal sources of information.

GENERAL TOXICOLOGY

The following references are essential and must be a part of every toxicology library if possible.

Doull, J., Klaassen, C. D., Amdur, M. O., Eds. *Casarett and Doull's Toxicology, The Basic Science of Poisons*. Macmillan, New York, 1980.

Gilman, A. G., Goodman, L. S., Rall, T. W., Murad, F., Eds. *The Pharmacological Basis of Therapeutics*. Macmillan, New York, 1985.

Gosselin, R. E., Hodge, H. C., Smith, R. P., Gleason, M. N. *Clinical Toxicology of Commercial Products*. Williams and Wilkins, Baltimore, 1976.

Hunter, D. *The Diseases of Occupations*, 4th ed. Little, Brown and Co., Boston, 1969.

Parmeggiani, L., Ed. *Encyclopedia of Occupational Health and Safety*, 3rd (rev.) ed, Vols. I and II. International Labor Office, Geneva, Switzerland, 1983.

Sax, N.I. *Dangerous Properties of Industrial Materials*, 6th ed. Van Nostrand Reinhold, New York, 1984.

ADDITIONAL SOURCES: ARRANGED ACCORDING TO GENERAL TOPICS

General

The citations listed below will give the reader an insight into the scope of the problem of chemical contamination of the environment and workplace.

Burrell, C. D., Sheps, C. G., Eds. *Primary Health Care in Industrialized Nations*, Vol. 310. Annals New York Academy of Sciences, 1978.

Claybrook, J. *Retreat from Safety*. Pantheon Books, New York, 1984.

Crone, H. D. *Chemicals and Society*. Cambridge University Press, Cambridge, 1986.

Finkel, A. J., Ed. *Hamilton and Hardy's Industrial Toxicology*. John Wright PSG, Inc., Boston, 1983.

Gilman, A. G., Goodman, L. S., Rall, T. W., Murad, F., Ed. *The Pharmacological Basis of Therapeutics*. Macmillan, New York, 1985.

Hammond, E. C., Selikoff, I. J., Eds. *Public Control of Environmental Health Hazards*, Vol. 329, Annals New York Academy of Sciences, 1979.

Levy, B., Wegman, D., *Occupational Health*. Little, Brown and Co., Boston, 1983.

Magnuson, W. G., Carper, J. *The Dark Side of the Market Place*. Prentice-Hall, Englewood Cliffs, N.J., 1968.

Nader, R., Brownstein, R., Richard, J. *Who's Poisoning America*. Sierra Club Books, San Francisco, 1981.

Rosenberg, B., Ed. *Papers in Science and Public Policy I*, Vol. 368, Annals New York Academy of Sciences, 1981.

Sax, N. I., Lewis, R. J., Sr. *Hazardous Chemicals Desk Reference*. Van Nostrand Reinhold, New York, 1987.

Sterrett, F. S., Ed. *Science and Public Policy*, Vol. 403, Annals New York Academy of Sciences, 1983.

Philosophy and Ethics

Although it may seem out of place to some to include references on philosophy and ethics in a book on chemicals and disease, it is essential that we have an appreciation of the elegance and precariousness of life. Lest we harm life's delicate balances, we may have to curtail our own actions and desires for the safeguard of our own and other species. Although they may not be current vogue, the past has shown that there are other worthy goals beside the maximization of profit and personal pleasure.

Asimov, I. *The Genetic Code*. Charles N. Potter, New York, 1962.

Bronowski, J. *Science and Human Values*. Harper and Row, New York, 1956, 1965.

Bronowski, J. *The Origins of Knowledge and Imagination*, Yale University Press, New Haven, Conn., 1978.

Crick, F. *Life Itself: Its Origin and Nature*. Touchstone Book, Simon and Schuster, New York, 1981.

Fuller, R. B. *Operating Manual for Spaceship Earth*. Simon and Schuster, New York, 1969.

Lovelock, J. E. *Gaia: A New Look at Life on Earth*. Oxford University Press, New York, 1979.

Rifkin, J. *Declaration of a Heretic*. Routledge and Kegan Paul, Boston, 1985.

Thomas, L. *The Lives of A Cell*. Bantam Books, New York, 1974.

Thomas, L. *The Medusa and the Snail*. Bantam Books, New York, 1974.

Thomas, L. *The Youngest Science*. Bantam Books, New York, 1983.

Environment

Bollier, D., Claybrook, J. *Freedom from Harm*. Public Citizen and Democracy Project, Washington, D.C., 1986.

Brodeur, P. *The Asbestos Hazard*. Annals New York Academy of Sciences, 1980.

Brown, M. *Laying Waste: The Poisoning of America by Toxic Chemicals*. Pocket Books, New York, 1981.

Chen, E. *PBB: An American Tragedy*. Prentice Hall, Englewood Cliffs, N.J., 1979.

Commoner, B. *The Closing Circle*. Alfred A. Knopf, New York, 1972.

DeBell, G., Ed. *The New Environmental Handbook*. Friends of the Earth, San Francisco, 1980.

Eckholm, E. P. *The Picture of Health: Environmental Sources of Disease*. W. W. Norton, New York, 1977.

Gorman, J. *Hazards to Your Health: The Problem of Environmental Disease*. Annals New York Academy of Sciences, 1979.

Hammond, P. B., Chairman. *Biologic Effects of Atmospheric Pollutants: Lead*. National Academy of Sciences, Washington, D.C., 1972.

Lederer, W. H. *Regulatory Chemicals of Health and Environmental Concern*. Van Nostrand Reinhold, New York, 1985.

National Research Council. *Decision Making for Regulating Chemicals in the Environment*. National Academy of Sciences, Washington, D.C., 1975.

Nelson, N., Chairman. *Principles for Evaluating Chemicals in the Environment*. National Academy of Sciences, Washington, D.C., 1975.

Nelson, N., Wittenberger, J. L., Chairmen. *Human Health and the Environment Some Research Needs*, Report of the Second Task Force for Research Planning in Environmental Health Science. U.S. Dept. Health, Education and Welfare, DHEW Pub. #NIH 77 1277, 1976.

Schiff, H. I., Chairman. *Stratospheric Ozone Depletion by Halocarbons: Chemistry and Transport*. National Academy of Sciences, Washington, D.C., 1979.

Schneider, M. J. *Persistent Poisons; Chemical Pollutants in the Environment*. Annals New York Academy of Sciences, 1979.

Tornton, R., Chairman. *Regulating Pesticides in Food: The Delaney Paradox*, Committee on Scientific and Regulatory Issues Underlying Pesticide Use Patterns and Agricultural Innovation. National Research Council, National Academy of Sciences, Washington, D.C., 1987.

Whiteside, T. *The Pendulum and the Toxic Cloud*. Yale University Press, New Haven, Conn., 1979.

Occupation

Ashford, N. A. *Crisis in the Workplace*. MIT Press, Cambridge, Mass., 1976.

Berman, D. M. *Death on the Job*. Monthly Review Press, New York, 1978.

Kazia, R., Grossman, R. L. *Fear at Work: Job Blackmail, Labor and the Environment*. Pilgrim Press, New York, 1982.

Kent, J. A., Ed. *Riegel's Handbook of Industrial Chemistry*, 8th ed. Van Nostrand Reinhold, New York, 1983.

Lee, J. S., Rom, W. N. *Legal and Ethical Dilemmas in Occupational Health*. Ann Arbor Science, Ann Arbor, Mich., 1982.

Mancuso, T. F. *Help for the Working Wounded*. International Assoc. of Machinists and Aerospace Workers, Washington, D.C., 1976.

Mendeloff, J. *Regulating Safety: An Economic and Political Analysis of Occupational and Safety Health Policy*. The MIT Press, Cambridge, Mass., 1979.

Polakoff, P. L. *Work and Health: It's Your Life*. Press Associates, Inc., Washington, D.C., 1984.

Proctor, N. H., Hughes, J. P. *Chemical Hazards in the Workplace*. Lippincott, Philadelphia, 1978.

Rosner, D., Markowitz, G., Eds. *Dying for Work: Workers' Safety and Health in Twentieth-Century America*. Indiana University Press, Bloomington, 1987.

Sax, N. I., Lewis, R. J., Sr. *Rapid Guide to Hazardous Chemicals in the Workplace*. Van Nostrand Reinhold, New York, 1986.

Stellman, J., Henifin, M. S. *Office Work Can Be Dangerous to Your Health*. Pantheon Books, New York, 1983.

Tucker, A. *The Toxic Metals*. Ballantine, New York, 1972.

Williams, P. L. *Industrial Technology: Safety and Health Applications in the Workplace*. Van Nostrand Reinhold, New York, 1985.

Xinteras, C., Johnson, B. L., deGroot, I. *Behavioral Toxicology: Early Detection of Occupational Hazards* U.S. Dept. Health, Education and Welfare, NIOSH Pub. #74-126, 1974.

Pesticides

Carson, R. *Silent Spring*. Houghton Mifflin Co., Boston, 1962; Fawcett Publications, Greenwich, Conn., 1962.

van den Bosch, R. *The Pesticide Conspiracy*. Doubleday, Garden City, N.Y., 1978.

Weir, D., Shapiro, M. *Circle of Poison*. Institute for Food and Development Policy, San Francisco, 1981.

Drugs

Berkowitz, R. L., Coustan, D. R., Mochizuki, T. K. *Handbook for Prescribing Medicines during Pregnancy*. Little, Brown and Co., Boston, 1981.

Bochner, F., Carruthers, G., Kampmann, J., Steiner, J. *Handbook of Clinical Pharmacology*. Little, Brown and Co., Boston, 1978.

Hirai, H., Alpert, E., Eds. *Carcinofetal Proteins: Biology and Chemistry*, Vol. 259. Annals New York Academy of Sciences, 1975.

Knightley, P, Evans, H., Potter, E., Wallace, M., The Insight Team of the Sunday *Times* of London. *Suffer the Children: the Story of Thalidomide*. Viking Press, New York, 1979.

Wolfe, S. M., Coley, C. M. *Pills That Don't Work*. Public Citizen's Health Research Group, Washington, D. C., 1981.

Cancer

Bardin, C. W., Sherins, R. J. *Cell Biology of the Testis*, Vol. 383. Annals New York Academy of Sciences, 1982.

Baserga, R. *Cell Proliferation, Cancer, and Cancer Therapy*, Vol. 397. Annals New York Academy of Sciences, 1982.

Boland, B., Ed. *Cancer and the Worker*. Annals New York Academy of Sciences, 1977.

Borek, C., Williams, G. M. *Differentiation and Carcinogenesis in Liver Cell Cultures*. Vol. 349. Annals New York Academy of Sciences, 1980.

Burnett, F. M. *Immunology, Aging and Cancer*. W. H. Freeman, San Francisco, 1970.

Epstein, S. S. *The Politics of Cancer*. Sierra Club Books, San Francisco, 1978.

Fraumeni, J. F. *Persons at High Risk of Cancer*. Academic Press, New York, 1975.

Friedman, H., Southam, C., Eds. *International Conference on Immunobiology of Cancer*, Vol. 276. Annals New York Academy of Sciences, 1976.

Hanna, M. G., Jr., Nettesheim, P., Gilbert, J. R., Eds. *Inhalation Carcinogenesis*. U.S. Atomic Energy Commission, NTIS, Springfield, Va., 1970.

Highland, J. H., Fine, M. E., Boyle, R. H. *Malignant Neglect*. Alfred A. Knopf, New York, 1979.

Hueper, W. C. *Occupational Tumors and Allied Diseases*. Thomas, Springfield, Ill., 1942.

Hueper, W. C. *Occupational and Environmental Cancers of the Urinary System*. Yale University Press, New Haven, Conn., 1969.

Kraybill, H. F., Dawe, C. J., Harshbarger, J. C., Tardiff, R. G. *Aquatic Pollutants and Biologic Effects with Emphasis on Neoplasia*, Vol. 298. Annals New York Academy of Sciences, 1977.

Mulvihill, J. J., Miller, R. W., Fraumeni, J. F., Jr., Eds. *Progress in Cancer Research and Therapy: Genetics of Human Cancer*. Raven Press, New York, 1977.

Nicholson, W. J., Moore, J. A., Eds. *Health Effects of Halogenated Aromatic Hydrocarbons*, Vol. 320. Annals New York Academy of Sciences, 1979.

Sax, N. I. *Cancer Causing Chemicals*. Van Nostrand Reinhold, New York, 1981.

Selikoff, I. J., Hammond, E. C., Eds. *Toxicity of Vinyl Chloride–Polyvinyl Chloride*, Vol. 246. Annals New York Academy of Sciences, 1975.

STATE-OF-THE-ART SAMPLE BIBLIOGRAPHIES

Appended to this chapter are bibliographies arranged in chronological order. This method allows the determination of what knowledge was available and when. It also facilitates an understanding of how knowledge about a particular chemical or industry has developed.

These particular bibliographies are not intended to be the last word or all-inclusive for each subject. Besides showing the existence of varied data from multiple sources, they represent a *method* of compilation. This method not only may be used retrospectively, in attempting to connect various disease states with chemical exposures, but may be used prospectively in regulatory and preventive efforts.

The following list shows the topics covered in the state-of-the-art bibliographies, in the order in which they are presented.

- Acrylics
- Arsenic and cancer
- Asbestos exposure and malignancies associated with immune dysfunction, including leukemias, lymphomas, sarcomas, and myelomas
- "Autoimmune" diseases and chemicals
- Birth defects and chemicals
- Chromium
- Epichlorhydrin
- Foundry industry and cancer
- Phenylbutazone, oxyphenylbutazone, and blood dyscrasias
- Phthalates
- Rubber workers
- Styrene
- 1,1,1-Trichloroethane
- Trichlorethylene

1. Halton, D. M. Computerized information resources in toxicology and industrial health—a review. *Toxicol. Ind. Health* 2(1): 113-25, 1986.
2. Corbett, P. K., Ifshin, S. L. Online retrieval of environmental and occupational health literature: a comparative study. *Med. Ref. Services Quart.* 2: 25-36, 1983.
3. Eddy, A. B., Rest, K. M., Miller, N. Clinical industrial toxicology: an approach to information retrieval. *Ann. Intern. Med.* 103(6): 967-72, 1985.

BIBLIOGRAPHIES

Acrylics

Deichmann, W. Toxicity of methyl, ethyl and N-butyl methacrylate. *J. Ind. Hyg. Toxicol.* 23(7): 343–51, 1941.

Spealman, C. R., Main, R. J., Haag, H. B. Monomeric methyl methacrylate. *Ind. Med.* 14: 292–98, Apr. 1945.

Laskin, D. M., Robinson, I. B., Weinmann, J. P. Experimental production of sarcomas by methyl methacrylate implants. *Proc. Soc. Exp. Biol. Med.* 87: 329–32, Nov. 1954.

Hamblin, D. O. The toxicity of acrylamide, a preliminary report, in *Hommage au doyen Rene Fabre.* Institut de Toxicologie a la Faculte de Pharmacie de Paris, Paris, 1956, p. 195.

Kupperman, S. S. Effects of acrylamide on the nervous system of the cat. *J. Pharmacol. Exp. Ther.* 123: 180, 1958.

Borzelleca, J. F., Larson, P. S., Hennigar, G. R., Jr., Huf, E. G., et al. Studies on the chronic oral toxicity of monomeric ethyl acrylate and methyl methacrylate. *Toxicol. Appl. Pharmacol.* 6: 29–36, 1963.

Stinson, N. E. The tissue reaction induced in rats and guinea-pigs by polymethyl-methacrylate (acrylic) and stainless steel *Brit. J. Exp. Pathol.* 45: 21–29, June 18, 1963.

McCollister, D. D., Oyen, F., Rowe, V. K. Toxicology of acrylamide. *Toxicol. Appl. Pharmacol.* 6: 172, 1964.

Fullerton, P. M., Barnes, J. M. Peripheral neuropathy produced in rats by acrylamide. *Brit. J. Ind. Med.* 23: 210, 1966.

Auld, R. B., Bedwell, S. F. Peripheral neuropathy with sympathetic overactivity from industrial contact with acrylamide. *Canad. Med. Assoc. J.* 96: 652–54, 1967.

Garland, T. O., Patterson, M. W. H. Six cases of acrylamide poisoning. *Brit. Med. J.* 4: 134–37, 1967.

Hopkins, A. D. The effect of acrylamide on the peripheral nervous system in the baboon *J. Neurol. Neurosurg.* 19: 1094–1100, 1969.

Leswing, R. J., Ribelin, W. E. Physiologic and pathologic changes in acrylamide neuropathy. *Arch. Environ. Health* 18: 22, 1969.

Pleasure, D. E., Mishler, K. C., Engle, W. K. Axonal transport of protein in experimental neuropathies. *Science* 166: 524–25, 1969.

Prineas, J. The pathogenesis of dying-back neuropathies. An ultrastructural study of experimental acrylamide intoxication in the cat. *J. Neuropathol. Exp. Neurol.* 28: 598–621, 1969.

Blagodatin, V. M., et al. Issues of industrial hygiene and occupational pathology in the manufacture of organic glass (Russ.). *Gig. Tr. Prof. Zabol* 14(8): 11–14, 1970.

Pleasure, D., Engle, W. K. Axonal flow in experimental neuropathies *J. Neuropathol. Exp. Neurol.* 29: 140, 1970.

Fessel, W. J. Fat disorders and peripheral neuropathy. *Brain* 94: 531–40, 1971.

Hopkins, A., Gilliatt, R. W. Motor and sensory nerve conduction velocity in the baboon: normal values and changes during acrylamide neuropathy. *J. Neurol. Neurosurg. Psychiat.* 34: 415, 1971.

Fowler, A. W. Methylmethacrylic cement and fat embolism. *Brit. Med. J.* 4: 108, 1972.

Singh, A. R., Lawrence, W. H., Autian, J. Embryo-fetal toxicity and teratogenic effects of a group of methacrylate esters in rats. *Toxicol. Appl. Pharmacol.* 22: 314–15, 1972.

Malloy, T. H., Stone, W. A., Piere, R. L. Potential hepatoxic effects of methyl methacrylate monomer. *Clin. Orthop.* 93: 366–68, 1973.

Hashimoto, K., Ando, K. Alterations of amino acid incorporation into proteins of the nervous system in vitro after administration of acrylamide to rats. *Biochem. Pharmacol.* 22: 1057, 1973.

Hashimoto, K., Aldridge, W. N. Biochemical studies on acrylamide. A neurotoxic agent. *Biochem. Pharmacol.* 19: 259, 1973.

Suzuki, K., Pfaff, L. D. Acrylamide neuropathy in rats. An electron microscopic study of degeneration and regeneration. *Acta Neuropathol.* 24: 197–213, 1973.

Spencer, P. S., Schaumberg, A. A review of acrylamide neurotoxicity, 1. Properties, uses and human exposure. *Canad. J. Neurol. Sci.* 1: 143, 1974.

Spencer, P. S., Schaumberg, A. Experimental animal neurotoxicity and pathologic mechanisms. *Canad. J. Neurol. Sci.* 1: 152, 1974.

Schaumberg, H.H., Wisniewski, M. D., Spencer, P. S. Ultrastructural studies of the dying-back process 1. Peripheral nerve terminal and anon degeneration in systemic acrylamide intoxication. *J. Neuropathol. Exp. Neurol.* 33: 260, 1974.

Tansy, M. F. The effects of methyl methacrylate vapor on gastric motor function. *J. Amer. Dent. Assoc.* 89: 372–76, 1974.

Autian, J. Structure–toxicity relationships of acrylic monomers. *Environ. Health Perspect.* 11: 141–52, 1975.

Bagramjan, S. B., et al. Mutagenic effect of small concentrations of volatile substances emitted from polychloroprene latex during their combined uptake by the animal. (Russ.). *Biol. Zh. Arm.* 29: 98–99, 1976.

Edwards, P. M. The insensitivity of the developing rat foetus to the toxic effects of acrylamide. *Chem. Biol. Interactions* 12: 13–18, 1976.

Cromer, J. C., Kronoveter, K. *A Study of Methyl Methacrylate Exposures and Employee Health.* National Institute for Occupational Safety and Health, Washington, D.C., 1976, pp. 11–12, 15.

Bratt, H., Hathway, D. E. Fate of methyl methacrylate in rats. *Brit. J. Cancer* 36: 114–19, 1977.

Kesson, C. M., Baird, A. W., Lawson, D. H. Acrylamide poisoning. *Postgrad. Med. J.* 53: 16–17, 1977.

Tansy, M. F., Kendall, F. M. Update on the toxicity of inhaled methyl methacrylate vapor. *Drug Chem. Toxicol.* 2(4): 315–30, 1979.

Nicholas, C. A., Lawrence, W. H., Autian, J. Embryotoxicity and fetotoxicity from maternal inhalation of methyl methacrylate monomer in rats. *Toxicol. Appl. Pharmacol.* 50: 451–58, 1979.

Goldstein, B. D. Functional correlates of toxic neuropathies. *Diss. Abstr. Int. B* 41(4): 1331–32, 1980; 1979, 150 pp.

Blum, Barbara. Response to Interagency Testing Committee Recommendations. *Federal Register* 44(94): 28095–97, May 14, 1979.

Innes, D. L., Tansy, M. F., Martin, J. S. Effects of acute methylmethacrylate inhalation on rat brain neuronal activity. *J. Dent. Res.* 58: 208, 1979.

Howland, R. D., Lowndes, H. E. Hepatic UDP-glucuronyltransferase activity in acrylamide neuropathy. *Experientia* 35(2): 248–49, 1979.

IARC. Methyl methacrylate and polymethyl methacrylate. *IARC Monographs on the Evaluation of the Carcinogenic Risk of Chemicals to Humans* 19: 187–211, 1979.

IARC. Acrylic acid, methyl acrylate, ethyl acrylate and polyacrylic acid *IARC Monographs on the Evaluation of the Carcinogenic Risk of Chemicals to Humans* 19: 47–71, 1979.

Innes, D. L., Tansy, M. F. Central nervous system effects of methyl methacrylate vapor. *Neurotoxicology* 2: 515–22, 1981.

Murray, J. S., Miller, R. R., Deacon, M. M., Hanley, T. R., Jr. et al. Teratological evaluation of inhaled ethyl acrylate in rats. *Toxicol. Appl. Pharmacol.* 60: 106–11, 1981.

Kuzelova, M., Kovarik, J., Fiedlerova, D., Popler, A. Acrylic compound and general health of exposed persons. *Pracovni Lekaratvi* 33(3): 95–99, 1981.

Boyes, W. K., Cooper, G. P. Acrylamide neurotoxicity: effects on far-field somatosensory evoked potentials in rats. *Neurobehav. Toxicol. Teratol.* 3(4): 487–90, 1981.

Cavanagh, J. B. Enhancement of sensitivity to acrylamide after nerve ligature. *Acta Neuropathol. (Berl)* 7(Supp.): 243–45, 1981.

Loeb, A., Anderson, R. J. Antagonism of acrylamide neurotoxicity by supplementation with vitamin B6. *Neurotoxicology* 2(4): 625–33, 1981.

Tilson, H. A. The neurotoxicity of acrylamide; an overview. *Neurobehav. Toxicol. Teratol.* 3(4): 445–61, 1981.

Von Burg, R., Penney, D. P., Conroy, P. J. Acrylamide neurotoxicity in the mouse: a behavioral, electrophysiological and morphological study. *J. Appl. Toxicol.* 1(4): 227–33, 1981.

Hashimoto, K., Sakamoto, J. Neurotoxicity of acrylamide and related compounds and their effects on male gonads in mice. *Arch Toxicol.* 47(3): 179–89, 1981.

Miller, R. A., Wynkoop, J. R., II, Allen, G. A., Lorton, L. Some effects of methyl methacrylate monomer upon renal function. *J. Dent. Res.* 61: 202, 1982.

Rafales, L. S., Bornschein, R. L., Caruso, V. Behavioral and pharmacological responses following acrylamide exposure in rats. *Neurobehav. Toxicol. Teratol.* 4: 355–64, 1982.

Korhonen, A., Hemminki, K., Vainio, H. Embryotoxic effects of acrolein, methacrylates, guanidines and resorcinal on three day chicken embryos. *Acta Pharmacol. Toxicol.* 52(2): 95–99, 1983.

IARC. AF-2 [2-(2-furyl)-3-(5-nitro-2-furyl) acrylamide: chemical and physical data. *IARC Monographs on the Evaluation of the Carcinogenic Risk of Chemicals to Humans* 31: 47–61, 1983.

Vainiotalo, S., Zitting, A., Jacobsson, S., Nickels, J., et al. Toxicity of polymethyl-methacrylate thermodegradation products. *Arch Toxicol.* 55(2): 137–42, 1984.

Seppalainen, A. M., Rajaniemi, R. Local neuropathy of methyl methacrylate among dental technicians. *Amer. J. Ind. Med.* 5(6): 471–77, 1984.

Miller, M. S., Spencer, P. S. The mechanisms of acrylamide axonopathy. *Ann. Rev. Pharmacol. Toxicol.* 25: 643–66, 1985.

Rajaniemi, R., Tola, S. Subjective symptoms among dental technicians exposed to the monomer methyl methacrylate. *Scand. J. Work Environ. Health* 11(4): 281–86, 1985.

Klaassen, C. D., Amdur, M. O., Doull, J., Eds. *Casarett and Doull's Toxicology*, 3rd ed. Macmillan, New York, 1986, p. 374.

Kanerva, L., Verkkala, E. Electron microscopy and immunohistochemistry of toxic and allergic effects of methylmethacrylate on the skin. *Arch. Toxicol. (Supp).* 9: 456–59, 1986.

Scolnick, B., Collins, J. Systemic reaction to methylmethacrylate in an operating room nurse. *J. Occup. Med.* 28(3): 196–98, 1986.

Levett, D. L., Woffindin, C., Bird, A. G., Hoenich, N. A., Ward, M. K., Kerr, D. N. Hemodialysis-induced activation of complement. Effects of different membranes. *Blood Purif.* 4(4): 185–93, 1986.

Arsenic and Cancer

It should be noted that the association between cancer and arsenic exposure has been known for the last 100 years. Carcinogenicity has been demonstrated in human epidemiological studies and animal studies following intentional and accidental exposure. Mutagenic and teratogenic studies of arsenic exposure have confirmed the fact that arsenic adversely affects the genetic material that we call life.

It has been postulated that Agrippina used arsenic to kill Claudius to make Nero the emperor of Rome, and that, later, Nero used the same material to poison Claudius' son, Britannicus (approx. A.D. 50).

Aricola. *de Re Metallica.* 1556.

Paris, J. A. *Pharmacologia*, 3rd Ed. W. Phillips, London, 1820, pp. 132–34.

Hutchinson, J. *Trans. Pathol. Soc. (London)* 39: 1888.

Haerting, F. H., Hesse, W. Der lungenkrebs, die bergkrankheit in den Schneeberger gruben. *Vrtljscher. gerichtl. Med.* 30: 296, 1879.

Morris, M., Oliver, T. *Dangerous Trades.* Murray, London, 1902.

Legge, T. M. *Annual Report of the Chief Inspector of Factories and Workshops for 1902.* London, 1903, p. 262.

Ehrlich, P., Bertheim, O. p-Aminophenylarsinic acid. *Ber. dtsch. chem. Ges.* 40: 3292–97 [*Chem. Abstr.* 1: 2715], 1907.

Pye-Smith, R. J. *Proc. Roy. Soc. Med.* 6: 229, 1913.

O'Donovan, W. J. *Report of the International Conference on Cancer*, 1928, p. 293.

Henry, S. A. *Industrial Maladies*, Legge, T. M., Ed. London, 1934, p. 83.

Prell, H. Die schadigung der tierwelt durch die fernwirkungen von industrieabgasen. *Arch. Gewerbepath. u. Gewerbehyg.* 7: 656–70, 1937.

Amor, A. J. Growths of the respiratory tract. *Report of the VIII International Congress for Industrial Accidents and Occupational Diseases (Leipzig)* 2: 941–62, 1939.

Webster, S. H. The lead and arsenic content of urines from 46 persons with no known exposure to lead or arsenic. *U.S. Pub. Health Serv. Rep.* 56: 1953–61, 1941.

Merewether, E. R. A. *Annual Review Chief Inspector of Factories*, London, 1944.

Peters, R. A., Stocken, L. A. Thompson, R. H. S. *Nature* 156: 616, 1945.

Hill, A. B., Faning, E. L. *Brit. J. Ind. Med.* 5: 2, 1948.

Perry, K., et al. Studies in the incidence of cancer in a factory handling inorganic compounds of arsenic-II. Clinical and environmental investigations. *Brit. J. Ind. Med.* 5: 6–15, 1948.

Snegireff, L. S., Lombard, O. M. Arsenic and cancer—observations in the metallurgic industry. *Arch. Ind. Hyg. Occup. Med.* 4: 199–205, 1951.

Sommers, S. C., McManus, R. G. Multiple arsenical cancers of the skin and internal organs. *Cancer* 6: 347–59, 1953.

Hueper, W. C. *A Quest into the Environmental Causes of Cancer of the Lung.* Public Health Monograph no. 36, U.S. Department of Health, Education and Welfare, Public Health Service, 1955, pp. 27–29.

Roth, F. Haemangioendothelioma of the liver after chronic arsenic intoxication (Ger.). *Zentralbl. allgemein, Pathol. pathol. Anat.* 93: 424–25, 1955.

Roth, F. Late consequencies of chronic arsenicism in Moselle vine-dressers (Ger.). *Dtsch. med. Wochenshr.* 82: 211–17, 1957.

Roth, F. Arsenic-liver-tumours (haemangioendothelioma) (Ger.). *Z. Krebsforsch.* 61: 468–503, 1957.

Roth, R. Concerning bronchial cancers in vine-growers injured by arsenic. *Virchows Arch. (Pathol. Anat.)* 331: 119–37, 1958.

Schrenk, H. H., Schreibeis, L., Jr. Urinary arsenic levels as an index of industrial exposure. *Amer. Ind. Hyg. Assoc. J.* 19: 225–28, 1958.

Vallee, B. L., Ulmer, D. D., Wacker, W. E. C. Arsenic toxicology and biochemistry. *Arch. Ind. Health* 21: 132–51, 1960.

Buchanan, M. D. *Toxicology of Arsenic Compounds.* Amsterdam, 1961.

Graham, J. H., Mazzanti, G. R., Helwigm, E. B. Chemistry of Bowen's disease, relationship to arsenic. *J. Invest. Dermatol.* 37: 317–32, 1961.

Halver, J. E. *Progress in Studies on Contaminated Trout Rations and Trout Hepatoma.* National Institute of Health Report, 1962.

Hueper, W. C., Payne, W. E. Experimental studies in metal carcinogeneses: chromium, nickel, iron and arsenic. *Arch. Environ. Health* 5: 445–562, 1962.

U.S. Public Health Service. *Public Health Service Drinking Water Standards.* U.S. Govt. Printing Office, Washington, D.C., 1962.

Pinto, S. S., Bennett, B. M. Effect of arsenic trioxide exposure on mortality. *Arch. Environ. Health* 7: 583–91, 1963.

Kraybill, H. F., Shimkin, M. G. Carcinogenesis related to food contaminated by processing and fungal metabolities. *Adv. Cancer Res.* 8: 191–248. Academic Press, New York, 1964.

Ginsburg, J. M. Renal mechanism for excretion and transformation of arsenic in the dog. *Amer. J. Physiol.* 208: 832–40, 1965.

Kyle, R. A., Oease, G. L. Hematologic aspects of arsenic intoxication. *New Engl. J. Med.* 273: 18–23, 1965.

Oppenheimer, J. J., Fishbein, W. N. Induction of chromosomal breaks on cultured normal human leucocytes by potassium arsenite, hydroxyurea and related compounds. *Cancer Res.* 25: 980–85, 1965.

Schroeder, H. A. The biological trace elements. *J. Chron. Dis.* 18: 217–28, 1965.

Schroeder, H. A., Balassa, J. J. Abnormal trace metals in man: arsenic. *J. Chron. Dis.* 19: 85–106, 1966.

Hunter, D. *The Disease of Occupations.* Little, Brown and Co., Boston, 1969.

Lee, A. M., Fraumeni, J. F., Jr. Arsenic and respiratory cancer in man—an occupational study. *J. Nat. Cancer Inst.* 43: 1045–52, 1969.

Morgan, G. G., Oxolins, G., and Tabor, E. C. Air pollution surveillance systems. *Science* 170: 289–96, 1970.

U.S. Public Health Service. *Community Water Supply Study: Analysis of National Survey*

Findings. U.S. Department of Health, Education and Welfare, Washington, D.C., 1970.

Osswald, H., Goerttler, K. Arsenic induced leukemia in mice after diaplacental and postnatal application. *Dtsch. Gesell. Pathol.* 55: 289–93, 1971.

Ferm, V. H. The teratogenic effect of metals on mammalian embryos. *Adv. Teratol.* 5: 51, 1972.

Hood, R. D., Bishop, S. L. Teratogenic effects of sodium arsenate in mice. *Arch. Environ. Health* 24: 62–65, 1972.

International Agency for Research on Cancer. *Evaluation of Carcinogenic Risk of Chemicals to Man.* Monograph 1, World Health Organization, Lyon, France, 1972.

Milham, S., Jr., Strong, T. Human arsenic exposure in relation to a copper smelter. *Environ. Res.* 7: 176–82, 1973.

Nieberle, K. Endemic cancer in the bone of sheep. *Z. Krebsforsch.* 49: 137, 1973.

NIOSH. *Criteria for a Recommended Standard: Occupational Exposure to Inorganic Arsenic.* U.S. Department of Health, Education and Welfare, Washington, D.C., 1973.

Konetzke, G. W. Die kanzerogene Wirkung von Arsen und Nickel. *Arch. Geschwulstforsch.* 44: 1, 1974.

Ott, M. G., Holder, B. B., Gordon, H. L. Respiratory cancer and occupational exposure to arsenicals. *Arch. Environ. Health* 29: 250–55, 1974.

Blot, W. J., Fraumeni, J. F. Arsenical air pollution and cancer. *Lancet*, July 26, 1975.

Casarett, L. J., Doull, J. *Toxicology, the Basic Science of Poisons.* Macmillan, New York, 1975.

OSHA, Department of Labor. Standard for exposure to inorganic arsenic: notice of proposed rule making. *Federal Register* 40: 3392–95. U.S. Govt. Printing Office, Washington, D.C., 1975.

Blejer, H. P., Wagner, W. Case study 4: Inorganic arsenic—ambient level approach to the control of occupational carcinogenic exposures. *Ann. N.Y. Acad. Sci.* 271: 179–86, 1976.

Ferguson, W. In *Health Effects of Occupational Lead and Arsenic Exposure*, Carnow, B., Ed. U.S. Department of Health, Education and Welfare, Public Health Service, 1976.

Hine, C. In *Health Effects of Occupational Lead and Arsenic Exposure*, Carnow, B., Ed. U.S. Department of Health, Education and Welfare, Public Health Service, 1976.

Milham, S., Jr. In *Health Effects of Occupational Lead and Arsenic Exposure*, Carnow, B., Ed. U.S. Department of Health, Education and Welfare, Public Health Service, 1976.

Sunderman, F. W. A review of the carcinogenicities of nickel, chromium and arsenic compounds in man and animals. *Prevent. Med.* 5: 279–94, 1976.

Wagoner, J. K. Occupational carcinogenesis, the two hundred years since Percivall Pott. *Ann. N.Y. Acad. Sci.* 271: 1–4, 1976.

Burgdorf, W., Kurvink, K., Cervenka, J. Elevated sister chromatid exchange rate in lymphocytes of subjects treated with arsenic. *Hum. Genet.* 36(1): 69–72, 1977.

Corbett, T. H. *Cancer and Chemicals.* Nelson-Hall, Chicago, 1977.

OSHA, Department of Labor. Occupational exposure to inorganic arsenic: final standard. *Federal Register* 19584–631, 1978.

Rebuttable presumption against registration for inorganic arsenicals. *Federal Register*, Part II, 48267–442, Oct. 18, 1978.

Meyers, F. H., Jawetz, E., Goldfein, A. *Review of Medical Pharmacology*, 6th ed. Lange Medical Pub., Los Altos, Calif., 1978.

Reymann, F., Moller, R., Nielsen, A. Relationship between arsenic intake and internal malignant neoplasms. *Arch. Dermatol.* 114(3): 378–81, 1978.

Axelson, O., Dahlgren, E., Jansson, C. D., Rehnlund, S. O. Arsenic exposure and mortality: a case-referent study from a Swedish copper smelter. *Brit. J. Ind. Med.* 35(1): 8–15, 1978.

Pinto, S. S., Henderson, V., Enterline, P. E. Mortality experience of arsenic-exposed workers. *Arch. Environ. Health* 33(6): 325–31, 1978.

Feussner, J. R., Shelburne, J. D., Bredehoeft, S., Cohen, H. J. Arsenic-induced bone marrow toxicity: ultrastructural and electron-probe analysis. *Blood* 53(5): 820–27, 1979.

Mabuchi, K., Lilienfeld, A. M., Snell, L. M. Lung cancer among pesticide workers exposed to inorganic arsenicals. *Arch. Environ. Health.* 34(5): 312–20, 1979.

Ivankovic, S., Eisenbrand, G., Preussmann, R. Lung cancer induction in BD rats after single intratracheal instillation of an arsenic-containing pesticide mixture formerly used in vineyards. *Int. J. Cancer* 24(6): 786–88, 1979.

Vainio, H., Sorsa, M. Chromosome aberrations and their relevance to metal carcinogenesis. *Environ. Health Perspect.* 40: 173–80, 1981.

Pershagen, G. The carcinogenicity of arsenic. *Environ. Health Perspect.* 40: 93–100, 1981.

Wicks, M. J., Archer, V. E., Auersbach, O., Kuschner, M. Arsenic exposure in a copper smelter as related to histological type of lung cancer. *Amer. J. Ind. Med.* 2(1): 25–31, 1981.

Landrigan, P. Arsenic—state of the art. *Amer. J. Ind. Med.* 2(1): 4–14, 1981.

Kasper, M. L., Schoenfield, L., Strom, R. L., Theologides, A. Hepatic angiosarcoma and bronchioloalveolar carcinoma induced by Fowler's solution. *J.A.M.A.* 252(24): 3407–8, 1984.

Lee, T. C., Lee, K. C., Tzeng, Y. J., Huang, R. Y., Jan, K. Y. Sodium arsenite potentiates the clastogenicity and mutagenicity of DNA crosslinking agents. *Environ. Mutagen.* 8(1): 119–28, 1986.

Peters, J. M., Thomas, D., Falk, H., Oberdorster, G., Smith, T. J. Contribution of metals to respiratory cancer. *Environ. Health Perspect.* 70: 71–83, 1986.

Wald, P. H., Becker, C. E. Toxic gases used in the microelectronics industry. *State Art Rev. Occup. Med.* 1(1): 105–17, 1986.

Lee-Feldstein, A. Cumulative exposure to arsenic and its relationship to respiratory cancer among copper smelter employees. *J. Occup. Med.* 28(4): 296–302, 1986.

Enterline, P. E., Henderson, V. L., Marsh, G. M. Exposure to arsenic and respiratory cancer. A reanalysis. *Amer. J. Epidemiol.* 125(6): 929–38, 1987.

Pershagen, G., Bergman, F., Klominek, J., Damber, L., Wall, S. Histological types of lung cancer among smelter workers exposed to arsenic. *Brit. J. Ind. Med.* 44(7): 454–58, 1987.

Asbestos Exposure and Malignancies Associated with Immune Dysfunction, Including Leukemias, Lymphomas, Sarcomas, and Myelomas

Gardiner, L. U., Cummings, D. E. Studies on experimental pneumoconiosis. Inhalation of asbestos dust; its effect upon primary tuberculosis infection. *J. Ind. Hyg.*, Mar. 1931.

Gloyne. The morbid anatomy and histology of asbestosis. *Tubercle* 14: 550, 1933.

Lynch, K. M., Smith, W. A. Pulmonary asbestosis III: carcinoma of lung in asbesto-silicosis. *Amer. J. Cancer*, 24:56–64, 1935.

Linzbach, A. J., Wedler, W. H. Beitrag zum Berufskrebs der Asbestarbeiter. *Virchow's Arch. Pathol. Anat. Physiol.*, 387–407, 1941.

Kuhn, A. Illness among dock workers at Willhelmshaven with special reference to industrial accidents and diseases (1930–1938). *Bull. Hyg.* 17(12): 819, 1942.

Mallory, T. B., Ed. Case reports of asbestosis. *New Engl. J. Med.* 236: 407, 1947.

Wyers, H. Asbestosis. *Postgrad. Med. J.* 631–38, Dec. 1949.

Merewether, E. P. A. The pneumoconioses: developments, doubts, difficulties. *Canad. Med. J.* 62:169–73, 1950.

Isselbacher, K. J., Hanna, K., Hardy, H. Asbestosis and bronchogenic carcinoma. *Amer. J. Med* 721–32, 1953.

Leicher, F. Primary peritoneal protective cell tumor in asbestosis. *Arch. Occup. Pathol.* 382–91, 1954.

Bonser, G. M., Fields, J. S., Stewart, M. J. Occupational cancer of the urinary bladder in dyestuffs operatives and of the lung in asbestos textile workers and iron-ore miners. *Amer. J. Clin. Pathol.* 25: 126–34, 1955.

Schepers, G. W. H., et al. The effects of inhaled commercial hydrous calcium silicate dust on animal tissues. *Arch. Indust. Health* 12: 348–60, 1955.

Merewether, E. R. A. Ed. *Indust. Med. Hyg.* Butterworth & Co., London, 1956.

Hueper, W. C. *A Quest into the Environmental Causes of Cancer of the Lung.* Public Health Monograph No. 36, 1956.

Van Der Shoot, H. C. M. Asbestosis en pleuragezwellen. *Nederl. Tylschrift von Geneeskunk. 102:* 1125–26, 1958.

Schmahl, D. Cancerogenous effect of asbestos in inplantation in rats. *J. Cancer Res.* 62: 1958.

Telischi, M., Rubenstone, A. I. Pulmonary asbestosis. *Arch. Pathol.* 72: 234–43, 1961.

Hervieux, J. Certain aspects of asbestosis. *Cahiers de medicine inter professionnelle.* 7: 2–3, 1962.

Buchanan. Asbestosis case references. *Int. Cong. Occup. Health* 14, Madrid. 1963.

Mancuso, T., Coulter, E. Methodology in industrial health studies. *Arch. Environ. Health* 6: 210–26, 1963.

Selikoff, I. J. Asbestosis exposure and neoplasia. *Amer. J. Med.* 188: 142–46, 1964.

Buchanan, W. D. Asbestosis and primary intrathoracic neoplasms. *Ann. N.Y. Acad. Sci.* 132: 507–18, 1965.

Lieben, J. Malignancies in asbestos workers. *Arch. Environ. Health* 13: 619–21, 1966.

Gerber, M. Asbestosis and neoplastic disorders of the hematopoietic system. *Amer. J. Clin. Pathol.* 53: 204–8, 1970.

Isobe, T., Osserman, E. F. Pathologic conditions associated with plasma cell dyscrasias, a study of 806 cases. *Ann. N.Y. Acad. Sci.* 190: 507, 1971.

September, B. A., Villet, W. T. Monoclonal gammopathy associated with malignant mesothelioma: case report. *South Afr. Med. J.* 46: 545–49, 1972.

DeCarvalho, S. Prodromal dysimmunity in cancer. *Oncology* 28: 1–34, 1973.

Parkes, W. R. Asbestos-related disorders. *Brit. J. Dis. Chest* 67: 261–300, 1973.

Potter, M. The developmental history of the neoplastic plasma cell in mice. A brief review of recent developments. *Semin. Hematol.* 10: 19, 1973.

Kang, K. Y., Sera, Y., Okochi, T., et al. T lymphocytes in asbestosis. *New Engl. J. Med.* 291: 735, 1974.

Lange, A., Smolik, R., Zatonski, W., et al. *Int. Arch. Arbeitsmed.* 32: 313, 1974.

Stansfield, D., Edge, J. R. Circulating rheumatoid factor and antinuclear antibodies in shipyard asbestos workers with pleural plaques. *Brit. J. Dis. Chest* 68: 166, 1974.

Becklake, M. R. Asbestos related diseases of the lung and other organs, their epidemiology and implications for clinical practice. *Amer. Rev. Respir. Dis.* 114: 187, 1976.

Dionne, G. P., Bieland, J. E., Wans, N. S. Letter: primary leiomyosarcoma of the diaphragm of an asbestos worker. *Arch. Pathol. Lab. Med.* 100(8): 398, 1976.

Kagan, E., Solomon, A., Cochrane, J. C., et al. Immunological studies of patients with asbestosis. I: Studies of cell-mediated immunity. *Clin. Exp. Immunol.* 28: 261, 1977.

Miller, K., Kagan, E. Immune adherence reactivity of of rat alveolar macrophages following inhalation of crocidolyte asbestos. *Clin. Exp. Immunol.* 29: 152, 1977.

Scully, R., Galdabini, J., McNeely, B. Case records of Mass. General Hospital. *New Engl. J. Med.* 297: 1211–28, 1977.

Haslam, P. L., Lukosek, A., Merchant, A., et al. Lymphocyte responses to phytohaemaglobin in patients with asbestosis and pleural mesothelioma. *Clin. Exp. Immunol.* 31: 178, 1978.

Kagan, E., Miller, K. Alveolar macrophage–splenic lymphocyte interactions following chronic asbestos inhalation in the rat. *J. Reticuloendothel. Soc.* 23 (Suppl.): 241, 1978.

Kagan, E., Solomon, A., Cochrane, J. C., et al. Cancer related to asbestos exposure. Immunological studies of patients at risk, in *Prevention and Detection of Cancer*, Part I, Vol. 2, Nieburgs, E. E., Ed. Marcel Dekker, New York, 1978, p. 1631.

Bigon, J., Monchaux, G., Sebastien, P., Hirsch, A., Lafuma, J. Human and experimental data on translocation of asbestos fibers through the respiratory system. *Ann. N.Y. Acad. Sci.* 330: 745–50, 1979.

Kagan, E., Jacobson, R., Yeung, K.-Y., Haidak, D. Asbestos-associated neoplasms of B cell lineage. *Amer. J. Med.* 67: 325–30, Aug. 1979.

Kagan, E., Miller, K. Alveolar macrophage–splenic lymphocyte interactions following chronic asbestos inhalation in the rat, in *Macrophages and Lymphocytes; Nature of Functions and Interactions*, Escobar, E. M., Friedman, H., Eds. New York, Plenum, 1979.

Robinson, C., Lemen, R. Mortality patterns, 1940–1975 among workers employed in an asbestos textile friction and packing products manufacturing facility, in *Dusts and Disease*. Society for Occupational and Environmental Health, Washington, D.C., 1979.

Lee, K. P. Barras, C. E., Griffith, F. D. Waritz, R. S., Pulmonary response and transmigration of inorganic fibers by inhalation exposure. *Amer. Assoc. Pathol.*, 102: 314–23, 1980.

Bianchi, C., Brollo, A., Bittesini, L. Mesothelioma da asbesto nel territorio di monfalcone. *Pathologica* 73: 649, 654–55, 1981.

Kagan, E., Miller, K. Asbestos inhalation and the induction of splenic lymphocytic proliferation in the rat. *Chest* 80: 11–12, 1981.

Maguire, F. W., Mills, R. C., Parker, F. P. Immunoblastic lymphadenopathy and asbestosis. *Cancer* 47: 791–97, 1981.

Chung, C. K., Stryker, J., Zaino, R., Sears, H. Liposarcoma after asbestos exposure. *Penn. Med.* 85(12): 47–48, 1982.

Ross, R., Nichols, P., Wright, W., Lukes, R., Dworsky, R., Paganini-Hill, A., Koss, M., Henderson, B. Asbestos exposure and lymphomas of the gastrointestinal tract and oral cavity. *Lancet* 1118–20, Nov. 20, 1982.

Rouhier, D., Andre, C., Allard, C., Gillon, J., Brette, R. Malignant alpha chain disease and exposure to asbestos. *Environ. Res.* 27: 222–25, 1982.

Kagan, E., Jacobson, R. Lymphoid and plasma cell malignancies: asbestos-related disorders of long latency. *Amer. J. Clin. Pathol.* 80(1): 14–20, 1983.

Anon. Asbestos and non-Hodgkin's lymphoma (letter). *Lancet* 1(8317): 189–90, 1983.

Olsson, H., Brandt, L. Asbestos exposure and non-Hodgkin's lymphoma (letter). *Lancet* 1(8324): 588, 1983.

Bozelka, B. E., Gaumer, H. R., Nordberg, J., Salvaggio, J. E. Asbestos-induced alterations of human lymphoid cell mitogenic responses. *Environ. Res.* 30(2): 281–90, 1983.

Tilkes, F., Beck, E. G. Macrophage functions after exposure to mineral fibers. *Environ. Health Perspect.* 51: 67–72, 1983.

Stevens, R. H., Cole, D. A., Cheng, H. F., Hodge, J. A., Will, L. A. Cell-mediated cytotoxicity expressed by lymphoid cells from rats with asbestos-induced peritoneal mesothelioma toward rat fetal cell. *Environ. Health Perspect.* 51: 91–96, 1983.

Kagan, E., Jacobson, R. J. Lymphoid and plasma cell malignancies: asbestos-related disorders of long latency. *Amer. J. Clin. Pathol.* 80(1): 14–20, 1983.

Ueki, A., Oka, T., Mochizuki, Y. Proliferation stimulating effects of chrysotile and crocidolite asbestos fibers on B lymphocyte cell lines. *Clin. Exp. Immunol.* 56(2): 425–30, 1984.

Hartmann, D. P., Georgian, M. M., Oghiso, Y., Kagan, E. Enhanced interleukin activity following asbestos inhalation. *Clin. Exp. Immunol.* 55(3): 643–50, 1984.

Efremidis, A. P., Waxman, J. S., Chahinian, A. P. Association of lymphocytic neoplasia and mesothelioma. *Cancer.* 55(5): 1056–59, 1985.

Lechner, J. F., Tokiwa, T., LaVeck, M., Benedict, W. F., Banks-Schlegel, S., Yeager, H., Jr., Banerjee, A., Harris, C. C. Asbestos-associated chromosomal changes in human mesothelial cells. *Proc. Nat. Acad. Sci. USA* 82(11): 3884–88, 1985.

White, K. L., Munson, A. E. Suppression of the in vitro humoral immune response by chrysotile asbestos. *Toxicol. App. Pharmacol.* 82(3): 493–504, 1986.

Lew, F., Tsang, P., Holland, J. F., Warner, N., Selikoff, I. J., Bekesi, J. G. High frequency of immune dysfunction in asbestos workers and in patients with malignant mesothelioma. *J. Clin. Immunol.* 6(3): 225–33, 1986.

Barbers, R. G., Oishi, J. Effects of in vitro asbestos exposure on natural killer and antibody-dependent cellular cytotoxicity. *Environ. Res.* 43(1): 217–26, 1987.

Hedenborg, M., Klockars, M. Production of reactive oxygen metabolites induced by asbestos fibers in human polymorphonuclear leukocytes. *J. Clin. Pathol.* 40(10): 1189–93, 1987.

"Autoimmune" Diseases and Chemicals

Bramwell, B. Diffuse scleroderma: its frequency, occurrence in stonemasons, treatment by fibrinolysin, elevations of temperature due to fibrinolysin injections. *Edinburgh Med. J.* 12: 387–401, 1914.

Rowe, V. K., McCollister, D. D., Spencer, H. C., Adama, E. M., Irish, D. D. Vapor toxicity of tetrachlorethylene for laboratory animals and human subjects. *Arch. Ind. Hyg. Occup. Med.* 5: 566–79, 1952.

Reinl, W. Schleroderma caused by trichlorethylene? *Bull. Hyg.* 32: 678–79, 1957.

Erasmus, L. D. Scleroderma in gold-miners on the Witwadersrand with particular reference to pulmonary manifestations. *South Afr. J. Labor Clin. Med.* 3: 209–31, 1957.

Kubota, J. Occupational diseases in synthetic resin and fibre industries. *Rodo Kagaku* 33: 1–22, 1957.

Suciu, I., Drejman, I., Valeskai, M. Investigation of the diseases caused by vinyl chloride. *Med. Interne* 15: 967–78, 1963.

Rumsby, M. G., Finean, J. B. The action of organic solvents on the myelin sheath of peripheral nerve tissue. III. Chlorinated hydrocarbons. *J. Neurochem.* 13: 1513–15, 1966.

Cordier, J., Fieviez, C., Lefevre, M. J., et al. Acro-osteolyse et lesions cutanees associees chez deux ouvriers affectes au nettoyage d'autoclaves. *Canada. Med. Travail* 4: 3, 1966.

Cordier, J., Fievez, M. J., Sevrin, A. Acro-osteolysis and exposure to vinyl chloride. *Canad. Med. Travail* 4: 14–19, 1966.

Harris, D. K., Adams, W. G. F. Acro-osteolysis occurring in men engaged in the polymerization of vinyl chloride. *Brit. Med. J.* 3: 712–14, 1967.

Wilson, R. H., McCormick, W. E., Tatum, C. F., et al. Occupational acroosteolysis: report of 31 cases. *J.A.M.A.* 201: 577–80, 1967.

Rodnan, G. P., Benedek, T. G., Medsger, T., et al. The association of progressive systemic sclerosis (scleroderma) with coal miners pneumoconiosis and other forms of silicosis. *Ann. Intern. Med.* 66: 323–34, 1967.

Gunther, G., Schuchardt, E. Silikose und progressive Sklerodermie. *Dtsch. Med. Wochenschr.* 95: 467–68, 1970.

O'Conner, R. B. A new occupational disease is born. *J. Occup. Med.* 12: 234, 1970.

Dinman, B. D., Cook, W. A., Whitehouse, W. M., Magnuson, H. J., Ditcheck, T. Occupational acroosteolysis. I. An epidemiological study. *Arch. Environ. Health* 22: 61–73, 1971.

Dodson, V. N., Dinman, B. D., Whitehouse, W. M., Nasr, A. N. M., Magnuson, H. J. Occupational acroosteolysis. III. A clinical study. *Arch. Environ. Health* 22: 83–91, 1971.

Medsger, T. A., Jr., Masi, A. T. Epidemiology of systemic sclerosis (scleroderma). *Ann. Intern. Med.* 74: 714–21, 1971.

Markowitz, S. S., McDonald, C. J., Fethiere, W., et al. Occupational acroosteolysis. *Arch. Dermatol.* 106: 219–23, 1972.

Takeuchi, Y., Mabuchi, C. A case of occupational acroosteolysis, presumably caused by vinyl chloride. *Sangyo Igaku* 15: 385–94, 1973.

Juhe, S., Lange, C. E., Stein, G., et al. The so called vinyl chloride disease. *Dtsch. Med. Wochenschr.* 98: 2034–37, 1973. English Translation via EPA, U.S. Dept. of Commerce TR75-6; PB 258838-T, 1975.

Bachner, U., Etzel, R., Lange, C. E., et al. Hamostaseologische Aspekte bei der Vinylchlorid-Krankheit. *Dtsch. Med. Wochenschr.* 99: 2409–10, 1974.

Lange, C. E., Juhe, S., Veltman, G. Ober das Auftreten von Angiosarkomen der Leber bei zwei Arbeitern der PVC-herstellenden Industrie. *Dtsch. Med. Wochenschr.* 99: 1598–99, 1974.

Lange, C. E., Juhe, S., Stein, G., Veltnam, G. Vinyl chloride disease. *Int. Arch. Occup. Health* 32: 1–32, 1974.

Moulin, P. G., Rety, J., Paliard, P., Vouillon, G., Guttin, G. Aspects sclerodermiques de l'acro-osteolyse professionnelle (polymerisation du chlorure de vinyle). *Ann. Derm. Syph.* 101: 33–44, 1974.

Lilis, R., Anderson, H., Nicholson, W. J., Daum, S., Fischbein, A. S., Selikoff, I. J. The prevalence of disease among vinyl chloride and polyvinyl chloride workers. *Ann. N.Y. Acad. Sci.* 246: 22–41, 1975.

Penin, H., Sargar, G., Lange, G. E., et al. *Neurologisch-psychiatrische und elektroen-zephalographische Befunde bei Patienten mit Vinylchlorid-Krankheit.* Verlag, Stuttgart, 1975, pp. 299–304.

Biersack, H. J., Lange, C. E., Ebinger, H., et al. Sequenzszintigraphische Untersuchungen von Leber und Milz bei Patienten mit Vinylchlorid-Krankheit. *Dtsch. Med. Wochenschr.* 100: 615–17, 1975.

Walker, A. E. A preliminary report of a vascular abnormality occurring in men engaged in the manufacture of polyvinyl chloride. *Brit. J. Dermatol.* 93(Supp. II): 22–23, 1975.

Veltman, G., Lange, C. E., Juhe, S., et al. Clinical manifestations and course of vinyl chloride disease. *Ann. N.Y. Acad. Sci.* 246: 6–17, 1975.

Ward, A. M., Udnoon, S., Watkins, J., et al. Immunological mechanisms in the pathogenesis of vinyl chloride disease. *Brit. Med. J.* 1: 936–38, 1976.

Walker, A. E. Clinical aspects of vinyl chloride disease: skin. *Proc. Roy. Soc. Med.* 69: 286–89, 1976.

Milford, W. A. Evidence of an immune complex disorder in vinyl chloride workers. *Proc. Roy. Soc. Med.* 69: 289, 1976.

Maricq, H. R., Johnson, M. N., Whetstone, C. L., et al. Capillary abnormalities in polyvinyl chloride production workers. Examination by in vivo microscopy. *J.A.M.A.* 236: 1368–71, 1976.

Jayson, M. I. V., Bailey, A. J., Black, C. Collagen studies in acro-osteolysis. *Proc. Roy. Soc. Med.* 69: 295–97, 1976.

Sparrow, G. P. A connective tissue disorder similar to vinyl chloride disease in a patient exposed to perchlorethylene. *Clin. Exp. Dermatol.* 2: 17–22, 1977.

Bauer, M., Rabens, S. F. Trichlorethylene toxicity. *Int. J. Dermatol.* 16: 113–16, 1977.

Medsger, T. A., Jr., Masi, A. T. The epidemiology of systemic sclerosis (scleroderma) among male U.S. veterans. *J. Chron. Dis.* 31: 73–85, 1978.

Thorgeirsson, A., Fregert, S., Ramnas O. Sensitization capacity of epoxy resin oligomers in the guinea pig. *Acta Derm. Venereol.* 58: 17–21, 1978.

Gama, C., Meira, J. B. B. Occupational acro-osteolysis. *J. Bone Joint Surg.* 60(A): 86–90, 1978.

Saihan, E. M., Burton, J. L., Heaton, K. W. A new syndrome with pigmentation, scleroderma, gynaecomastia, Raynaud's phenomenon and peripheral neuropathy. *Brit. J. Dermatol.* 99: 437–40, 1979.

Bretza, J., Goldman, J. A. Scleroderma simulating vinyl chloride disease. *J. Occup. Med.* 21(6): 436–38, 1979.

Stachow, A., Jablonska, S., Skiendzielewska, A. Biogenic amines derived from tryptophan in systemic and cutaneous scleroderma. *Acta Derm. Venereol.* 59: 1–5, 1979.

Kumagai, Y., Abe, C., Shiokawa, Y. Scleroderma after cosmetic surgery. *Arthritis Rheum.* 22: 532–37, 1979.

Sternberg, E. M., Van Woert, M., Young, S. N., et al. Development of a scleroderma-like illness during therapy with L-5-hydroxytryptophan and carbidopa. *New Engl. J. Med.* 303: 782–87, 1980.

Yamakagae, A., Ishikawa, H., Saito, Y., Hattori, A. Occupational scleroderma-like disorder occurring in men engaged in the polymerization of epoxy resins. *Dermatologica* 161: 33–44, 1980.

Bernstein, R., Prinsloo, I., Zwi, S., et al. Chromosomal aberrations in occupation-associated progressive systemic sclerosis. *South Afr. Med. J.* 58: 235–37, 1980.

Veltman, G. Klinische Befunde und arbeitsmedizinische Aspekte der Vinylchlorid-Krankheit. *Dermatol. Monatsschr.* 166: 705–12, 1980.

Tabeunca, J. M. Toxic-allergic syndrome caused by ingestion of rapeseed oil denatured with aniline. *Lancet* 2: 567–68, 1981.

Haustein, U. F., Ziegler, V., Zschunke, E., et al. The coincidence of progressive systemic sclerosis with silicosis in the GDR: an epidemiologic study. Int. Conf. Progressive Systemic Sclerosis, Austin, Tex., Oct. 1981.

Kondo, H. Progressive systemic sclerosis following cosmetic surgery ("adjuvant disease"). Int. Conf. Progressive Systemic Sclerosis, Austin, Tex., Oct. 1981.

Maricq H. R. Vinyl chloride disease. Int. Conf. Progressive Systemic Sclerosis, Austin, Tex., Oct. 1981.

Pfluger, H., Stachelberger, H. Kann eine Siliziumdiffusion aus Silikonkautschukkathetern Harnrohrenstrikturen verursachen? *Z. Urol. Nephrol.* 75: 817–20, 1982.

Ziegler, V., Pampel, W., Zschunke, E., et al. Kristalliner Quarz: eine Ursache der progressiven Sklerodermie? *Dermatol. Monatsschr.* 168: 398–401, 1982.

Yamakage, A., Ishikawa, H. Generalized morphea-like scleroderma occurring in people exposed to organic solvents. *Dermatologica* 165: 186–93, 1982.

Kohanka, V. Epidemiologische Untersuchungen der Immunitat vom Spattyp an Arbeitern unter Vinylchlorid-Monomer. *Borgyog. Venerol. Szemle.* 58: 145–49, 1982.

Destouet, J. M., Murphy, W. A. Acquired acroosteolysis and acronecrosis. *Arthritis Rheum.* 26: 1150–54, 1983.

Walder, B. K. Do solvents cause scleroderma? *Int. J. Dermatol.* 22: 157–58, 1983.

Black, C. M., Walker, A. E., Catoggio, L. J., et al. Genetic susceptibility to scleroderma-like syndrome induced by vinyl chloride. *Lancet* 1: 53–55, Jan. 1983.

Mateo, I. M., Izquierdo, M., Fernandez-Dapica, M. P., et al. Toxic epidemic syndrome: musculoskeletal manifestations. *J. Rheumatol.* 11: 333–38, 1984.

Rush, P. J., Bell, M. J., Fam, A. G. Toxic oil syndrome (Spanish Oil Disease) and chemically induced scleroderma-like conditions (editorial). *J. Rheumatol.* 11: 3, 1984.

Haustein, U. F. and Ziegler, V. Environmentally induced systemic sclerosis-like disorders. *Int. J. Dermatol.* 24: 148–51, 1985.

Jaffe, I. A. Adverse effects profile of sulfhydryl compounds in man. *Amer. J. Med.* 80(3): 471–76, 1986.

Pereyo, N. Hydrazine derivatives and induction of systemic lupus erythematosus (letter). *J. Amer. Acad. Dermatol.* 14(3): 514–15, 1986.

Lambert, J., Schepens, P., Janssens, J., Dockx, P. Skin lesions as a sign of subacute pentachlorophenol intoxication. *Acta Derm. Venereol. (Stockh.)* 66(2): 170–72, 1986.

Magnavita, N., Bergamaschi, A., Garcovich, A., Giuliano, G. Vasculitis purpura in vinyl chloride disease: a case report. *Angiology* 37(5): 382–88, 1986.

Sergott, T. J., Limoli, J. P., Baldwin, C. M., Laub, D. R. Human adjuvant disease, possible autoimmune disease after silicone implantation: a review of the literature, case studies, and speculation for the future. *Plast. Reconstr. Surg.* 78(1): 104–14, 1986.

Ingalls, T. H. Endemic clustering of multiple sclerosis in time and place, 1934–1984. Confirmation of a hypothesis. *Amer. J. Forensic Med. Pathol.* 7(1): 3–8, 1986.

Albrecht, W. N., Boiano, J. M., Smith, R. D. Glomerulonephritis associated with hydrocarbon solvents (letter). *J. Soc. Occup. Med.* 36(4): 145, 1986.

The multiple reports of immunological, blood, collagen, and skin diseases induced by pharmaceutical agents have not been included, as they are readily available from the standard literature search programs. The chemicals and diseases cited in this section are encountered in the context of occupational and/or environmental exposures.

Birth Defects and Chemicals

Wiesner, B. P. The post-natal development of the genital organs in the albino rat. *J. Obstet. Gynecol.* 41: 867–922, 1934.

Willier, B. H., Gallagher, T. F., Koch, F. C. Sex-modification in the chick embryo resulting from injections of male and female hormones. *Proc. Nat. Acad. Sci.* 21: 625–31, 1935.

Burns, R. K. The effects of crystalline sex hormones on sex differentiation in amblystoma. *Anatom. Record.* 73–93, 1939.

Greene, R. R., Burrill, M. W., Ivy, A. C. Experimental intersexuality: modification of sexual development of the white rate with synthetic estrogen. *Proc. Soc. Exper. Biol. Med.* 41: 169–70, 1939.

Greene, R. R., Burrill, M. W., Ivy, A. C. Experimental intersexuality: the effects of estrogens on the antenatal sexual development of the rat. *Amer. J. Anat.* 67: 305–45, 1940.

Larsen, C. D. Pulmonary tumor induction by transplacental exposure to urethane. *J. Nat. Cancer Inst.* 8(2): 63–70, 1947.

McOmie, W. A. Comparative toxicology of methacrylonitrile and acrylonitrile. *J. Indust. Hyg. Toxicol.* 31: 113–16, 1949.

Dickey, F. H., Cleland, G. H., Lotz, C. The role of organic peroxides in the induction of mutations. *Nat. Acad. Sci. Proc.* 35: 581–86, 1949.

Auerbach, C. The mutagenic action of formalin. *Science* 110: 419–20, 1949.

Auerbach, C. Mutation tests on drosophila melanogaster with aqueous solutions of formaldehyde. *Amer. Nat.* 86, 330–32, 1952.

Burdette, W. J. Tumor incidence and lethal mutation rate in a tumor strain of drosophila treated with formaldehyde. *Cancer Res.* 11: 555–558, 1961.

Campbell, G. D. Chlorpropamide and fetal damage. *Brit. Med. J.* 1: 59–60, 1963.

Lenz, W. Malformations caused by drugs in pregnancy. *Amer. J. Dis. Child.* 112: 99–106, 1966.

Herbst, A. L., Scully, R. E. Adenocarcinoma of the vagina in adolescence. *Cancer.* 25: 745–57, 1970.

Grasso, C., Bernardi, G., Martinelli, F. Significance of the presence of chlorinated insecticides in the blood of newborn children and of their mothers. *Ig. Mod.* 66: 362–71, 1973.

Ames, B. N., McCann, J., Yamasaki, E. Methods for detecting carcinogens and mutagens with the salmonella/mammalian-microsome mutagenicity test. *Mut. Res.* 31: 347–63, 1975.

Autian, J. Structure-toxicity relationships of acrylic monomers. *Environ. Health Perspect.* 11: 141–52, 1975.

Herbst, A. L., Poskanzer, D. C., Robboy, S. J., Friedlander, L., Scully, R. E. Prenatal exposure to stilbesterol. *New Engl. J. Med.* 292(7): 334–39, 1975.

Dowty, B. J., Laseter, J. L., Storer, J. The transplacental migration and accumulation in blood of volatile organic constituents. *Pediatr. Res.* 10(7): 696–701, 1976.

Edwards, P. M. The insensitivity of the developing rat foetus to the toxic effects of acrylamide. *Chem. Biol. Interact.* 12(1): 13–8, 1976.

Gordon, J., Meinhardt, T. J. Ethylene oxide–spontaneous abortions (Letter to the Editor). *Brit. Med. J.* 286: 1976–77, 1976.

Hanson, J. W., Smith, D. W. Fetal hydantoin syndrome. *Lancet.* 1: 692, 1976.

Auerbach, C., Moutschen-Dahmen, M., and Moutachen, J. Genetic and cytogenetical effects of formaldehyde and related compounds. *Mutat. Res.* 39: 317–362, 1977.

Blattner, W. A., Henson, D. E., Young, R. C., Fraumeni, J. F., Jr. Malignant mesenchymoma and birth defects. *J. A. M. A.* 238(4): 334–35, 1977.

Heinonen, O. P., Sloane, D., Shapiro, S. *Birth Defects and Drugs in Pregnancy.* Publishing Sciences Group, Inc., Littleton, MA., 1977.

Choi, B. H., Lapham, L. W., Amin-Zaki, L., Saleem, T. Abnormal neural migration, deranged cerebral cortical organization, and diffuse white matter astrocytosis of human fetal brain: a major effect of methylmercury poisoning in utero. *J. Neuropath. Exper. Neurol.* 37(6): 719–33, 1978.

Fujita, M., Fujimoto, T., Hirata, S. Embryotoxic effects of methylmercuric chloride administered to mice and rats during organogenesis. *Teratol.* 18: 353–66, 1978.

Murray, F. J., Schwets, B. A., Nitschke, K. D., John, J. A.; Norris, J., *et al.* Teratogenicity of acrylonitrile given to rats by gavage or by inhalation. *J. Cosmet. Toxicol.*, 16: 547–51, 1978.

O'Brien, T. E., McManus, C. E. Drugs and the fetus: A consumer's guide by generic and brand name. *Birth Family J.* 5(2): 58–86, 1978.

Scott, R. Reproductive hazards. *Job Saf. Health* 6(5): 7–13, 1978.

Spyker, J. M., Avery, D. L. Neurobehavioral effects of prenatal exposure to the organophosphate diazinon in mice. *J. Toxicol. Environ. Health.* 4(5-6): 989–1002, 1978.

Cooper, P. Genetic effects of formaldehyde. *Food Cosmet. Toxicol.* 17(3): 300–1, 1979.

Elovaara, E., Hemminki, K., Vainio, H. Effects of methylene chloride, trichloroethane, trichlorethylene, tetrachloroethylene and toluene on the development of chick embryos. *Toxicology*, 12: 111–19, 1979.

Garry, V. F., Hozier, J., Jacobs, D., Wade, R. L., Gary, P. G. Ethylene oxide: evidence of human chromosome effects. *Environ. Mutagen.* 1: 375–382, 1979.

Lucier, G. W., McDaniel, O. S. Development toxicology of the halogenated aromatics: effects on enzyme development. *Ann. N. Y. Acad. Sci.* 449–57, 1979.

Nicholas, C., Lawrence, W. H., Autian, J. Embryotoxicity and fetotoxicity from maternal inhalation of methyl methacrylate monomer in rats. *Toxicol. Appl. Pharmacol.* 50: 451–458, 1979.

Stellman, J. M. The effects of toxic agents on reproduction. *Occup. Health. Saf.* 1–8, Apr. 1979.

Baden, J. M., Simmon, V. F. Mutational effects of inhalational anesthetics. *Mutat. Res.* 75: 169–180, 1980.

Druga, A., Nyitray, M., Szaszovszky, E. Experimental teratogenicity of structurally similar compounds with or without piperazine-ring: a preliminary report. *Pol. J. Pharmacol.* 32: 199–204, 1980.

Hashimoto, K. The toxicity of acrylamide. *Jap. J. Ind. Health*, 22: 233, 1980.

LaBorde, J. B., Kimmel, C. A. The teratogenicity of ethylene oxide administered intravenously to mice *Toxicol. Pharmacol.* 56: 16–22, 1980.

Clement Associates, Inc. Council on Environmental Quality *Chemical Hazards to Human Reproduction.* pp. 1–44 and appendices. Jan. 1981.

Ferm, V. H. Cyanide teratogenicity in the Syrian golden hamster (SGH). *Pharmacologist* 23(3): 214, 1981.

Ford, W. D. A., Little, K. E. T. Fetal ovarian dysplasia possibly associated with clomiphene. *Lancet.* 2: 1107, 1981.

Laing, I. A., Steer, C. R., Dudgeon, J., Brown, J. K. Clomiphene and congenital retinopathy. *Lancet* 2: 1107–08, 1981.

Murray, J. S., Miller, R. R., Deacon, M. M., Hanley, T. R., Hayes, W. C. Teratological evaluation of inhaled ethyl acrylate in rats. *Toxicol. Appl. Pharmacol.* 60: 106–111, 1981.

National Institute for Occupational Safety and Health (NIOSH). Current intelligence bulletin: ethylene oxide (EtO) DHHS (NIOSH) Publication No. 81-130, Cincinnati, OH., 1981.

Whitehead, E. D., Leeiter, E. Genital abnormalities and abnormal semen analysis in male patients exposed to diethylstilbesterol in utero. *J. Urol.* 125: 47–50, 1981.

Willhite, C. C., Ferm, V. H., Smith, R. P. Teratogenic effects of aliphatic nitriles teratology 23: 317–23, 1981.

Willhite, C. C., Marin-Padilla, M., Ferm, V. H., Smith, R. P. Morphogenesis of axial skeletal (dysraphic) disorders induced by aliphatic nitriles. *Teratology* 23: 325–33, 1981.

Brix, K. Environmental and occupational hazards to the fetus. *J. Reprod. Med.* 577–83, 1982.

Chvapil, M., Droegemueller, W., Earnest, D. Liver function tests in women using intravaginal spermicide nonoxynol-9. *Fertil. Steril.* 37: 281–82, 1982.

Eskenazi, B., Bracken, M. B. Bendectin (Debendox) as a risk factor for pyloric stenosis. *Amer. J. Obstet. Gynecol.* 144(8): 919–24, 1982.

Hemminki, R., Mutanen, P., Saloniemi, I., Niemi, M. L., Vainio, H. Spontaneous abortions in hospital staff engaged in sterilizing instruments with chemical agents. *Brit. Med. J.* 285: 1461–63, 1982.

Lynch, D. W., Lewis, T. R., Moorman, W. J. Chronic inhalation toxicity of ethylene oxide and propylene oxide in rats and monkeys—a preliminary report *Toxicologist* 2: 11–18, 1982.

Pero, R. W., Bryngelsson, T., Widegren, G., Godstedt, B., Welinder, H. A reduced capacity for unscheduled DNA synthesis from individuals exposed to propylene oxide and ethylene oxide. *Mutat. Res.* 104: 193–200, 1982.

Perocco, P., Pane, G., Bolognesi, S., Zannotti, M. Increase of sister chromatid exchange and unscheduled synthesis of deoxyribonucleic acid by acrylonitrile in human lymphocytes in vitro. *Scand. Work Environ. Health.* 8(4): 290–93, 1982.

Shapiro, S., Slone, D., Heinonen, O., Kaufman, D., Rosenberg, L., et al. Birth defects and vaginal spermicides. *J.A.M.A.* 247(17): 2381–84, 1982.

Singh, A. R., Lawrence, W. H., Autian, J. Embryonic-fetal toxicity and teratogenic effects of a group of methacrylate esters in rats. *J. Dental Res.* 51(6): 1632–38, 1982.

Yager, J. W., Benz, R. D. Sister chromatoid exchanges induced in rabbit lymphocytes by ethylene oxide after inhalation exposure. *Environ. Mutagen.* 4: 121–134, 1982.

Anon. A vaginal contraceptive sponge. *Med. Lett. Drugs Therap.* 25(642): 78–80, 1983.

Hardin, B., Niemeier, R., Sikov, M., Hackett, P. Reproductive-toxicologic assessment of the epoxides etheylene oxide, proplyene oxide, butylene oxide, and styrene oxide. *Scand. J. Work Environ. Health*, 9: 94–102, 1983.

Yager, J. W., Hines, C. J., Spear, R. C. Exposure to ethylene oxide at work increases sister chromatoid exchanges in peripheral lymphocytes. *Science* 219: 1221–23, 1983.

Hogstedt, B., Gullberg, B., Hedner, K., Kolnig, A., Mitelman, F. Chromosomal aberrations and micronuclei in bone marrow cells and peripheral blood lymphocytes in humans exposed to ethylene oxide. *Hereditas.* 98: 105–13, 1983.

Korhonen, A., Hemminki, K., Vainio, H. Embryotoxic effects of acolein, methacrylates, guanidines and resorcinol on three day chicken embryos. *Acta Pharmacol. Toxicol.* 52(2): 95–9, 1983.

National Institute of Environmental Health Sciences (NIEHS) Teratologic evaluation of ethylene oxide (CAS No. 75-21-8) in New Zealand white rabbits. NIEHS Contract No. PR 259231, Apr. 1983.

Occupational Safety and Health Administration Occupational exposure to ethylene oxide: proposed rule. Federal Register 48: 17284–319 Apr. 21, 1983.

Rafales, L. S., Bornschein, R. L., Caruso, Y. Behavioral and pharmacological responses following acrylamide exposure in rats Neurobehav. Toxicol. Teratol. 4: 355–64, 1983.

Snellings, W. M., Maronpot, R. R., Zelanek, J. P., Lafoon, C. P. Teratology study in Fischer 344 rats exposed to ethylene oxide by inhalation. Toxicol. Appl. Pharmacol. 64: 476–81, 1983.

Strong, C. Defective infants and their impact on families: ethical and legal considerations. Law Med. Health Care, 168–81, 1983.

Yager, J. W., Hines, C. J., Spear, R. C. Exposure to ethylene oxide at work increases sister chromatoid exchanges in peripheral lymphocytes. Science 219: 1221–23, 1983.

Yunis, J. J. The chromosomal basis of human neoplasia Science 221: 227–35, 1983.

Devereux, M. E. Equal employment opportunity under Title VII and the exclusion of fertile women from the toxic workplace. Law, Med. Health Care. 164–72, Sept. 1984.

Hemminki, K., Lindbohm, M. L., Hemminki, T., Vainio, H. Reproductive hazards and plastics industry. Prog. Clin. Biol. Res. 141: 79–87, 1984.

Hoffman, D. J., Albers, P. H. Evaluation of potential embryotoxicity and teratogenicity of 42 herbicides, insecticides and petroleum contaminants to mallard eggs. Arch. Environ. Contam. Toxicol. 13(1): 15–28, 1984.

Lambotte-Vandepaer, M., Duverger-Van Bogaert, M. Genotoxic properties of acrylonitrile. Mutat. Res. 134(1): 49–59, 1984.

Landrigan, P., Meinhardt, T., Gordon, J., Lipscomb, J., Burg, J. R. Ethylene oxide: A review of toxicologic and epidemiologic research. Amer. J. Indust. Med. 6: 103–15, 1984.

Sarto, F. Toxicity of ethylene oxide and its importance for man. Med. Lav. 75(4): 254–63, 1984.

Norwood, C. Terata. Mother Jones. 15–21, Jan. 1985.

Dixon, R. L. Toxic responses of the reproductive system. In: Casarett and Doull's Toxicology. 3rd Edition. Eds: Klaassen, C. D., Amdur, M. O., Doull, J. Macmillan Publishing Co., New York. Chapter 16, 1986.

Tabacova, S. Maternal exposure to environmental chemicals. Neurotoxicology 7(2): 421–40, 1986.

Yasuda, Y., Konishi, H., Matuso, T., Tanimura, T. Accelerated differentiation of seminiferous tubules of fetal mice prenatally exposed to ethinyl estradiol. Anat. Embryol. 174: 289–99, 1986.

State of California, Health and Welfare Agency. Safe Drinking Water and Toxic Enforcement Act of 1986. (Proposition 65). Chemicals known to cause cancer or reproductive toxicity. Sacramento, CA, Jan. 1, 1988.

Chromium

Hueper, W. C. Occupational Tumors and Allied Diseases. Charles C. Thomas, Springfield, Ill., 1942.

Gross, E., Kolsch, F. Lung cancer in the chromate dye industry. Arch. Gewerbepathol. Gewerbehyg. 12: 164–70, 1943.

Machle, W., Gregorius, F. Cancer of the respiratory system in the United States chrome-producing industry. U.S. Public Health Bull. 63: 1114–27, 1948.

Baetjer, A. M. Pulmonary carcinoma in chromate workers; a review of the literature and report of cases. *Arch. Ind. Hyg. Occup. Med.* 2: 487–504, 1950.

Baetjer, A. M. Pulmonary carcinoma in chromate workers, II. Incidence on basis of hospital records. *Arch. Ind. Hyg. Occup. Med.* 2: 505–16, 1950.

Mancuso, T. F., Hueper, W. C. Occupational cancer and other health hazards in a chromate plant: a medical appraisal, I. Lung cancer in chromate workers. *Ind. Med. Surg.* 20: 358–63, 1951.

Mancuso, T. F. Occupational cancer and other health hazards in a chromate plant, medical approach, clinical and toxicological aspects. *Ind. Med. Surg.* 20(9): 393–407, 1951.

Gafawer, W. M. *Health of Workers in Chromate Producing Industry.* Pub. Health Service Bulletin #192, Washington, D.C., 1953.

Hueper, W. C. Experimental studies in experimental carcinogenesis: VII. Tissue reactions to parenterally introduced powdered metallic chromium and chromite ore. *J. Nat. Cancer Inst.* 16: 447–69, 1955.

Bidstrup, P. L., Case, R. A. M. Carcinoma of the lung in workmen in the bichromates-producing industry in Great Britain. *Brit. J. Ind. Med.* L2: 260–64, 1956.

Hueper, W. C. Experimental studies in metal carcinogenesis. (10) Carcinogenic effects of chromite ore roast deposited in muscle, tissue and pleural cavity of rats. *Arch. Ind. Health* 18: 284, 1958.

Hueper, W. C., Payne, W. M. Experimental cancers in rats produced by chromium compounds and their significance to industry and public health. *Amer. Ind. Hyg. Assoc. J.* 20: 274–80, 1959.

Baetjer, A. M., Damron, C., Budacz, V. The distribution and retention of chromium in men and animals. *Arch. Ind. Health* 20: 136–50, 1959.

Payne, W. W. The role of roasted chromite ore in the production of cancer. *Arch. Environ. Health* 1: 20–26, 1960.

Payne, W. W. Production of cancers in mice and rats by chromium compounds. *Arch. Ind. Health* 21: 530–35, 1960.

Hueper, W. C., Payne, W. W. Experimental studies in metal carcinogenesis. *Arch. Environ. Health* 5: 445–62, 1962.

Hanslian, L., Kadlec, K., Barborik, M. *Chrom. Prac. Lek.* 15: 7–10, 1963.

Roe, F. F. C., Carter, R. L. Chromium carcinogenesis: calcium chromate as a potent carcinogen from the subcutaneous tissues of the rat. *Brit. J. Cancer* 23: 172–76, 1969.

Williams, C. D. Asthma related to chromium compounds. *N.C. Med. J.* 30: 482–90, 1969.

Barborik, M. The problem of harmful exposure to chromium compounds. *Ind. Med.* 39: 45–48, 1970.

Enterline, P. E. Respiratory cancer among chromate workers. *J. Occup. Med.* 16(8): 523–26, 1974.

Waterhouse, J. A. H. Cancer among chromium platers. *Brit. J. Cancer* 32: 262, 1975.

Young, M. A. Health hazards of electroplating. *J. Occup. Med.* 7: 348–52, 1975.

Mancuso, T. F. Consideration of chromium as an industrial carcinogen. Int. Conf. on Heavy Metals in the Environment, Toronto, Ontario, Canada, 1975.

Markel, H. L., Jr., Lucas, J. B. Health hazard evaluation report 72-106-44, No. American Rockwell Automotive Division, Grenada, Miss. Hazard Evaluation Service Branch, NIOSH, Cincinnati, Ohio, Report No. 72-106-44, 1975, 13 pp.

Langard, S., Norseth, T. A cohort study of bronchial carcinomas in workers producing chromate pigments. *Brit. J. Ind. Med.* 32: 62–65, 1975.

Dankman, H. S. Industrial exposure to chromates in New York State. *Arch. Ind. Hyg. Occup. Med.* 5: 228-31, 1975.

Royle, H. Toxicity of chromic acid in the chromium plating industry. *Environ. Res.* 10(1): 39-53, 1975.

Raffetto, G., Parodi, S., De Ferrari, M., Troiano, R., Brambilla, G. Direct interaction with cellular targets as the mechanism for chromium-carcinogenesis. *Tumori* 63(6): 503-12, 1977.

Davies, J. M. Lung cancer mortality of workers making chrome pigments. *Lancet* 1: 384, 1978.

Ohsaki, Y, Abe, S., Kimura, K., Tsuneta, Y., Mikami, H., Murao, M. Lung cancer in Japanese chromate workers. *Thorax* 33: 372-74, 1978.

Heuper, W. C. Some comments on the history and experimental explorations of metal carcinogens and cancers. *J. Nat. Cancer Inst.* 62(4): 723-25, 1979.

Dalager, N. A., Mason, T. J., Fraumeni, J. J., Jr., Hoover, R., Payne, W. W. Cancer mortality among workers exposed to zinc chromate paints. *J. Occup. Med.* 22: 25-29, 1980.

IARC. Chromium and chromium compounds. *Monographs in the Evaluation of the Carcinogenic Risk of Chemicals to Humans* 23: 205-323, 1980.

Norseth, T. Asbestos and metals as carcinogens. *J. Toxicol. Environ. Health* 6(5-6): 1021-28, 1980.

Norseth, T. Cancer hazards caused by nickel and chromium exposure. *J. Toxicol. Environ. Health* 6(5-6): 1219-27, 1980.

Tsuneta, Y., Ohsaki, Y., Kimura, K., Mikami, H., Abe, S., Murao, M. Chromium content of lungs of chromate workers with lung cancer. *Thorax* 35(4): 294-97, 1980.

Vainio, H., Sorsa, M. Chromosome aberrations and their relevance to metal carcinogenesis. *Environ. Health Perspect.* 40: 173-80, 1981.

Bencko, V. Chromium: a review of environmental and occupational toxicology. *J. Hyg. Epidemiol. Microbiol. Immunol.* 29(1): 37-46, 1985.

Kim, S., Iwai, Y., Fujino, M., Furumoto, M., Sumino, K., Miyasaki, K. Chromium-induced pulmonary cancer. Report of a case and a review of the literature. *Acta Pathol. Jpn.* 35(3): 643-54, 1985.

Glaser, U., Hochrainer, D., Kloppel, H., Oldiges, H. Carcinogenicity of sodium dichromate and chromium (VI/III) oxide aerosols inhaled by male Wistar rats. *Toxicology* 42(2-3): 219-32, 1986.

Hamilton, J. W., Wetterhahn, K. E. Chromium (VI)-induced DNA damage in chick embryo liver and blood cells in vivo. *Carcinogenesis* 7(12): 2085-88, 1986.

Hertel, R. F. Sources of exposure and biological effects of chromium. *IARC Sci. Publ.* 71:63-77, 1986.

Kawanishi, S., Inoue, S., Sano, S. Mechanism of DNA cleavage induced by sodium chromate (VI) in the presence of hydrogen peroxide. *J. Biol. Chem.* 262(13): 5952-58, 1986.

Levy, L. S., Martin, P. A., Bidstrup, P. L. Investigation of the potential carcinogenicity of a range of chromium containing materials on the rat lung. *Brit. J. Ind. Med.* 43(4): 243-56, 1986.

Levy, L. S., Venitt, S. Carcinogenicity and mutagenicity of chromium compounds: the association between bronchial metaplasia and neoplasia. *Carcinogenesis* 7(5): 831-35, 1986.

Newton, M. F., Lilly, L. J. Tissue specific clastogenic effects of chromium and selenium salts in vivo. *Mut. Res.* 169(1-2): 61-69, 1986.

Reichrtovia, E., Takiaio, L., Kranerovia, J., Bencko, V., Sulicovia, L., Holuisa, R. Biomonitoring of environmental pollution hazards from a nickel smelter waste dump. *J. Hyg. Epidemiol. Microbiol. Immunol.* 30(4): 359–64, 1986.

Sen, P., Costa, M. Incidence and localization of sister chromatid exchanges induced by nickel and chromium compounds. *Carcinogenesis* 7(9): 1527–33, 1986.

Sky-Peck, H. H. Trace metals and neoplasia. *Clin. Physiol. Biochem.* 4(1): 99–111, 1986.

van der Wal, J. F. Further studies on the exposure of welders to fumes. Chromium, nickel and gases in Dutch industries; plasma welding and cutting stainless steel. *Ann. Occup. Hyg.* 30(2): 153–61, 1986.

Friedman, J., Shabtai, F., Levy, L. S., Djaldetti, M. Chromium chloride induces chromosomal aberrations in human lymphocytes via indirect action. *Mut. Res.* 191(3–4): 207–10, 1987.

Kishi, R., Tarumi, T., Uchino, E., Miyake, H. Chromium content of organs of chromate workers with lung cancer. *Amer. J. Ind. Med.* 11(1): 67–74, 1987.

Sjorgren, B., Gustavsson, A., Hedstrom, L. Mortality in two cohorts of welders exposed to high and low levels of hexavalent chromium. *Scand. J. Work Environ. Health* 13(3): 247–51, 1987.

Epichlorhydrin

Koss. Veber das epichlorhydrin. *Pharmacodyn.* 4: 351–359, 1898.

Freuder, E., Leake, C. D. The toxicology of epichlorhydrin. *University of California Pub. Pharmacol.* 2: 69–78, 1941.

Carpenter, C. P., Smith, H. F., Jr., Pozzani, U. C. The assay of acute vapor toxicity and the grading and interpretation of results on 96 chemical compounds. *J. Ind. Hyg. Toxicol.* 31: 343–46, 1949.

Kingsley, W. H. Health hazards and control of an epoxy resin operation. *Amer. Ind. Hyg. Assoc. J.* 19: 258–60, 1958.

Gage, J. G. The toxicity of epichlorhydrin vapor. *Brit. J. Ind. Med.* 16: 11–14, 1959.

Kremneva, S. N. Investigation of toxicity of epichlorhydrin and dichlorhydrin *Chem. Abstracts* 57: 6267c, 1962 (orig. pub. in Russian, 1960).

Kotin, P., Falk, H. L. Organic peroxides, hydrogen peroxide, epoxides and neoplasia. *Radiat. Res.* (Suppl. 3): 193–211, 1963.

Soloimskaya, E. A. Biochemical changes occurring in acute epichlorhydrin poisoning. *Chem. Abstracts* 67: 80777e, 1967 (orig. pub. in Russian, 1967).

Fedyania, V. N. Sanitary toxicological investigations of epichlorhydrin as a basis for the establishment of its standard concentration in bodies of water. *Hyg. Sanitat.* 33: 52–57, 1968.

Fomin, A. P. Effect of small concentrations of epichlorhydrin vapors on animals. *Chem. Abstracts* 70: 108963A, 1969 (orig. pub. in Russian, 1968).

Hahn, J. D. Post-testicular antifertility effects of epichlorhydrin and 2,3 epoxypropanol. *Nature* 226: 87, 1970.

Anon. Epichlorhydrin. Toxicology Studies, Union Carbide Corp., New York, 1971.

Van Duuren, B. L., Katz, C., Goldschmidt, B. M., Direct-acting alkylating carcinogens. Chloro ethers and related compounds. #65 Epichlorhydrin. *Toxicol. Appl. Pharmacol.* 22: 279–80, 1972.

Lawrence, W. H., Malik, M., Turner, J. E., Autian, J. Toxicity profile of epichlorhydrin. *J. Pharm. Sci.* 61: 1712–17, 1972.

Van Duuren, B. L., Goldschmidt, B. M., Katz, C., Seidman, I., Paul, J. S. Carcinogenic activity of alkylating agents. *J. Nat. Cancer Inst.* 53: 695–700, 1974.

Cooper, E. R. A., Jones, A. R., Jackson, H. Effects of α-chlorhydrin and related compounds on the reproductive organs and fertility of the male rat. *J. Reprod. Fertil.* 38: 379–86, 1974.

Enterline, P. E. Updated mortality in workers exposed to epichlorhydrin. Submission to OSHA Docket #H-100, File #4, 1978.

Picciano, D. Cytogenetic investigation of occupational exposure to epichlorhydrin. *Mut. Res.* 66: 169–73, 1979.

White, A. D. In vitro induction of SCE in human lymphocytes by epichlorhydrin with and without metabolic activation. *Mut. Res.* 78: 171–76, 1980.

Konishi, Y., Kawabata, Denda, A., Ikeda, T. Forestomach tumors induced by orally administered epichlorhydrin in male wistar rats. *Gann* 71: 922–23, 1980.

Apfeldorf, R., Infante, P. F. Review of epidemiologic study results of vinyl chloride-related compounds. *Environ. Health Perspect.* 41: 221–26, Oct. 1981.

Chemical Hygiene Fellowship. *Epichlorhydrin.* Special Report 33-41, Mellon Institute, Carnegie-Mellon University, Jan. 28, 1983.

Perocco, P., Rocchi, P., Ferreri, A. M., Capucci, A. Toxic. DNA-damaging and mutagenic activity of epichlorohydrin on human cells cultured in vitro. *Tumori* 69(3): 191–94, 1983.

Sriam, R. J., Landa, L., Samkovia, I. Effect of occupational exposure to epichlorohydrin on the frequency of chromosome aberrations in peripheral lymphocytes. *Mut. Res.* 122(1): 59–64, 1983.

Henschler, D., Elaiasser, H., Romen, W., Eder, E. Carcinogenicity study of trichloroethylene, with and without epoxide stabilizers, in mice. *J. Cancer Res. Clin. Oncol.* 107(3): 149–56, 1984.

Wester, P. W., van der Heijden, C. A., Bisschop, A., van Esch, G. J. Carcinogenicity study with epichlorohydrin by gavage in rats. *Toxicology* 36(4): 325–39, 1985.

Prens, E. P., de Jong, G., van Joost, T. Sensitization to epichlorohydrin and epoxy system components. *Contact Dermatitis* 15(2): 85–90, 1986.

Foundry Industry and Cancer

Turner, H. M., Grace, H. G. An investigation into cancer mortality among males in certain Sheffield trades. *J. Hyg.* 38: 90–103, 1938.

McLaughlin, A. I. G., Harding, H. E. Pneumoconiosis and other causes of death in iron and steel foundry workers. *Arch. Ind. Health* 14: 350–78, 1956.

Tubich, G. E. New materials and processes create new liabilities for the foundry. *Ind. Med. Surg.* 79–85, 1964.

Boddey, R. F. The use of chemicals in the modern foundry. *Ann. Occup. Hyg.* 10: 231–39, 1967.

Tanimura, H. Benzo (α) pyrene in an iron and steel works. *Arch. Environ. Health* 17: 172–77, 1968.

Anon. Health risks in the foundry. *Mich. Occup. Health* 12: 2–5, 1968.

Bingham, E., Falk, H. L. Environmental carcinogens. *Arch. Environ. Health* 19: 779, 1969.

Garcia, W. H., Feldman, R., Lijanski, W., Shubik, P. Tumorigenesis in mice by petroleum asphalts and coal-tar pitches of known polynuclear aromatic hydrocarbon content. *Toxicol. Appl. Pharmacol.* 18: 41–42, 1971.

Eckartt, R. E. Recent developments in industrial carcinogen. *J. Occup. Med.* 15: 904–7, 1973.

Plamenac, P., Nikulin, A., Pikula, B. Cytologic changes in the respiratory epithelium in iron foundry workers. *Acta Cytol.* 18: 34–40, 1973.

Bigelow-Sherman, J., Wolfe, S., Hricko, A., Mets, M. A health research group study on disease among workers in the auto industry. Public Citizen, Inc., 1973.

Orr, R. Foundry men urge twenty-five year retirements, *Detroit Free Press*, Feb. 9, 1973.

Bates, C., Scheel, L. Processing emissions and occupational health in the ferrous foundry industry. *Amer. Ind. Hyg. Assoc. J.* 452–62, Aug. 1974.

Euokarava, R. Mortality, disability and turnover among foundry workers. *Duodecim* 90: 311–14, 1974.

Mason, T. J., McKay, F. W., Hoover, R., Blot, W. J., Fraumeni, J. F. *Atlas of Cancer Mortality for U.S. Counties: 1950–1969*. U.S. Department of Health, Education and Welfare Pub. #NIH-75-780. 1975.

Bierbaum, P. J. Personal correspondence with author, Jan. 29, 1975.

Clark, M., Lord, M. Cancer's hot spots, *Newsweek*, July 7, 1975.

Anon. *Cancer Questions and Answers about Rates and Risks*. U.S. Department of Health, Education and Welfare, Public Health Service, National Institutes of Health, Pub. #NIH-76-1040, 1975.

Fallentien, B., Wilhardt, P. Occurrence of, measures against, and control with new, dangerous substances and foundries. IMF Mechanical Engineering and Foundry Conference, Vienna, May, 1975.

Mastromatteo, M. Air pollution in foundries. IMF Mechanical Engineering and Foundry Conference, Vienna, May, 1975.

McCann, J., Choi, E., Yamasaki, E., Ames, B. N. Detection of carcinogens as mutagens in the salmonella microsome test: assay of 300 chemicals. *Proc. Nat. Acad. Sci.* 72: 5135–39, 1975.

Taylor, D. M., Davies, J. C. A. Six cases of ferro-alloy workers' disease. *Cent. Afr. J. Med.* 4: 67–71, 1975.

Henderson, B. E., Louie, E., Jing, J. S., Buell, P., Gardiner, M. B. Risk factors associated with nasal pharyngeal carcinoma. *New Engl. J. Med.* 295: 1101–6, 1976.

Maltoni, C. Occupational chemical carcinogenesis: new facts, priorities and perspectives. *Inserm* (IARC Scientific Publications) 52: 127–50, 1976.

Riitta-Sisko, K., Hernberg, S., Karava, R., Jarvinen, E., Nurminen, M. A mortality study of foundry workers. *Scand. J. Work Environ. Health* 1: 73–89, 1976.

West, C. Working steel. *Labor Occup. Health Prog. Mon.* 3: 1–2, May 1976.

Usery, W. J. (Secretary of Labor). *Job Safety and Health*. U.S. Department of Labor, Occupational Safety and Health Administration, Apr. 1976.

Tossavainen, A. Metal fumes in foundries. *Scand. J. Work Environ. Health* 1: 42–49, 1976.

Virtamo, M., Tossavainen, A. Gases formed from furan binding agents. *Scand. J. Work Environ. Health* 2: 50–53, 1976.

Doll, R. Strategy for detection of cancer hazards to man. *Nature* 265: 589–96, 1977.

Egan, B., Waxweiler, R. J., Wolfe, J., Blade, L., Wagoner, J. K. A preliminary report of mortality patterns among foundry workers. Society for Occupational and Environmental Health Meeting, Washington, D.C., Dec. 5–7, 1977.

Scott, W. D., Bates, C. E., James, R. H. Chemical emissions from foundry molds. *A.F.S. Trans.* 203–8, 1977.

Train. R. E. Environmental Cancer. *Science* 195: editorial page, Feb. 4, 1977.

U.S. Department of Health, Education and Welfare, NIOSH. Melting and pouring department: health hazards in a foundry. 1977.

U.S. Department of Health, Education and Welfare, NIOSH. Shakeout, cleaning, grinding and inspections departments: health hazards in a foundry. 1977.

U.S. Department of Health, Education and Welfare, NIOSH. Pattern shop, core room, molding shop and sand handling department: health hazards in a foundry. 1977.

Bridbord, K., Decoufle, P., Fraumeni, J. F., Hoel, D. G., Hoover, R. N., Rall, D. P., Saffiotti, U., Schneiderman, M. A., Upton, A. C. *Estimates of the Fraction of Cancer in the United States Related to Occupational Factors.* National Cancer Institute, National Institute of Environmental Health Sciences, National Institute for Occupational Safety and Health, Washington, D.C., Sept. 15, 1978.

Lloyd, D. L. Respiratory cancer clustering associated with localised industrial air pollution. *Lancet* 1(8059): 318–20, 1978.

Metropolitan Life Insurance Company. Mortality from cancer by socio-economic status, *Statistical Bulletin*, pp. 2–5, Jan.–Mar. 1978.

Decoufle, P., Wood, D. J. Mortality patterns among workers in a gray iron foundry. *Amer. J. Epidemiol.* 109(6): 667–75, 1979.

Egan, B., Waxweiler, R. J., Blade, L., Wolfe, J., Wagoner, J. K. A preliminary report of mortality patterns among foundry workers. *J. Environ. Pathol. Toxicol.* 2(5): 259–72, 1979.

Tola, S., Koskela, R. S., Hernberg, S., Jiarvinen, E. Lung cancer mortality among iron foundry workers. *J. Occup. Med.* 21(11): 753–59, 1979.

Tola, S. Epidemiology of lung cancer in foundries. *J. Toxicol. Environ. Health* 6(5–6): 1195–1200, 1980.

Palmer, W. G., Scott, W. D. Lung cancer in ferrous foundry workers: a review. *Amer. Ind. Hyg. Assoc. J.* 42(5): 329–40, 1981.

Goldsmith, D. F., Guidotti, T. L., Johnston, D. R. Does occupational exposure to silica cause lung cancer? *Amer. J. Ind. Med.* 3(4): 423–40, 1982.

Blot, W. J., Brown, L. M., Pottern, L. M., Stone, B. J., Fraumeni, J. F., Jr. Lung cancer among long-term steel workers. *Amer. J. Epidemiol.* 117(6): 706–16, 1983.

Gibson, E. S., McCalla, D. R., Kaiser-Farrell, C., Kerr, A. A., Lockington, J. N., Hertzman, C., Rosenfeld, J. M. Lung cancer in a steel foundry: a search for causation. *J. Occup. Med.* 25(8): 573–78, 1983.

Fletcher, A. C., Ades, A. Lung cancer mortality in a cohort of English foundry workers. *Scand. J. Work Environ. Health* 10(1): 7–16, 1984.

Lloyd, O. L., Smith, G., Lloyd, M. M., Holland, Y., Gailey, F. Raised mortality from lung cancer and high sex ratios of births associated with industrial pollution. *Brit. J. Ind. Med.* 42(7): 475–80, 1985.

Lloyd, O. L., Williams, F. L., Gailey, F. A. Is the Armadale epidemic over? Air pollution and mortality from lung cancer and other diseases, 1961–82. *Brit. J. Ind. Med.* 42(12): 815–23, 1985.

Vena, J. E., Sultz, H. A., Fiedler, R. D., Barnes, R. E. Mortality of workers in an automobile engine and parts manufacturing complex. *Brit. J. Ind. Med.* 42(2): 85–93, 1985.

Silverstein, M., Maizlish, N., Park, R., Silverstein, B., Brodsky, L., Mirer, F. Mortality among ferrous foundry workers. *Am. J. Ind. Med.* 10(1): 27–43, 1986.

Westerholm, P., Ahlmark, A., Maasing, R., Segelberg, I. Silicosis and risk of lung cancer or lung tuberculosis: a cohort study. *Environ. Res.* 41(1): 339–50, 1986.

Phenylbutazone, Oxyphenylbutazone, and Blood Dyscrasias

Bean, R. H. D. Phenylbutazone and leukemia. *Brit. Med. J.* 2: 1552–55, 1950.

Steinberg, C. L., Bohrod, M. G., Roodenburg, A. I. Agranulocytosis following phenylbutazone (Butazolidin) therapy. *J.A.M.A.* 152: 33–36, 1953.

Engleman, E. P., Krupp, M. A., Rinehart, J. F., Jones, R. C., Gibson, J. R. Hepatitis following the ingestion of phenylbutazone. *J.A.M.A.* 156: 98–101, 1954.

O'Brien, D. J., Storey, G. Death from hypersensitivity to phenylbutazone. *Brit. Med. J.* 1: 797–94, 1954.

Burns, J. J., Rose, R. K., Goodwin, D., Reichenthal, J., Horning, E. C., Brodie, B. B. The metabolic fate of phenylbutazone (Butazolidin) in man. *J. Pharm. Exp. Ther.* 113: 481–89, 1955.

MacCarthy, J. M. Hepatic necrosis and other visceral lesions associated with phenylbutazone therapy. *Brit. Med. J.* 2: 240–42, 1955.

Mauer, E. F. The toxic effects of phenylbutazone (Butazolidin) *New Engl. J. Med.* 253(10): 404–10, 1955.

Girdwood, R. H., Lenman, J. A. R. Megaloblastic anemia occurring during primidone therapy. *Brit. Med. J.* 1: 146–47, 1956.

Hodge, P. R., Lawrence, J. R. Two cases of myocarditis associated with phenylbutazone therapy. *Med. J. Australia* 1: 640–41, 1957.

Robson, H. N., Lawrence, J. R. Megaloblastic anemia induced by phenylbutazone. *Brit. Med. J.* 475–77, 1959.

Scheuer-Karpin, R. Myeloblastic leukemia after analgesic drugs. *Hematologicia Polonica* 3: 33–45, 1959.

Burns, J. J., Yu, T. F., Dayton, P. G., Gutman, A. B., Brodie, B. B. Biochemical pharmacological considerations of phenylbutazone and its analogues. *Ann. N.Y. Acad. Sci.* 86: 253–62, 1960.

Cast, I. P. Phenylbutazone and leukemia. *Brit. Med. J.* 2: 1569–70, 1961.

Clinicopathologic Conference. Agranulocytosis and anuria during phenylbutazone therapy for arthritis. *Amer. J. Med.* 30: 268–80, 1961.

Garrett, J. V. Phenylbutazone and leukemia. *Brit. Med. J.* 1: 53, 1961.

Cadman, E. F. B., Linmont, W. Phenylbutazone and leukemia. *Brit. Med. J.* 1: 798, 1962.

Best, W. R. Drug-associated blood dyscrasias. *J.A.M.A.* 185: 140–44, 1963.

Randall, L. O. Non-narcotic analgesics, in *Physiological Pharmacology*, Vol. 1: *The Nervous System*, Part A: Central Nervous System Drugs, Root, W. S., Hoffman, F. G., Eds. Academic Press, New York, 1963, pp. 313–416.

Chalmers, T. M., McCarthy, D. D. Phenylbutazone therapy associated with leukemia. *Brit. Med. J.* 1: 747, 1964.

Chatterjea, J. B. Leukemia and phenylbutazone. *Brit. Med. J.* 2: 875, 1964.

Hart, G. D. Acute leukemia following phenylbutazone therapy. *Canad. Med. Assoc. J.* 91: 449–50, 1964.

Sen, S., Siddique, K. K. H. Phenylbutazone and leukemia. *Bull. Inst. Postgrad. Med. Educ. Res.* 6: 23–24, 1964.

Thorpe, G. J. Leukemia and phenylbutazone. *Brit. Med. J.* 1: 1707, June 1964.

Woodliff, H. J., Dougan, L. Acute leukemia associated with phenylbutazone treatment. *Brit. Med. J.* 1: 744–46, 1964.

Dameshek, W. Sideroblastic anemia: is this a malignancy? *Brit. J. Haematol.* 11: 52–58, 1965.

DeGowin, R. L. Preleukemic phase of acute and chronic myelocytic leukemia. *Clin. Med.* 72: 1135–43, 1965.

Dougan, L., Woodliff, H. J. Acute leukemia associated with phenylbutazone treatment. A review of the literature and report of a further case. *Med. J. Aust.* 1: 217–19, 1965.

Golding, J. R., Hamilton, M. G., Moody, H. E. Monocytic leukemia and phenylbutazone. *Brit. Med. J.* 1: 1673, 1965.

Jensen, M. K., Roll, K. Phenylbutazone and leukemia. *Acta Med. Scand.* 178: 505–13, 1965.

Lewis, S. M. Course and prognosis in aplastic anemia. *Brit. Med. J.* 1: 1027, 1965.

Sjoberg, S. G., Peters, D. Leukemia after oxyphenylbutazone *Lancet* 441, Aug. 1965.

Bloom, G. E., et al. Chromosome abnormalities in constitutional aplastic anemia. *New Engl. J. Med.* 274: 8–14, Jan. 1966.

Dvorak, K., Blazkova. *Vnitr. Lek.* 11: 1000, 1965; per: *Int. Pharm. Abst.* 3: 1222, 1966.

Huguley, C. M., Jr. Heamatological reactions. *J.A.M.A.* 196: 408–10, May 1966.

Perers, D., Sjoberg, S. G. Akut leukemia, leukemoid reaktion och leukocytos after behandling med oxifenylbutazon. *Sv. Lakartidn* 63: 53–56, Jan. 1966.

Fraumeni, J. F. Bone marrow depression induced by chloramphenicol or phenylbutazone. *J.A.M.A.* 201: 828–34, 1967.

Fraumeni, J. F., Jr., Miller, R. W. Epidemiology of human leukemia: Recent observations. *J. Nat. Cancer Inst.* 38: 593–605, 1967.

Coer, P. Phenylbutazone, adjuvant in treatment of malignant hemopathy and especially Hodgkin's disease. *Therapeutique* 45: 780–85, 1969.

Leavesley, G. M., Stenhouse, N. S., Dougan, L., Woodliff, H. J. Phenylbutazone and leukemia: is there a relationship? *Med. J. Aust.* 2: 963, 1969.

Wallerstein, R. O., Condit, P. K., Kasper, C. K., Brown, J. W., Morrison, F. R. *J.A.M.A.* 208: 2045, 1969.

Yunis, A. A. Drug-induced bone marrow injury. *Adv. Intern. Med.* 15: 357, 1969.

Breull, W., Karzel, K. Action of phenylbutazone in cytostatic concentrations on cell volume, protein and DNA content of Ehrlich ascites tumor cells culture in vitro. *Arch. Int. Pharmacodyn.* 184: 317–27, 1970.

Woodbury, D. M. Analgesic-antipyretics, anti-inflammatory agents and inhibitors of uric acid syntheses, in *The Pharmacological Basis of Therapeutics*, Goodman, L. S., Gilman, A., Eds. Macmillan, New York, 1970, pp. 335–37.

Stevenson, S. L., Hill, A. G. S., Hill, H. F. H. Chromosomal studies in patients taking phenylbutazone. *Ann. Rheum. Dis.* 30: 487, 1971.

Bottiger, L. E., Westerholm, B. Drug induced blood dyscrasias in Sweden. *Brit. Med. J.* 339–42, Aug. 1973.

Davidson, C., Manohitharajah, S. M. Drug-induced antiplatelet antibodies. *Brit. Med. J.* 3: 545, 1973.

Pisciotta, A. V. Immune and toxic mechanisms in drug-induced agranulocytosis. *Semin. Hematol.* 10: 279–310, 1973.

Williams, D., Lynch, R., Cartwright, G. Drug induced aplastic anemia. *Semin. Hematol.* 10: 195, 1973.

Cuthbert, M. F. Adverse reactions to non-steroidal antirheumatic drugs. *Curr. Med. Res. Opin.* 2(9): 600–609, 1974.

Linman, J. W., Saarini, M. I. The preleukemia state. *Semin. Hematol.* 11: 93, 1974.

Meyers, F. H., Jawetz, E., Goldfieln, A. *Review of Medical Pharmacology.* Lange Medical Publications, Los Altos, Calif., 1974: 4th ed., pp. 281–83; 1976: 5th ed., pp. 285–87; 1978: 6th ed., pp. 290–92; 1980: 7th ed., pp. 289–91.

Casarett, L. J., Doull, J. *Toxicology; the Basic Science of Poisons.* Macmillan, New York, 1975: 1st ed., p. 226; 1980: 2nd ed., p. 313.

Inman, W. H. Study of fatal bone marrow depression with special reference to phenylbutazone and oxyphenylbutazone. *Brit. Med. J.* 1(6075): 1500–5, 1977.

Smith, C. S., Chinn, S., Watts, R. W. The sensitivity of human bone marrow granulocyte/monocyte precursor cells to phenylbutazone, oxyphenylbutazone and gammahydroxyphenylbutazone in vitro, with observations on the bone marrow colony formation in phenylbutazone-induced granulocytopenia. *Biochem. Pharmacol.* 26(9): 847–52, 1977.

Bottiger, L. E. Prevalence and etiology of aplastic anemia in Sweden, in *Aplastic Anemia*, Hibind, S., Ed. University Park Press, Baltimore, 1978, pp. 171–81.

Firkin, F. C., Moore, M. A. S. Atypical phenylbutazone sensitivity of marrow colony forming units in phenylbutazone-induced aplastic anemia, in *Aplastic Anemia*, Hibind, S., Ed. University Park Press, Baltimore, 1978.

Najean, T., Pecking, A., Dresch, C. A prospective analysis of 350 cases of aplastic anemia, *Aplastic Anemia*, Hibind, S. Ed. University Park Press, Baltimore, 1978, pp. 181–84.

Schiffman, F. J., Uehara, Y., Fisher, J. M., Rabinovita, M. Potentiation of chlorambucil activity by phenylbutazone. *Cancer Letters* 4: 211–16, 1978.

Sidi, Y., Douer, D., Pinkhas, J. Phenylbutazone and acute leukemia. *Rev. Roum. Med. Int.* 15: 413–15, 1978.

Timonen, T. T., Ilvonen, M. Contact with hospital, drugs and chemicals as aetiological factors in leukemia. *Lancet* 1(8060): 350–52, 1978.

Haak, H. L. Experimental drug-induced aplastic anemia. *Clin. Haematol.* 9(3): 621–39, 1980.

Heimpel, H., Heit, W. Drug-induced aplastic anemia: clinical aspects. *Clin. Haematol.* 9(3): 641–62, 1980.

Venning, G. R. Identification of adverse reactions to new drugs. II (continued): How were 19 important adverse reactions discovered and with what delays? *Brit. Med. J.* 286(6362): 365–68, 1983.

Anon. Phenylbutazone and oxyphenylbutazone: time to call a halt. *Drug Ther. Bull.* 22(2): 5–6, 1984.

Anon. Phenylbutazone, oxyphenylbutazone labeling revised. *FDA Drug Bull.* 14(3): 23–24, 1984.

Dear Doctor Letter from Ciba Pharmaceuticals, Summit, N.J., Apr. 4, 1985.

Biron, P. Withdrawal of oxyphenylbutazone: what about phenylbutazone? (editorial). *Canad. Med. Assoc. J.* 134(10): 1119–20, 1986.

Meakawa, A., Onodera, H., Tanigawa, H., Furota, K., Kanno, J., Matsuoka, C., Ogiu, T., Hayashi, Y. Long-term studies on carcinogenicity and promoting effect of phenylbutazone in DONRYU rats. *J. Nat. Cancer Inst.* 79(3): 577–84, 1987.

Ohyashiki, K., Ohyashiki, J. H., Raza, A., Preisler, H. D., Sandberg, A. A. Phenylbutazone-induced myelodysplastic syndrome with Philadelphia translocation. *Cancer Genet. Cytogenet.* 26(2): 213–16, 1987.

Phthalates

Mallette, F. S., Von Hamm, E. Studies on the toxicity and skin effects of compounds used in the rubber and plastics industries, II. Plasticizers. *Arch. Ind. Hyg. Occup. Med.* 6: 231–36, 1952.

Calley, D., Autian, J., Guess, W. L. Toxicology of a series of phthalate esters. *J. Pharm. Sci.* 55: 158–62, 1966.

Dillingham, E. O., Autian, J. Teratogenicity, mutagenicity, and cellular toxicity of phthalate esters. *Environ. Health Perspect.* 3: 81–89, 1973.

Miklov, L. E., Aldyreva, M. V., Popova, T. B., et al. Health status of workers exposed to phthalate plasticizers in the manufacture of artificial leather and films. *Environ. Health Perspect.* 3: 175–78, 1973.

Statsek, N. K. Hygienic investigations of certain esters of phthalic acid and of polyvinylchloride materials plastificated thereby. *Gig. Sanit.* 6: 25, 1974.

Timofievskaya, L. A., Aldyreva, M. V., Kazbekov, I. M. Experimental research of the influence of groups of plastic phthalate plasticizers on the organism. *Gig. Sanit.* 12: 26, 1974.

Yamaguchi, T., Kaneahima, H., Okui, T., Ogawa, H., Yamagishi, T. Studies on the effect of phthalate esters on the biological system (1). Distribution of 14-C-dibutyl phthalate in mice. *Jap. J. Hyg.* 31: 331–36, 1976.

Kawano, M. Toxicological studies on phthalate esters 2. Metabolism, accumulation and excretion of phthalate esters in rats. *Japan. J. Hyg.* 35(4): 693–701, 1980.

Shiota, K., Chou, M. J., Nishimura, H. Embryotoxic effects of di-2-ethylhexyl phthalate (DEHP) and di-n-butyl phthalate (DBP) in mice. *Environ. Res.* 22: 245–53, 1980.

Kasuya, M. Toxicity of butylbenzyl phthalate (BBP) and other phthalate esters to nervous tissue in culture. *Toxicol. Letters* 6: 373–78, 1980.

Shukla, P. R., Kamal, R. Verma, N. Screening of metal complexes for biological activity; Part II Pharm. investigations on metal complexes capable of affecting central nervous system. *Indian J. Hosp. Pharm.* 17(1): 14–16, 1980.

Bell, F. P. Effects of phthalate esters on lipid metabolism in various tissues, cells and organelles in mammals. *Environ. Health Perspect.* 45: 41–50, 1982.

Kluwe, W. M. Overview of phthalate ester pharmacokinetics in mammalian species. *Environ. Health Perspect.* 45: 3–10, 1982.

Parkhie, M. R., Webb, M., Norcross, M. A. Dimethoxyethyl phthalate: embryopathy, teratogenicity, fetal metabolism and the role of zinc in the rat. *Environ. Health Perspect.* 45: 87–97, 1982.

Shiota, K., Nishimura, H. Teratogenicity of di(2-ethylhexyl) phthalate (DEHP) and di-n-butyl phthalate (DBP) in mice. *Environ. Health Perspect.* 45: 65–70, 1982.

Donner, M., Hytonen, S., Sorsa, M. Drosophila in a rubber factory. *Mut. Res.* 113: 247–48, 1983.

Turbin, E. V., Aldyreva, M. V., Milkov, L. E. Changes in various indices of the nervous system in workers exposed to phthalate plasticizers. *Gig. Tr. Prof. Zabol (USSR)* (4): 46–48, Apr. 1983.

Moller, D. R., Gallagher, J. S., Bernstein, D. I., Wilcox, T. G., Burroughs, H. E., Bernstein, I. L. Detection of IgE-mediated respiratory sensitization in workers exposed to hexahydrophthalic anhydride. *J. Allergy Clin. Immunol.* 75(6): 663–72, 1985.

Patterson, R., Harris, K. E. Responses of human airways to inhaled chemicals. *New Engl. Reg. Allergy Proc.* 6(3): 238–40, 1985.

Ritter, E. J., Scott, W. J., Randall, J. L., Ritter, J. M. Teratogenicity of dimethoxyethyl phthalate and its metabolites methoxyethanol and methoxyacetic acid in the rat. *Teratology* 32(1): 25–31, 1985.

Sjoberg, P. O., Bondesson, U. G., Sedin, E. G., Gustafsson, J. P. Exposure of newborn infants to plasticizers. Plasma levels of di-(2-ethylhexyl) phthalate and mono-(2-ethylhexyl) phthalate during exchange transfusion. *Transfusion* 25(5): 424–28, 1985.

Smith, J. J. Chemical hazard information profile: draft report (dimethoxyethyl phthalate). Chemical Effects Information Task Group, Oak Ridge National Laboratory, Aug. 21, 1985.

Gray, T. J. Testicular toxicity in vitro: Sertoli-germ cell co-cultures as a model system. *Food Chem. Toxicol.* 24(6–7): 601–5, 1986.

Gray, T. J., Gangolli, S. D. Aspects of the testicular toxicity of phthalate esters. *Environ. Health Perspect.* 65: 229–35, 1986.

Hinton, R. H., Mitchell, F. E., Mann, A., Chescoe, D., Price, S. C., Nunn, A., Grasso, P., Bridges, J. W. Effects of phthalic acid esters on the liver and thyroid. *Environ. Health Perspect.* 70: 195–210, 1986.

Kluwe, W. M. Carcinogenic potential of phthalic acid esters and related compounds: structure–activity relationships. *Environ. Health Perspect.* 65: 271–78, 1986.

Komitowski, D., Schmezer, P., Schmitt, B., Ehemann, V., Muto, S. Quantitative analysis of the early changes of hepatocyte nuclei after treating Syrian golden hamsters with di(2-ethylhexyl) phthalate. *J. Cancer Res. Clin. Oncol.* 111(2): 103–7, 1986.

Lake, B. G., Gray, T. J., Gangolli, S. D. Hepatic effects of phthalate esters and related compounds—in vivo and in vitro correlations. *Environ. Health Perspect.* 67: 283–90, 1986.

Mitchell, F. E., Bridges, J. W., Hinton, R. H. Effects of mono (2-ethylhexyl) phthalate and its straight chain analogues mono-*n*-hexylphthalate and mono-*n*-octyl phthalate on lipid metabolism in isolated hepatocytes. *Biochem. Pharmacol.* 35(17): 2941–47, 1986.

Oishi, S. Testicular atrophy induced by di(2-ethylhexyl) phthalate: changes in histology, cell specific enzyme activities and zinc concentrations in rat testis. *Arch. Toxicol.* 59(4): 290–95, 1986.

Pfaffli, P. Phthalic acid excretion as an indicator of exposure to phthalic anhydride in the work atmosphere. *Int. Arch. Occup. Environ. Health* 58(3): 209–16, 1986.

Sjoberg, P., Lindquist, N. G., Ploen, L. Age-dependent response of the rat testes to di(2-ethylhexyl) phthalate. *Environ. Health Perspect.* 65: 237–42, 1986.

Tomaszewski, K. E., Agarwal, D. K., Melnick, R. L. In vitro steady-state levels of hydrogen peroxide after exposure of male F344 rats and female B6C3F1 mice to hepatic peroxisome proliferators. *Carcinogenesis* 7(11): 871–76, 1986.

Ganning, A. E., Brunk, U., Edlund, C., Elhammer, A., Dallner, G. Effects of prolonged administration of phthalate ester on the liver. *Environ. Health Perspect.* 73; 251–58, 1987.

Melnick, R. L., Morrissey, R. E., Tomaszewski, K. E. Studies by the National Toxicology Program on di(2-ethylhexyl) phthalate. *Toxicol. Ind. Health* 3(2): 99–118, 1987.

Nassberger, L., Arbin, A., Ostelius, J. Exposure of patients to phthalates from polyvinyl chloride tubes and bags during dialysis. *Nephron* 45(4): 286–90, 1987.

Reddy, J. K., Rao, M. S. Xenobiotic-induced peroxisome proliferation: role of tissue specificity and species differences in response in the evaluation of the implications for human health. *Arch. Toxicol.* 10: 43–53, 1987.

Rock, G., Labow, R. S., Franklin, C., Burnett, R., Tocchi, M. Hypotension and cardiac arrest in rats after infusion of mono (2-ethylhexyl) phthalate (MEHP), a contaminant of stored blood. *N. Engl. J. Med.* 316(19): 1218–19, 1987.

Smith-Oliver, T., Butterworth, B. E. Correlation of the carcinogenic potential of di(2-ethylhexyl) phthalate (DEHP) with induced hyperplasia rather than with genotoxic activity. *Mut. Res.* 188(1): 21–28, 1987.

Rubber Workers—Toxicology

Arlidge, J. T. *The Hygiene of Diseases and Mortality of Occupations.* Percival Co., London, 1892, pp. 483, 487.

Butlin, Henry T. Cancer of the scrotum in chimney sweeps and others. *Brit. Med. J.* 1–6, 1892.

Butlin, Henry T. Cancer of the scrotum in chimney sweeps and others. *Brit. Med. J.* 41–47, 1892.

Butlin, Henry T. Cancer of the scrotum in chimney sweeps and others. *Brit. Med. J.* 66–71, 1892.

Rehn, L. *Arch. Klin. Chir.* 50: 588, 1892.

Williams, W. R. *Brit. Med. J.* 103, 1892.

Selling, L. Preliminary report of some cases of purpura hemorrhagica due to benzol poisoning. *Bull. Johns Hopkins Hosp.* 33–37, 1910.

Krantz, G. D. Control and prevention of a rash among rubber factory employees. *India Rubber Wld.* 57: 145–46, 1917.

Earle, R. D. Report of the committee on poisonous nature of some accelerators, and precautions regarding their use. *India Rubber Wld.* 82, 1918.

Leitch, A. Paraffin cancer and its experimental production *Brit. Med. J.* 2: 1104–6, 1922.

Hamilton, A. *Industrial Poisons in the United States.* Macmillian Co., New York, 1925.

Kennaway, E. L. Experiments on cancer-producing substances. *Brit. Med. J.* 2: 1–4, 1925.

Davis, P. A. Toxic substances in the rubber industry: VI—hexamethylenetetramine. *Rubber Age.* 25: 199–200, 1929.

Davis, P. A. Toxic substances in the rubber industry: benzene. *Rubber Age* 25: 367–68, 1929.

Davis, P. A. Toxic substances in the rubber industry: III—aniline. *Rubber Age* 25: 611–12, 1929.

Davis, P. A. Toxic substances in the rubber industry: XVI—guanidine compounds. *Rubber Age* 26: 143–44, 1930.

Merewether, E. R. A. The occurrence of pulmonary fibrosis and other pulmonary afflictions in asbestos workers. *J. Ind. Hyg.* 12: 239–59, 1930.

Wilson, F. The very least an employer should know about dust and fume diseases. *Saf. Eng.* 62(5): 317–18, 1931.

Anon. Occupational diseases are not generally understood. *Nat. Underwriter* 3: 28, Dec. 15, 1932.

Davis, P. A. Toxic substances in the rubber industry: XXIV—methylenediphenyldiamine. *Rubber Age* 30: 313, 1932.

General Register Office. Occupations, Table 2, in *Census of England and Wales, 1931.* H.M.S.O., London, 1934.

Hamilton, A. The lessening menace of benzol poisoning in American industry. *J. Ind. Hyg.* 10: 227–30, 1934.

Anon. Compounding materials used in the rubber industry. *Nat. Saf. Counc. J.*, Chicago, 1–12, 1938.

Gall, E. A., Benzene poisoning with bizarre extramedullary hematopoiesis. *Arch. Path.* 25: 316–26, 1938.

Hueper, W. C., Wiley, F. H., Wolfe, H. D. Experimental production of bladder tumors in dogs by administration of beta-naphthylamine. *J. Ind. Hyg.* 20: 46–84, 1938.

Hueper, W. C. "Aniline tumors" of the bladder. *Arch. Path.* 25: 856–99, 1938.

Registrar General. Part II-a: Occupational mortality, in *Registrar General's Decennial Supplement, England and Wales.* H.M.S.O., London, 1938.

Hunter, F. T. Chronic exposure to benzene (Benzol). II. The clinical effects. *J. Ind. Hyg. Toxicol.* 21(8): 331–54, 1939.

Mallory, T. B., Gall, E. A., Brickley, W. Chronic exposure to benzene (Benzol). III. The pathologic results. *J. Ind. Hyg.* 21(8): 355–77, 1939.

Salter, W. T. Benzol and leukemia. *New Engl. J. Med.* 222(4): 146–47, 1940.

Schrenk, H. H., Yant, W. P. Comparative physiological effects of pure, commercial and crude benzenes. *J. Ind. Hyg. Toxicol.* 22(2): 53–63, 1940.

Dublin, I. L., Vane, R. J. Occupational hazards and diagnostic signs, a guide to impairments to be looked for in hazardous occupations. *U.S. Dept. Labor Bull.* 41: 22, 25, 1941.

von Oettingen, W. F. The aromatic amino and nitro compounds: their toxicity and potential dangers. *Pub. Health Bull.* # 271, 1941.

Anon. Occupational hazards and diagnostic signs. *U.S. Dept. Labor Bull.* 41: 99, 1942.

Dudley, H. C., Neal, P. A. Toxicology of acrylonitrile (vinyl cyanide). *J. Ind. Hyg. Toxicol.* 24(2): 27–36, 1942.

Hueper, W. C. *Occupational Tumors and Allied Diseases.* Charles C. Thomas, Springfield, Ill., 1942, pp. 399–405, 816–50.

Davis, P. A. Carbon tetrachloride as an industrial hazard *J.A.M.A.* 123: 962, 1943.

Hamilton, A. *Exploring the Dangerous Trades, the Autobiography of Alice Hamilton.* Little Brown and Co., Boston, 1943, pp. 294–96, 388.

Hueper, W. C. Cancer in its relation to occupation and environment. *Amer. Soc. Control Cancer Bull.* 63–69, 1943.

Mallette, F. S. Industrial hygiene in synthetic rubber manufacture. *Ind. Med.* 112: 495–99, 1943.

Wilson, R. H. Toluene poisoning. *J.A.M.A.* 123: 1106–8, 1943.

Anon. Environmental cancer. *J.A.M.A.* 124: 836, 1944.

Carpenter, C. P., Schaffer, C. B., Weill, C. S., Smyth, H. F., Jr. Studies of inhalation of 1:3 butadiene with a comparison of its narcotic effect with benzol, toluol, and styrene, and a note on the elimination of styrene in the human. *J. Ind. Hyg. Toxicol.* 26: 69–78, 1944.

Wilson, R. H. Health hazards encountered in the manufacture of synthetic rubber. *J.A.M.A.* 124: 701–3, 1944.

Cook, W. A. Maximum allowable concentrations of industrial atmospheric contaminants. *Ind. Med.* 936–46, 1945.

Hueper, W. C. Significance of industrial cancer in the problem of cancer. *J. Occup. Med.* 2(1): 190–200, 1946.

Hueper, W. C. Industrial management and occupational cancer. *J.A.M.A.* 131: 731–41, 1946.

Wilson, R. H., Hough, G. V., McCormick, W. E. Medical problems encountered in the manufacture of American-made rubber. *Ind. Med.* 17: 199–207, 1948.

Yant, W. P. Industrial hygiene codes and regulations. Ind. Hyg. Found. Amer., Inc., Nov. 18, 1948, pp. 48–61.

Conklin, G. Cancer and environment. *Sci. Amer.* 180(1): 11–15, 1949.

Goldblatt, M. W. Vesical tumors induced by chemical compounds. *Brit. J. Ind. Med.* 6: 65–81, 1949.

Hamilton, A., Hardy, H. L. *Industrial Toxicology*, 2nd ed. Paul B. Hoeber, Inc., New York, 1949.

Hueper, W. C. Environmental and occupational cancer. *Pub. Health Rep.* Suppl. 209, pp. 1–14. Govt. Printing Office, Washington, D.C., 1949.

Mancuso, T. F. Occupational cancer survey in Ohio. *Proc. Pub. Health Cancer Assoc. Amer.* 56–70, 1949.

Frost, J. Three cases of asbestosis. *Doctor's Weekly*, Copenhagen, 112(37): 1284–89, 1950.

Merewether, E. R. A. The pneumokonioses: developments, doubts, differences. *Canad. Med. Assoc. J.* 62: 169–73, 1950.

Srbova, J., Teisinger, J., Skramovsky, S. Absorption and elimination of inhaled benzene in man. *Arch. Ind. Hyg. Occup. Med.* 2(1): 1–8, 1950.

Bonser, G. M., Clayson, D. B., Jull, J. W. An experimental inquiry into the cause of industrial bladder cancer. *Lancet* 2: 286–88, 1951.

Falk, H. L., Steiner, P. E., Goldfein, S., Breslow, A., Hykes, R. Carcinogenic hydrocarbons and related compounds in processed rubber. *Cancer Res.* 11: 318–24, 1951.

Hueper, W. C. Environmental lung cancer. *Ind. Med. Surg.* 20 (2): 49–62, 1951.

Mallette, F. S., von Hamm, E. Studies on the toxicity and skin effects of compounds used in the rubber and plastics industries, I, accelerators. activators and antioxidants. *Arch. Ind. Hyg. Occup. Med.* 5: 311–17, 1952.

von Hamm, E., Mallette, F. S. Studies on the toxicity and skin effects of compounds used in the rubber and plastics industries III. Carcinogenicity of carbon black extracts. *Arch. Ind. Hyg. Occup. Med.* 6: 237–42, 1952.

Case, R. A. M., Hosker, M. E. Tumour of the urinary bladder as an occupational disease in the rubber industry in England and Wales. *Brit. J. Prev. Soc. Med.* 8: 39–50, 1954.

Wilson, R. H. McCormick, W. E. Toxicology of plastics and rubber—plastomers and monomers. *Ind. Med. Sur.* 23: 479–86, 1954.

Anon. Conference on environmental carcinogens and the medical aspects. *Consumer Reports* 20: 66, 1955.

Breslow, L. Industrial aspects of bronchogenic neoplasms. *Dis. Chest.* 28: 421, 1955.

Hueper, W. C. Editorial: silicosis, asbestosis and cancer of the lung. *Amer. J. Clin. Path.* 25: 1955.

Druckrey, H. Schmahl, D., Merckle, R., Jr. The cancerogenic effect of rubber after implantation into rats. *Z. Krebsforschung* 61(1): 55–64, 1956.

Hueper, W. C. A quest into the environmental causes of cancer of the lung. *Pub. Health Monograph #36*, pp. 1–46. USPHS, 1956.

Mancuso, T. F. Advances in industrial toxicology for the year 1955. *Arch. Ind. Health* 14: 206–12, 1956.

Fairhall, L. R. *Industrial Toxicology*, 2nd ed. Hafner Publishing Co., New York, 1957, pp. 148–49, 159–60, 161–68, 343–46.

Boyland, E. The biochemistry of cancer of the bladder. *Brit. Med. Bull.* 14(2): 153–58, 1958.

Nau, C. A., Neal, J., Stembridge, V. A study of the physiological effects of carbon black. *Arch. Ind. Health* 17: 21–28, 1958.

Williams, M. H. C., Walpole, A. L. Aromatic amines as carcinogens in industry. *Brit. Med. Bull.* 14(2): 141–45, 1958.

Harris, D. K. Some hazards in the manufacture and use of plastics. *Brit. J. Ind. Med.* 16: 221–29, 1959.

Nau, C. A., Neal, J. M. S., Stembridge, V. A. A study of the physiological effects of carbon black. *Arch. Environ. Health* 59: 513–33, 1960.

Pagnotto, L. D., Elkins, H. B., Grusch, H., Walkley, J. Industrial benzene exposure from petroleum naphtha: I. Rubber coating industry. *Ind. Hyg. J.* 417–21, 1961.

Zapp, J. A. Toxic and health effects of plastic and resins. *Arch. Environ. Health* 4: 125–36, 1962.

Mancuso, T. F. Tumors of the central nervous system. Industrial considerations. *Acta Union Int. Control Cancer* 19: 488–89, 1963.

Anon. Cancer research. *Lancet.* 2: 25–26, 1964.

Boyland, E., Haddow, A. Letters to the editor: Cancer research *Lancet* 2: 527, 1964.

Case, R. A. M. Letters to the editor: Cancer research. *Lancet* 309–10, 1964.

Hueper, W. C., Conway, W. D. *Chemical Carcinogenesis and Cancers*, 3rd ed. Charles C. Thomas, Springfield, Ill., 1964, pp. 88–89, 156.

Parkes, G. Letters to the editor: Cancer research. *Lancet* 254, 414, 1964.

Wallace, D. Letters to the editor: Cancer research. *Lancet* 365, 1964.

Anon. Occupational bladder tumors and the control of carcinogens. *Lancet* 306–7, 1965.

Anon. Industrial cancer of the bladder. *Lancet* 328, 1965.

Anon. Bladder tumours in industry. *Lancet* 627–28, 1965.

Anon. Medicine and the law: Death of a rubber worker. *Lancet* 635, 1965.

Anon. Bladder cancer in the rubber industry. *Brit. Med. J.* 1: 329–30, 1965.

Anon. Medicine and the law. Inquest on a former rubber worker. *Lancet* 1259, 1966.

Mancuso, T. J. Environmental study of cancer of the nervous system. Summary progress report of Grant #CA 07588-01, 1966.

Kipling, M. D., Waterhouse, J. A. H. Cadmium and prostatic cancers. *Lancet* 1: 730, 1967.

U. S. Department of Health, Education and Welfare. Occupational characteristics of disabled workers, by disabling condition. Disability insurance benefit awards made in 1959–1962 to men under age 65. *Pub. Health Serv. Bull.* #1531, U. S. Govt. Printing Office, 1967.

Boyland, E., Carter, R. L., Garrod, J. W., Roe, F. J. C. Carcinogenic properties of certain rubber additives. *Eur. J. Cancer* 4: 232–39, 1968.

Fristedt, G., Mattsson, S. B., Schutz, A. Talcosis by exposure to granular talc in a rubber industry. *Nord. Hyg. Tidskr.* 49: 66–71, 1968.

Mancuso, T. F., Ciocco, A., El-Attar, A. An epidemiological approach to the rubber industry. *J. Occup. Med.* 10(5): 213–31, 1968.

Sax, I. N. *Dangerous Properties of Industrial Materials*, 3rd ed. Van Nostrand Reinhold, New York, 1968.

Hunter, D. *The Diseases of Occupations.* Little Brown and Co., Boston, 1969, pp. 363, 591, 593, 613.

Lijinski, W., Epstein, S. S. Nitrosamines as environmental carcinogens. *Nature* 225: 21–23, 1970.

Mancuso, T. J., Brennan, M. J. Epidemiological considerations of cancer of the gallbladder, bile ducts and salivary glands in the rubber industry. *J. Occup. Med.* 12(9): 333–41, 1970.

Anon. Medicolegal: damages for rubber workers. *Brit. Med. J.* 2: 412–13, 1971.

Adelstein, A. M. Occupational mortality: cancer. *Ann. Occup. Hyg.* 15: 53–57, 1972.

Hueper, W. C. Letter to J. S. Kelman, Mar. 27, 1972.

Noweir, M., El-Dakhakhny, A. A. Exposure to chemical agents in rubber industry. *J. Egyptian Pub. Health Assoc.* 47(3): 182–201, 1972.

Osman, H., Wahdan, M., Noweir, M. Health programs resulting from prolonged exposure to chemical agents in rubber industry. *J. Egyptian Pub. Health* 47(5): 189–99, 1972.

Gaubatz, E., Gaubatz-Trott, H. Pneumoconiosis in the workers of the rubber processing plant. *Prax. Pneumol.* 27: 740–42, 1973.

Fox, A. J., Lindars, D. C., Owen, R. A survey of occupational cancer in the rubber and cablemaking industries: results of five year analysis, 1967–71. *Brit. J. Ind. Med.* 31: 140–51, 1974.

McMichael, A. J., Spirtas, R., Kupper, L. L. An epidemiologic study of mortality within a cohort of rubber workers, 1964–72. *J. Occup. Med.* 16: 458–64, 1974.

Wagner, J. C., Berry, G., Skidmore, J. W., Timbrell, V. The effects of inhalation of asbestos in rats. *Brit. J. Cancer* 29: 252–69, 1974.

do Pico, G., Rankin, J., Chosy, L., Reddan, W., Barbee, R., Gee, B., Dickie, H. Respiratory tract disease from thermosetting resins. *J. Intern. Med.* 83: 177–84, 1975.

Gold, A. Carbon black adsorbates: separation and identification of a carcinogen and some oxygenated polyaromatics. *Analyt. Chem.* 47(8): 1469–72, 1975.

Mancuso, T. Epidemiological investigation of occupational cancers in the rubber industry, in *New Multinational Health Hazards*, Levinson, C., Ed. ICF, Switzerland, 1975, pp. 80–136.

McMichael, A. J. Letter to P. F. Infante, Feb. 4, 1975.

McMichael, A. J., Andjelkovic, D. A., Tyroler, H. A. Cancer mortality among rubber workers: an epidemiologic study. New York Academy of Science, Conference on Occupational Carcinogenesis, New York, Mar. 24–28, 1975.

McMichael, A. J., Spirtas, R., Kupper, L. L., Gamble, J. F. Solvent exposure and leukemia among rubber workers: an epidemiologic study. *Occup. Med.* 17(4): 234–39, 1975.

Selikoff, I. J., Hammond, E. C. Toxicity of vinyl chloride–polyvinyl chloride. *Ann. N.Y. Acad. Sci.* 246: 5–337, 1975.

Andjelkovic, D., Taulbee, J., Symons, M. Mortality experience of a cohort of rubber workers, 1964–73. *J. Occup. Med.* 18(6): 387–94, 1976.

Fine, L. J., Peters, J. M. Respiratory morbidity in rubber workers. *Arch. Environ. Health* 31(1): 5–14, 1976.

Fine, L. J., Peters, J. M., Burgess, W., Bernardinis, L. Studies of respiratory morbidity in rubber workers. *Arch. Environ. Health* 31: 195–200, 1976.

Fox, J., Collier, P. F. A survey of occupational cancer in the rubber and cablemaking industries: analysis of deaths occurring in 1972–74. *Brit. J. Ind. Med.* 33: 249–64, 1976.

Gamble, J. F., McMichael, A., William, T., Battigelli, M. Respiratory function and symptoms: an environmental-epidemiological study of rubber workers exposed to a phenolformaldehyde type resin. *Amer. Ind. Hyg. J.* 499–513, 1976.

Gamble, J., Spirtas, R. Job classification and utilization of complete work histories in occupational epidemiology. *J. Occup. Med.* 18(6): 399–404, 1976.

Lee, M. L., Hites, R. Characterization of sulfur-containing polycyclic aromatic compounds in carbon blacks. *Analyt. Chem.* 48(13): 1890–93, 1976.

Mancuso, T. F. Problems and perspective in epidemiological study of occupational health hazards in the rubber industry. *Environ. Health Perspect.* 17: 21–30, 1976.

McMichael, A. J. Standardized mortality ratios and the "healthy worker effect": scratching beneath the surface. *J. Occup. Med.* 18(3): 165–68, 1976.

McMichael, A. J., Andjelkovic, D. A., Tyroler, H. A. Cancer mortality among rubber workers: an epidemiologic study *Ann. N. Y. Acad. of Sci.* 271: 125–37, 1976.

McMichael, J. A., Gerber, W. S., Gamble, J. F., Lednar, W. M. Chronic respiratory symptoms and job type within the rubber industry. *J. Occup. Med.* 18(9): 611–17, 1976.

Monson, R. R., Nakano, K. K. Mortality among rubber workers and other employees. *Amer. J. Epidemiol.* 103(3): 297–303, 1976.

Nutt, A. Measurement of some potentially hazardous materials in the atmosphere of rubber factories. *Environ. Health Perspect.* 17: 117–23, 1976.

Selikoff, I. Priorities in the investigation of human health hazards in the plastics and synthetic rubber industries. *Environ. Health Perspect.* 17: 5–11, 1976.

Tyroler, H. A., Andjelkovic, D., Harris, R., Lednar, W., McMichael, A., Symons, M. Chronic diseases in the rubber industry. *Environ. Health Perspect.* 17: 13–20, 1976.

van Duuren, B. L. Tumor-promoting and co-carcinogenic agents in chemical carcinogenesis, in *Chemical Carcinogens* (ACS Monograph #173), Searle, C. E., Ed. American Chemical Society, Washington, D.C., 1976, pp. 24–51.

Andjelkovick, D., Taulbee, J., Symons, M., Williams, T. Mortality of rubber workers with reference to work experience. *J. Occup. Med.* 19(6): 397–405, 1977.

Boland, B., Ed. *Cancer and the Worker*, New York Academy of Science, New York, 1977.

Lednar, W. M., Tyroler, H. A., McMichaels, A. J., Shy, C. M. The occupational determinants of chronic disabling pulmonary disease in rubber workers. *J. Occup. Med.* 19(4): 263–68, 1977.

NIOSH. *Carbon Black, Criteria Document.* U.S. Department of Health, Education and Welfare. Washington, D.C., 1977.

Sedlander, J. R. Letter to W. L. Smelser, Manager, Kelly Springfield Tire Co. (containing survey of the Cumberland, MD plant, done by Traveler's Insurance Co.), Aug. 4, 1977.

Andjelkovich, D., Taulbee, J., Blum, S. Mortality of female workers in a rubber manufacturing plant. *J. Occup. Med.* 20(6): 409–13, 1978.

Epstein, S. S. *The Politics of Cancer.* Sierra Club Books, San Francisco, 1978.

U.S. Environmental Protection Agency. Assessment of health effects of benzene germane to low level exposure. EPA-600/1-78-061, 1978.

Urban Environment Conference, Inc. *Environmental Cancer, Causes, Victims, Solutions*, proceedings of a conference held Mar. 21–22, 1977, Washington, D.C., published 1978.

Fajen, J. M., Caron, G. A. N-Nitrosamines in the rubber and tire industry. *Science* 205: 1262–64, 1979.

Hammond, E. C., Selikoff, I. J., Eds. Public control of environmental health hazards. *Ann. N.Y. Acad. Sci.* 329: 1–405, 1979.

Pagnotto, L., Elkins, H., Brugsch, H. Benzene exposure in the rubber coating industry— a follow-up. *Amer. Ind. Hyg. Assoc. J.* 40:137–46, 1979.

U.S. Environmental Protection Agency. Carcinogen Assessment Group's final report on population risk to ambient-benzene exposures. EPA-450/5-80-004, 1979.

Baxter, P. J. Mortality in the British rubber industries, 1967–76. *Health and Safety Executive.* H. M. Stationery Office, London, pp. 1–48, 1980.

Goldsmith, D. F., Smith, A. H., McMichael, A. A case-control study of prostate cancer within a cohort of rubber and tire workers. *J. Occup. Med.* 22: 533–41, 1980.

Laskin, L., Sellakumar, A., Kuschner, M., Nelson, N., Mendola, S., Rusch, G., Katz, G., Dulak, N., Albert, R. Inhalation carcinogenicity of epichlorohydrin in noninbred Sprague-Dawley rats. *J. Nat. Cancer Inst.* 65(4): 571–57, 1980.

U.S. Environmental Protection Agency. Support document health effects test rule: chlorinated benzenes. Assess. Div. Office of Pesticides and Toxic Substances, Washington, D.C., U.S. EPA 560/11-80-014, 1980.

Van Ert, M. D., Arp, E. W., Harris, R. L., Symons, M. J., Williams, T. M. Worker exposures to chemical agents in the manufacture of rubber tires: solvent vapor studies. *Amer. Ind. Hyg. Assoc. J.* 41: 212–18, 1980.

Williams, T., Harris, R., Arp, E., Symons, M., Van Ert, M. Worker exposure to chemical agents in the manufacture of rubber tires and tubes: particulates. *Amer. Ind. Hyg. Assoc. J.* 41: 204–11, 1980.

Delzell, E., Monson, R. Mortality among rubber workers III. Cause-specific mortality, 1940–78. *J. Occup. Med.* 23(10): 677–84, 1981.

Hedenstedt, A., Ramel, C., Wachtmeister, C. A. Mutagenicity of rubber vulcanization gases in *Salmonella typhimurium*. *J. Toxicol. Environ. Health* 8: 805–14, 1981.

Nicholson, W. J., Ed. Management of assessed risk for carcinogens. *Ann. N.Y. Acad. Sci.* 363: 1–301, 1981.

Vainio, H., Sorsa, M., Rantanen, J., Hemminiki, C., Aitio, A. Biological monitoring in the identification of the cancer risk of individuals exposed to chemical carcinogens. *Scand. J. Work Environ. Health* 7:241–51, 1981.

Weeks, J., Peters, J., Monson, R. Screening for occupational health hazards in the rubber industry. Part I. *Amer. J. Ind. Med.* 2: 125–41, 1981.

Alexander, V., Leffingwell, S. S., Lloyd, W. J., Waxweiler, R. J., Miller, R. L. Investigation of an apparent increased prevalence of brain tumors in a U.S. petrochemical plant. *Ann. N.Y. Acad. Sci.* 381: 97–107, 1982.

Delzell, E., Monson, R. Mortality among rubber workers: V. processing workers. *J. Occup. Med.* 24: 539–45, 1982.

IARC. Monographs on the Evaluation of the Carcinogenic Risk of Chemicals to Humans: The Rubber Industry, 28: 1–454, WHO, Lyon, 1982.

Maltoni, C., Ciliberti, A., Carretti, D. Experimental contributions in identifying brain potential carcinogens in the petrochemical industry. *Ann. N.Y. Acad. Sci.* 281: 216–49, 1982.

Mancuso, T. F. Epidemiological study of tumors of the central nervous system in Ohio. *Ann. N.Y. Acad. of Sci.* 381: 17–39, 1982.

Rall, D. National toxicology program. *NTP Tech. Bull.* 6: 1–12, 1982.

Zimmerman, H. M. Production of brain tumors with hydrocarbons. *Ann. N.Y. Acad. Sci.* 381: 320–24, 1982.

Arp, E. W., Jr., Wolf, P. H., Checkoway, H. Lymphocytic leukemia and exposures to benzene and other solvents in the rubber industry. *J. Occup. Med.* 25(8): 598–602, 1983.

Falck, K. Biological monitoring of occupational exposure to mutagenic chemicals in the rubber industry. Use of the bacterial urinary mutagenicity assay. *Scand. J. Work Environ. Health* 9(2): 39–42, 1983.

Schroeder, E. E. Rubber, in *Reigel's Handbook of Industrial Chemistry,* Kent, J. A., Ed. Van Nostrand Reinhold, New York, 1983, pp. 281–310.

Korhonen, A., Hemminki, K., Vainio, H. Embryotoxic effects of phthalic acid derivatives, phosphates and aromatic oils used in the manufacturing of rubber in three day chicken embryos. *Drug Chem. Tox.* 6(2): 191–207, 1983.

Michaels, D. Occupational cancer in the black population: the health effects of job discrimination. *J. Nat. Med. Assoc.* 75(10): 1014–18, 1983.

Norell, S., Ahlbom, A., Lipping, H., Osterblom, L. Oesophageal cancer and vulcanization work. *Lancet* 1(8322): 462–63, 1983.

Nutt, A. Rubber work and cancer—past, present and perspectives. *Scand. J. Work Environ. Health* 9(2): 49–57, 1983.

Riggan, W., Acquavella, J., Beaubier, J., Mason, T. J. *U.S. Cancer Mortality Rates and Trends, 1950–79*, Vols. I, II, III and various tables. NCI/EPA Interagency Agreement on Environmental Carcinogenesis. EPA 600/1-83-015a, 1983.

Sorsa, M., Maki-Paakkanen, J., Vainio, H. A chromosome study among worker groups in the rubber industry. *Scand. J. Work Environ. Health* 9(2): 43–47, 1983.

Benigni, R., Calcagnile, A., Fabri, G., Giuliani, A., Leopardi, P., Paoletti, A. Biological monitoring of workers in the rubber industry. II. UV-induced unscheduled DNA synthesis in the lymphocytes of vulcanizers. *Mut. Res.* 138(1): 105–11, 1984.

Checkoway, H., Wilcosky, T., Wolf, P., Tyroler, H. An evaluation of the association of leukemia and rubber industry solvent exposures. *Amer. J. Ind. Med.* 5(3): 239–49, 1984.

Delzell, E., Monson, R. R. Mortality among rubber workers: VIII. Industrial products workers. *Amer. J. Ind. Med.* 6(4): 273–79, 1984.

NIOSH. Health hazard evaluation report, HETA 79-109-1538. Kelly Springfield Tire Company, Cumberland, Md., 1984.

Oliver, L. C., Weber, R. P. Chest pain in rubber chemical workers exposed to carbon disulfide and methemoglobin formers. *Brit. J. Ind. Med.* 41(3): 296–304, 1984.

Vainio, H., Sorsa, M., Falck, K. Bacterial urinary assay in monitoring exposure to mutgens and carcinogens. *IARC Sci. Pub.* 59: 247–58, 1984.

Delzell, E., Monson, R. R. Mortality among rubber workers: X. Reclaim workers. *Amer. J. Ind. Med.* 7(4): 307–13, 1985.

Delzell, E., Monson, R. R. Mortality among rubber workers: IX. Curing workers. *Amer. J. Ind. Med.* 8(6): 537–44, 1985.

Sorahan, T., Parkes, H. G., Veys, C. A., Waterhouse, J. A. Cancer mortality in the British rubber industry: 1946–80. *Brit. J. Ind. Med.* 43(6): 363–73, 1986.

Thomas, R. J., Bascom, R., Yang, W. N., Fisher, J. F., Baser, M. E., Greenhut, J., Baker, J. H. Peripheral eosinophilia and respiratory symptoms in rubber injection press operators: a case-control study. *Amer. J. Ind. Med.* 9(6): 551–59, 1986.

Bernardinelli, L., de Marco, R., Tinelli, C. Cancer mortality in an Italian rubber factory. *Brit. J. Ind. Med.* 44(3): 187–91, 1987.

Bourguet, C. C., Checkoway, H., Hulka, B. S. A case-control study of skin cancer in the tire and rubber manufacturing industry. *Amer. J. Ind. Med.* 11(4): 461–73, 1987.

Goldsmith, D. F. Calculating cancer latency using data from a nested case-control study of prostatic cancer. *J. Chron. Dis.* 40(2): 1198–238, 1987.

Rinsky, R. A., Smith, A. B., Hornung, R., Filloon, T. G., Young, R. J., Okun, A. H., Landrigan, P. J. Benzene and leukemia. An epidemiologic risk assessment. *New Engl. J. Med.* 316(17): 1044–50, 1987.

Styrene

Spencer, H. C., Irish, D. D., et al. The response of laboratory animals to monomeric styrene. *J. Ind. Hyg. Toxicol.* 24: 295–301, 1942.

Wilson, R. H., Hough, G. V., McCormick, W. E. Medical problems encountered in the manufacture of American-made rubber. *Ind. Med.* 17: 199–207, 1948.

Wolf, M. A., Rowe, V. K., et al. Toxicological studies of certain alkylated benzenes and bezene. *Arch. Ind. Health* 14: 387–98, 1956.

Oppenheimer, B. S., Oppenheimer, E. T., et al. The latent period in carcinogenesis by plastics in rats and its relation to the presarcomatous stage. *Cancer* 11: 204–13, 1958.

Zielhuis, R. L. Systemic toxicity from exposure to epoxy resins, hardeners, and styrene. *J. Occup. Med.* 3: 25–29, 1961.

Pratt-Johnson, J. A. Case report: retrobulbar neuritis following exposure to vinyl benzene (styrene). *Canad. Med. Assoc. J.* 90: 975–77, 1964.

Fiserova-Bergerova, V., Teisinger, J. Pulmonary styene vapor retention. *Ind. Med. Surg.* 34: 620–22, 1965.

Dutkiewicz, T., Tyras, H. Skin absorption of toluene, styrene, and xylene by man. *Brit. J. Ind. Med.* 25: 243, 1968.

Leibman, K. C., Ortiz, E. Styrene epoxide—an intermediate in microsomal oxidation of styrene to its glycol. *Pharmacol.* 10: 203, 1968.

Stewart, R. D., Dodd, H. C., et al. Human exposure to styrene vapor. *Arch. Environ. Health* 16: 656–62, 1968.

Leibman, K. C., Ortiz, E. Oxidation of styrene in liver microsomes. *Biochem. Pharmacol.* 18: 552–54, 1969.

van Duuren, B. L. Carcinogenic epoxides, lactones and halo-ethers and their mode of action. *Ann. N.Y. Acad. Sci.* 163: 633, 1969.

Rudd, J. F. Styrene polymers, physical properties—general, in *Encyclopedia of Polymer Science and Technology, Plastics, Resins, Rubbers, Fibers,* Bikales, N. M., Ed., 13: 243–44, 251, 1970.

Araki, S., Abe, A., Ushiro, K., Fujino, M. A case of skin atrophy, neurogenic muscular atrophy and anxiety reaction following long exposure to styrene. *Jap. J. Ind. Health* 13: 427–31, 1971.

Gotell, P., Axelson, O., Lindelof, B. Field studies on human styrene exposure. *Work. Environ. Health* 9: 76–83, 1972.

Klimkova-Deutschova, E., Dandova, D., Salamonova, Z., Schwarzova, K., Titman, O. Recent advances concerning the clinical picture of professional styrene exposure. *Cs. Neurol.* 36: 20–25, 1973.

Astrand, I., Kilbom, A., Ovrum, P., Wahlberg, I., Vesterberg, O. Exposure to styrene. I. Concentration in alveolar air and blood at rest and during exercise and metabolism. *Work Environ. Health* 11: 69–85, 1974.

van Rees, H. The partition coefficients of styrene between blood and air and between oil and blood. *Int. Arch. Arbeitsmed.* 33: 39–47, 1974.

Astrand, I. Uptake of solvents in the blood and tissues of man. A review. *Scand. J. Work Environ. Health* 1: 199–218, 1975.

Leibman, K. C. Metabolism and toxicity of styrene. *Environ. Health Perspect.* 11: 115–19, 1975.

Block, J. B. A Kentucky study: 1950–1975, in *Proceedings of NIOSH Styrene-Butadiene Briefing, Covington, Kentucky,* Ede, L., Ed., HEW Publ. (NIOSH) 77-129: 28–32, 1976.

Chielewski, J., Renke, W. Clinical and experimental research into the pathogenesis of toxic effect of styrene. *Bull. Inst. Mar. Trop. Med. Gdynia* 27: 63–68, 1976.

Gosselin, R. E., Smith, R. P., Hodge, H. C. *Clinical Toxicology of Commercial Products,* 5th ed. Williams and Wilkins, Baltimore, 1976, pp. D-104-5.

Lemen, R. A., Young, R. Investigations of health hazards in styrene butadiene rubber facilities, in *Proceedings of NIOSH Styrene-Butadiene Briefing,* Ede, L., Ed., HEW Publ. (NIOSH) 77-129: 3–8, 1976.

Lilis, R., Nicholson, W. J. Cancer experience among workers in a chemical plant producing styrene monomers, in *Proceedings of NIOSH Styrene-Butadiene Briefing,* Ede, L., Ed., HEW Publ. (NIOSH) 77-129: 22–27, 1976.

Loprieno, N., Abbondandolo, A., et al. Mutagenicity of industrial compounds: styrene and its possible metabolite styrene oxide. *Mut. Res.* 40: 317–24, 1976.

Lorimer, W. V., Lilis, R., et al. Clinical studies of styrene workers: preliminary report, in *Proceedings of NIOSH Styrene-Butadiene Briefing*, Ede, L., Ed., HEW Publ. (NIOSH) 77-129: 163–69, 1976.

Milvy, P., Garro, A. J. Mutagenic activity of styrene oxide (1,2-epoxyethylbenzene), a presumed styrene metabolite. *Mut. Res.* 40: 15–18, 1976.

Ryan, A. J., James, M. O., et al. Hepatic and extrahepatic metabolism of 14C-styrene oxide. *Environ. Health Perspect.* 17: 135–44, 1976.

Vainio, H., Paakkonen, R., et al. A study on the mutagenic activity of styrene and styrene oxide. *Scand. J. Work. Environ. Health* 3: 147–51, 1976.

Wolff, M. S. Evidence for existence in human tissues of monomers for plastics and rubber manufacture. *Environ. Health Perspect.* 17: 183–87, 1976.

de Meester, C., Poncelet, F., et al. Mutagenicity of styrene and styrene oxide. *Mut. Res.* 56: 147–52, 1977.

Meretoja, T., Vainio, H., et al. Occupational styrene exposure and chromosomal aberrations. *Mut. Res.* 56: 193–97, 1977.

Savolainen, H. Vainio, H. Organ distribution and nervous system binding of styrene and styrene oxide. *Toxicol.* 8: 135–41, 1977.

Smith, A. H., Ellis, L. Styrene butadine rubber synthetic plants and leukemia. *J. Occup. Med.* 19: 441, 1977.

Vainio, H., Hemminki, K. et al. Toxicity of styrene and styrene oxide on chick embryos. *Toxicol.* 8: 319–25, 1977.

Vainio, H., Makinen, A. Styrene and acrylonitrile induced depression of hepatic nonprotein sulfhydryl content in various rodent species. *Res. Commun. Chem. Pathol. Pharmacol.* 17: 115–24, 1977.

Wolff, M. S., Daum, S. M., et al. Styrene and related hydrocarbons in subcutaneous fat from polymerization workers. *J. Toxicol. Environ. Health* 2: 997–1005, 1977.

Anon. MCA reports preliminary results of teratology, inhalation tests. *Chem. Regul. Reporter*, pp. 1759–62, Feb. 17, 1978.

IARC. *Information Bulletin on the Survey of Chemicals Being Tested for Carcinogenicity* 7: 35, 238, 256, 337, 382, 1978.

IARC. *Directory of Ongoing Research in Cancer Epidemiology* 26: 330, 1978.

Lilis, R., Lorimer, W. V., Diamond, S., Selikoff, I. J. Neurotoxicity of styrene production and polymerization workers. *Environ. Res.* 15: 133–38, 1978.

Meretoja, R., Vainio, H. Clastogenic effects of styrene exposure on bone marrow cells of rat. *Toxicol. Letters* 1: 315–18, 1978.

Fjeldstad, P. E., Thorud, S., Wannag, A. Letter to the editor: Styrene oxide in the manufacture of reinforced polyester plastics. *Scand. J. Work Environ. Health.* 5: 162–63, 1979.

IARC Monographs on the Evaluation of the Carcinogenic Risk of Chemicals to Humans 19: 73–114, 231–74, 1979.

Maltoni, C., Failla, G., Kassapidis, C. First experimental demonstration of the carcinogenic effects of styrene oxide. *Med. Lavoro* 10(5): 358–62, 1979.

Nicholson, W. J., Selikoff, I. J., Seidman, H. Mortality experience of styrene-polystyrene polymerization workers. Initial findings. *Scand. J. Work. Environ. Health* 4 (Suppl. 2): 247–52, 1978.

Ponomarkov, V., Tomatis, L. Effects of long-term oral administration of styrene to mice and rats. *Scand. J. Environ. Health* 4(2): 127–35, 1979.

Withey, J. R., Collins, P. G. The distribution and pharmacokinetics of styrene monomer in rats by the pulmonary route. *J. Environ. Pathol. Toxicol.* 2: 1329–42, 1979.

Skoog, K. O., Nilsson, S. E. G. Changes in the c-wave of the electroretinogram and in the standing potential of the eye after small doses of toluene and styrene. *Acta Opthamol. (Denmark)* 59(1): 71–79, 1981.

Pacifici, G. M., Rane, A. Metabolism of styrene oxide in different human fetal tissues. *Drug Metab. Dis.* 10(4): 302–5, 1982.

Samimi, B., Falbo, L. Monitoring of workers exposure to low levels of airborne monomers in a polystyrene production plant. *Amer. Ind. Hyg. Assoc. J.* 43: 858–62, 1982.

Camurri, L., Codeluppi, S., Pedroni, C., Scarduelli, L. Chromosomal aberrations and sister chromatid exchanges in workers exposed to styrene. *Mut. Res. (Netherlands)* 119: 361–69, 1983.

Chakrabarti, S., Brodeur, J. The nephrotoxic potential of styrene in Sprague-Dawley rats. *Toxicol. Letters* 18(3): 315–22, 1983.

Hardin, B. D., Niemeier, R. W., Sikov, M. R., Hackett, P. L. Reproductive-toxicologic assessment of the epoxides ethylene oxide, propyene exide, butylene oxide, and styrene oxide. *Scand. J. Work Environ. Health* 9(2): 94–102, 1983.

NIOSH. *Criteria for a Recommended Standard: Occupational Exposure to Styrene.* U.S. Department of Health and Human Services, Washington, D.C., Sept. 1983, pp. 1–250.

Norppa, H., Vainio, H. Genetic toxicity of styrene and some of its derivatives. *Scand. J. Work Environ. Health (Finland)* 9(2): 108–14, 1983.

Norppa, H., Vainio, H. Induction of sister-chromatid exchanges by styrene analogues in cultured human lymphocytes. *Mut. Res. (Netherlands)* 116(3–4): 379–87, 1983.

Edling, C., Tagesson, C. Raised serum bile acid concentrations after occupational exposure to styrene: a possible sign of hepatotoxicity. *Brit. J. Ind. Med.* 41(2): 257–59, 1984.

Mutti, A., Falzoi, M., Romanelli, A., Framchini, I. Regional alterations of brain catecholamines by styrene exposure in rabbits. *Arch. Toxicol.* 55(3): 173–77, 1984.

Mutti, A., Mazzucchi, A., Rustichelli, P., Frigeri, G., Arfini, G., Franchini, I. Exposure-effect and exposure-responsible relationships between occupational exposure to styrene and neuropsychological functions. *Amer. J. Ind. Med.* 5(4): 275–86, 1984.

Mutti, A., Vescovi, P. P., Falzoi, M., Arfini, G., Valenti, G., Franchini, I. Neuroendocrine effects of styrene on occupationally exposed workers. *Scand. J. Work Environ. Health* 10(4): 225–28, 1984.

Sax, N. I. *Dangerous Properties of Industrial Materials.* Van Nostrand Reinhold, New York, 1984, pp. 554, 2473–74.

Vainio, H., Norppa, H., Belvedere, G. Metabolism and mutagenicity of styrene and styrene oxide. *Prog. Clin. Biol. Res.* 141: 215–25, 1984.

Baker, E. L., Smith, T. J., Landrigan, P. J. The neurotoxicity of industrial solvents: a review of the literature. *Amer. J. Ind. Med.* 8(3): 207–17, 1985.

Crandall, M. S., Hartle, R. W. An analysis of exposure to styrene in the reinforced plastic boat-making industry. *Amer. J. Ind. Med.* 8(3): 183–92, 1985.

Lemasters, G. K., Carson, A., Samuels, S. J. Occupational styrene exposure for twelve product categories in the reinforced-plastics industry. *Amer. Ind. Hyg. Assoc. J.* 46(8): 434–41, 1985.

Jantunen, K., Maki-Paakkanen, J., Norppa, H. Induction of chromosome aberrations by styrene and vinylacetate in cultured human lymphocytes: dependence on erythrocytes. *Mut. Res.* 159(1–2): 109–16, 1986.

Chakrabarti, S. K., Labelle, L., Tuchweber, B. Studies on the subchronic nephrotoxic potential of styrene in Sprague-Dawley rats. *Toxicology* 44(3): 355–65, 1987.

Pryor, G. T., Rebert, C. S., Howd, R. A. Hearing loss in rats caused by inhalation of mixed xylenes and styrene. *J. Appl. Toxicol.* 7(1): 55–61, 1987.

1,1,1-Trichloroethane

Adams, E. M., Spencer, H. C., Rowe, V. K., Irish, D. D. Vapor toxicity of 1,1,1,-trichloroethane (methylchloroform) determined by experiments on laboratory animals. *Arch. Ind. Hyg. Occup. Med.*, Feb. 1950.

Torkelson, R. T., Oyen, F., McCollister, D. D., Rowe, V. K. Toxicity of 1,1,1-trichloroethane as determined on laboratory animals and human subjects. *Amer. Ind. Hyg. Assoc. J.*, Oct. 1958.

Hake, C. L., Waggoner, T. B., Robertson, D. N., Rowe, V. K. The metabolism of 1,1,1-trichloroethane by the rat. *Arch. Environ. Health*, 1: 101–05, 1960.

1,1,1-Trichloroethane (Methyl Chloroform). American Industrial Hygiene Assoc., 1961.

Stewart, R., Gay, H., Erley, D., Hake, C., Schaeffer, A. Human exposure to 1,1,1-trichloroethane vapor: relationship of expired air and blood concentrations to exposure and toxicity. *Amer. Ind. Hyg. Assoc. J.*, 22(4): 253–62, 1961.

Steward, R., Dodd, H. Absorption of carbon tetrachloride, trichloroethylene, tetrachloroethylene, methylene chloride, and 1,1,1-trichloroethane through the human skin. *Amer. Ind. Hyg. Assoc. J.*, 25: 439–46, Sept–Oct. 1964.

Biochemical Research Laboratory, Dow Chemical Co. *Summary of Toxicological Information and Practical Considerations in the Handling and Use of Chlorothene NU (Inhibited 1,1,1-Trichloroethane)*. Dow Chemical Co., Midland, Mich., Dec. 1961.

Manufacturing Chemists Association. Chemical Safety Data Sheet SD-90, Properties and essential information for safe handling and use of 1,1,1-trichloroethane. Manufacturing Chemists Assoc., 1825 Connecticut Avenue, N.W., Washington, D.C., adopted 1965.

Stewart, R. D., and Andrews, J. T. Acute intoxication with methylchloroform. *J.A.M.A.*, 195: 904–906, 1966.

Hall, F. B., Hine, C. H. Trichloroethane intoxication: a report of two cases. *J. Forensic Sci.*, 11(3): 404–13, July 1966.

Stewart, R. D., Dodd, H. C., Erley, D., and Holder, B. Diagnosis of solvent poisoning. *J.A.M.A.* 193: 1097–1100, Sept. 27, 1965.

Klassen, C., Plaa, G. Relative effects of various chlorinated hydrocarbons on liver and kidney function in mice. *Toxicol. Appl. Pharmacol.* 9: 139–51, 1966.

Biochemical Research Laboratory, Dow Chemical Co. *Toxicity, Health Hazards and Safe Handling of Chlorothene NU Solvent (Inhibited 1,1,1-Trichloroethane)*. Dow Chemical Co., Midland, Mich., Jan. 5, 1967.

Prendergast, J. A., Jones, R. A., Jenkins, L. J., Jr., Siegel, J. Effects on experimental animals of long-term inhalation of trichloroethylene, carbon tetrachloride, 1,1,1-trichloroethane, dichlorodifluoromethane etc. *Toxicol. Appl. Pharmacol.* 10: 2, 1967.

Klaassen, C. D., Plaa, G. L. Relative effects of various chlorinated hydrocarbons on liver and kidney function in dogs. *Toxicol. Appl. Pharmacol.* 10: 119–31, 1967.

Stewart, R. D. The toxicology of 1,1,1-trichloroethane. *Ann. Occup. Hyg.* 2: 71–79, 1968.

Klaassen, C. D., Plaa, G. L. Comparison of biochemical alterations in livers from rats treated with carbon tetrachloride, chloroform, 1,1,2-trichloroethane, 1,1,1-trichloroethane. *Biochem. Pharmacol.* 18(8): 2019–27, Aug. 1969.

Stewart, R., Gay, H., Schaffer, A., Erley, D., Rowe, V. K. Experimental human exposure to methyl chloroform vapor. *Arch. Environ. Health* 19: 467–72, Oct. 1969.

Dow Chemical Co. *Chlorothene NU Solvent—The One with a Thousand and One Uses.* Dow Chemical Co., Midland, Mich., 1970.

Dow Chemical Co. *Chlorothene VG Cleaning Solvent . . . Why the Best Costs Less.* Dow Chemical Co., Midland, Mich., 1970.

American National Standards Institute, Inc. American National Standard acceptable concentrations of methyl chloroform (1,1,1 trichloroethane). *Amer. Ind. Hyg. Assoc.*, Jan. 27, 1970.

Fuller, G. C., Olshan, A., Puri, S. K., Lal, H. Induction of hepatic drug metabolism in rats by methylchloroform inhalation. *J. Pharmacol. Exp. Therm.* 175(2): 311–17, Nov. 1970.

Salvani, M., Binaschi, S., Riva, M. Evaluation of the psychophysiological functions in humans exposed to the 'threshold limit value' of 1,1,1-trichloroethane. *Brit. J. Ind. Med.* 28: 286–92, 1971.

Dow Chemical Co. *Properties, Health Hazards and Precautions for Safe Handling of Chlorothene Brand Solvents (Chlorothene Industrial, Chlorothene NU, Chlorothene VG).* Dow Chemical Co., Midland, Mich., Apr. 16, 1973.

Dow Chemical Co. *Subjective Responses to 1,1,1-Trichloroethane, Perchloroethylene, Trichloroethylene and Methylene Chloride.* Dow Chemical Co., Midland, Mich., May 1, 1973.

McNutt, N. S., Amster, R. L., McConnell, E. E., Morris, F. Hepatic lesions in mice after continuous inhalation exposure to 1,1,1-trichloroethane. *Lab Invest.* 32(5): 642–54, May 1975.

Caplan, Y., Backer, R., Whitaker, J. 1,1,1-Trichloroethane: report of a fatal intoxication. *Clin. Toxicol.* 9(1): 69–74, 1976.

Gosselin, R., Hodge, H., Smith, R., Gleason, M. *Clinical Toxicology of Commercial Products.* Williams and Wilkins Co., Baltimore, 1976.

U.S. Department of Health, Education and Welfare. *Recommended Standard for Occupational Exposure to 1,1,1-Trichloroethane (Methyl Chloroform).* NIOSH, July 1, 1976.

Ando, K., Hayashi, M., Kosaka, H., Tabuchi, T. Toxicological studies of methylchloroform. II. Distribution of methylchloroform in rats. *Osaka-furitsu Koshu Eisei Kenkyusho Kenkyu Holoku, Rodo Eisei Hen*, Vol. 15, 1977.

Bonventre, J., Brennan, O., Jason, D., Henderson, A., Bastos, M. L. Two deaths following accidental inhalation of dichloromethane and 1,1,1-trichloroethane. *J. Anal. Toxicol.* 1(4): 158–61, 1977.

Fukabori, S., Nakaaki, K., Yonemoto, J., Tada, O. On the cutaneous absorption of 1,1,1-trichloroethane (2). *J. Sci. Labour*, 53(1): 89–95, 1977.

Holmberg, B. O., Jakobson, I., Sigvardsson, K. A study on the distribution of methylchloroform and *n*-octane in the mouse during and after inhalation. *Scand. J. Work Environ. Health* 3(1): 43–52, 1977.

Key, M., Henschel, A. F., Butler, J., Tabershaw, I. *Occupational Diseases, a Guide to Their Recognition.* U.S. Department of Health, Education and Welfare, June 1977.

Maroni, M., Bulgheroni, C., Cassitto, M., Merluzzi, F., Gilioli, R., Foa, V. A clinical, neurophysiological and behavioral study of female workers exposed to 1,1,1-trichloroethane. *Scand. J. Work Environ. Health* 3: 16–22, 1977.

Proctor, N., Hughes, J. *Chemical Hazards of the Workplace.* J. B. Lippincott Co., Philadelphia, 1977.

Savolainen, H., Pfaffli, P., Tengen, M., Vainio, H. Trichloroethylene and 1,1,1-trichloroethane: effects on brain and liver after five days intermittent inhalation. *Arch. Toxicol.* 38(3): 229–37, 1977.

U.S. Department of Health, Education and Welfare *Chloroethanes: Review of Toxicity*. Current Intelligence Bull. 27, NIOSH, Aug. 21, 1978.

U.S. Department of Health and Human Services and U.S. Department of Labor, *Occupational Health Guidelines for Methyl Chloroform*. Washington, D.C., Sept. 1978.

Vainio, H., Savolainen, H., Pfaffli, P. 1,1,1-Trichloroethane and trichloroethylene on rat liver and brain. *Xenobiotica* 8(3): 191–96, 1978.

Nathan, A. W., Toseland, P. A. Goodpasture's syndrome and trichloroethane intoxication. *Brit. J. Clin. Pharmacol.* 8(3): 28406, Sept. 1979.

Halvey, J., Pitlik, S., Rosenfeld, J., Eitan, B. D. 1,1,1-Trichloroethane intoxication: a case report with transient liver and renal damage. *Clin. Toxicol.* 16(4): 467–72, 1980.

Garriott, J., Petty, C. S. Death from inhalant abuse: toxicological and pathological evaluation of 34 cases. *Clin. Toxicol.* 16(3): 305–15, 1980.

Nestmann, E. R., Lee, G. G. H., Matula, T. I., Douglas, G. R., Mueller, J. C. Mutagenicity of constituents identified in pulp and paper mill effluents using the salmonella/mammalian-microsome assay. *Mut. Res.* 79(3): 203–12, 1980.

Carlson, G. P. Effect of alterations in drug metabolism on epinephrine-induced cardiac arrhythmias in rabbits exposed to methylchloroform. *Toxicol. Letters* 9(4): 307–13, 1981.

Ivanetich, K. M., Van den Honert, L. H. Chloroethanes: their metabolism by hepatic cytochrome P-450 in vitro. *Carcinogenesis* 2(8): 697–702, 1981.

Moody, Y. D. E., James, J. L., Clawson, G. A., Smuckler, E. A. Correlations among the changes in hepatic microsomal components after intoxication with alkyl halides and other hepatotoxins. *Mol. Pharmacol.* 20(3): 685–93, 1981.

Northfield, R. R. Avoidable deaths due to acute exposure to 1,1,1-trichloroethane. *J. Soc. Occup. Med.* 31: 1981.

Viola, A., Signon, M., Pittoni, G., Sarto, F., Pennelli, N. Serum enzyme activities and histological changes after percutaneous application of methylchloroform. *Med. Lavoro* 72(5): 410–15, 1981.

Caperos, J. R., Droz, P. O., Hake, C. L., Humbert, B. E., Jacot-Guillarmod, A. 1,1,1-Trichloroethane exposure, biologic monitoring by breath and urine analyses. *Int. Arch. Occup. Environ. Health* 29(3–4): 293–303, 1982.

Guillemin, M., Guberan, E. Value of the simultaneous determination of PCO_2 in monitoring exposure to 1,1,1-trichloroethane by breath analysis. *Brit. J. Ind. Med.* 39(2): 161–68, 1982.

Thiele, D. L., Eigenbrodt, E. H., Ware, A. J. Cirrhosis after repeated trichloroethylene and 1,1,1-trichloroethane exposure. *Gastroenterology* 83(4): 926–29, 1982.

Hobara, T., Kobayashi, H., Higashihara, E., Kawamoto, T., Iwamoto, S., Sakai, T. Changes in hematologic parameters with acute exposure to 1,1,1-trichloroethane. *Ind. Health* 21(4): 255–61, 1983.

Hobara, T., Kobayashi, H., Higashihara, E., Kawamoto, T., Sakai, T. Acute effects of 1,1,1-trichloroethane, trichloroethylene, and toluene on the hematologic parameters in dogs. *Arch. Environ. Contam. Toxicol.* 13(5): 589–93, 1984.

Keogh, A. M., Ibels, L. S., Allen, D. H., Isbister, J. P., Kennedy, M. C. Exacerbation of Goodpasture's syndrome after inadvertent exposure to hydrocarbon fumes. *Brit. Med. J.* 288(6412): 188, 1984.

Moser, V. C., Balster, R. L. Acute motor and lethal effects of inhaled toluene, 1,1,1-trichloroethane, halothane, and ethanol in mice: effects of exposure duration. *Toxicol. Appl. Pharmacol.* 77(2): 285–91, 1985.

Kobayashi, H., Hobara, T., Satoh, T., Sakai, T. Respiratory disorders following 1,1,1-

trichloroethane inhalation: a role of reflex mechanism arising from lungs. *Arch. Environ. Health* 41(3): 149–54, 1986.

Nelson, B. K. Development neurotoxicology of in utero exposure to industrial solvents in experimental animals. *Neurotoxicology* 7(2): 441–47, 1986.

Rudolph, L., Swan, S. H. Reproductive hazards in the microelectronics industry. *State Art Rev. Occup. Med.* 1(1): 135–43, 1986.

Trichloroethylene

Barsoum, C. S., Saad, K. Relative toxicology of certain chlorine derivatives of the alaphatic series. *Quart. J. Pharm.* 7:205, 1934.

von Oettien, W. F. The halogenated hydrocarbons: their toxicity and potential dangers. *J. Ind. Hyg.* 19: 349, 1937.

Barrett, H. M., Johnston, J. H. The fate of trichloroethylene in the organism. *J. Biol. Chem.* 127: 765–70, 1939.

Waters, R. M., Orth, O. S., Gillespie, N. A. Trichloroethylene anesthesia and cardiac rhythm. *Anesthesiology* 4: 1, 1943.

Barnes, C. G., Ives, J. EKG changes during trilene anesthesia. *Proc. Roy. Soc. Med.* 37: 528, 1944.

Seifter, J. Liver injury in dogs exposed to trichloroethylene. *J. Ind. Hyg. Toxicol.* 26: 250–52, 1944.

Powell, J. F. Trichloroethylene: absorption, elimination and metabolism. *Brit. J. Ind. Med.* 2: 142–45, 1945.

Butler, T. C. Metabolic transformations of trichloroethylene. *J. Pharmacol. Exp. Ther.* 97: 84–92, 1949.

Hunter, A. R. The toxicity of trichlorethylene. *Brit. J. Pharmacol.* 4: 177–80, 1949.

Adams, E. M., Spencer, H. C., Rowe, V. K., McCollister, D. D. Vapor toxicity of trichloroethylene determined in laboratory animals. *Arch. Ind. Hyg. Occup. Med.* 4: 469–81, 1951.

Browning, E. *Toxicology of Industrial Solvents*. Chemical Pub. Co., New York, 1953, p. 169.

Ostlere, G. *Trichloroethylene Anesthesia*. E. and S. Livingstone, Ltd., Edinburgh and London, 1953, p. 26.

Klinefield, M., Tabershaw, I. R. Trichlorethylene toxicity, report of 5 fatal cases. *Arch. Ind. Hyg.* 10: 134–41, 1954.

McBirney, R. S. Trichlorethylene and dichlorethylene poisoning. *Arch. Ind. Hyg.* 10: 130–33, 1954.

Nowill, W. K., Stephen, R. R., Margolis, G. The chronic toxicity of trichloroethylene— a study. *Anaesthesiology* 15: 462–65, 1954.

Grandjean, E., Munchinger, R., Turrian, V., Haas, A., Knoepel, H. K., Rosenmund, H. Investigations into the effects of exposure to trichloroethylene in mechanical engineering. *Brit. J. Ind. Med.* 12: 131, 1955.

Joron, G. E., Cameron, D. G., Halpenny, G. W. Massive necrosis of the liver due to trichloroethylene. *Canad. Med. Assoc. J.* 73: 890–91, 1955.

von Oettigen, W. F. *The Halogenated Hydrocarbons: Toxicity and Potential Dangers*. Public Health Service Bull. #414, U.S. Govt. Printing Office, Washington, D.C., 1955.

Bardodej, Z., Vyskocil, J. The problems of trichloroethylene in occupational medicine. *Arch. Ind. Health* 13: 581–92, 1956.

Soucek, B., Vlachova, D. Excretion of trichloroethylene metabolites in human urine. *Brit. J. Ind. Med.* 17: 60–64, 1960.

DeFalque, R. J. Pharmacology and toxicology of trichlorethylene. *Clin. Pharmacol. Ther.* 2: 665, 1961.

Wirtschafter, Z. T., Cronyn, M. W. Relative hepatotoxicity—pentane, trichloroethylene, benzene, carbon tetrachloride. *Arch. Environ. Health* 9: 180–85, 1964.

Gutch, C. F., Tomhave, W. G., Stevens, S. C. Acute renal failure due to inhalation of trichloroethylene. *Ann. Intern. Med.* 63: 128–34, 1965.

Kylin, B., Sumegi, I., Yllner, S. Hepatotoxicity of inhaled trichloroethylene and tetrachloroethylene—long term exposure. *Acta Pharmacol. Toxicol.* 22: 379–85, 1965.

Stewart, R., Dodd, H., Duncan, E., Holder, B. Diagnosis of solvent poisoning. *J.A.M.A.* 193: 1097–1100, 1965.

Klaassen, C., Plaa, G. Relative effects of various chlorinated hydrocarbons on liver and kidney functions in mice. *Toxicol. Appl. Pharmacol.* 9: 139–51, 1966.

Leibman, K. C., McAllister, W. J., Jr. Metabolism of trichloroethylene in liver microsomes—III. Induction of the enzymic activity and its effect on excretion of metabolites. *J. Pharmacol. Exp. Ther.* 157: 574–80, 1967.

Mazza, V., Brancaccio, A. Characteristics of the formed elements of the blood and bone marrow in experimental trichloroethylene intoxication. *Folia Med.* 50: 318–24, 1967.

Prendergast, J. A., Jones, R. A., Jenkins, L. J., Jr., Siegel, J. Effects on experimental animals of long-term inhalation of trichloroethylene, carbon tetrachloride, 1,1,1-trichloroethane, dichlorodifluoromethane, etc. *Toxicol. Appl. Pharmacol.* 10: 2, 1967.

Friborska, A. The phosphates of peripheral white blood cells in workers exposed to trichloroethylene and perchloroethylene. *Brit. J. Ind. Med.* 26: 159–61, 1969.

Lilis, R., Stanescu, D., Muica, N., Roventa, A. Chronic effects of trichloroethylene exposure. *Med. Lavoro* 60: 595–601, 1969.

Ogata, M., Takatsuak, Y., Tomokuni, K. Excretion of organic chlorine compounds in the urine of persons exposed to vapours of trichloroethylene and tetrachloroethylene. *Brit. J. Ind. Med.* 28: 386–91, 1971.

Seage, A. J., Burns, M. W. Pulmonary oedema following exposure to trichlorethylene. *Med. J. Aust.* 2: 484–86, 1971.

Correspondence Re: TCE. David P. Brown, NIOSH, to Janette D. Sherman, M.D., July 25, 1972.

Dow Chemical Co. Subjective responses to 1,1,1-trichloroethane, perchloroethylene, trichloroethylene and methylene chloride. Dow Chemical Co., Midland, Mich., 1973.

U.S. Department of Health, Education and Welfare. *Occupational Exposure to Trichloroethylene.* Public Health Service, NIOSH, 1973.

Plaa, G. L., Traiser, G. J., Hanasono, G. K., Witschi, H. Effect of alcohols on various forms of chemically induced liver injury. *Alcohol. Liver Pathol.* (Proc. Int. Symp. Alcohol Drug Res.) 225–44, 1975.

Carcinogenesis Technical Report #2. *Carcinogenesis Bioassay of Trichloroethylene.* DHEW Pub. #(NIH) 76-802, U.S. Govt. Printing Office, 1976.

Correspondence Re: TCE. David P. Brown, Epidemiologist, NIOSH to Frank Mirer, UAW, Mar. 24, 1976; to Gene Kortsha, General Motors, Mar. 24, 1976; to Janette D. Sherman, M.D., Apr. 5, 1976.

Gosselin, R., Hodge, H., Smith, R., Gleason, M. *Clinical Toxicology of Commercial Products: Acute Poisoning.* Williams and Wilkins Co., Baltimore, 1976, p. 112.

NEW News. *Bioassay of Trichloroethylene (TCE).* U.S. Department of Health, Education and Welfare, 1976.

Anon. Trichloroethylene. I. An Overview. *J. Toxicol. Environ. Health* 2(3): 671–707, 1977.

Savolainen, H., Pfaffli, P., Tengen, M., Vainio, H. Trichloroethylene and 1,1,1-trichloroethane: effects on brain and liver after five days intermittent inhalation. *Arch. Toxicol.* 38(3): 229–37, 1977.

NIOSH. *Chloroethanes: Review of Toxicity.* U.S. Department of Health, Education and Welfare, 1978.

Vainio, H., Savolainen, H., Pfaffli, P. Biochemical and toxicological effects of combined exposure to 1,1,1-trichloroethane and trichloroethylene on rat liver and brain. *Xenobiotica* 8(3): 191–96, 1978.

Garriott, J., Petty, C. S. Death from inhalant abuse: toxicological and pathological evaluation of 34 cases. *Clin. Toxicol.* 16(3): 305–15, 1980.

Annau, Z. The neurobehavioral toxicity of trichloroethylene. *Neurobehav. Toxicol. Teratol.* 3(4): 91–94, 1981.

Haglid, K. G., Briving, C., Hansson, H. A., Rosengren, L., Kjellstrand, P., Stavron, D., Swedin, U., Wronski, A. Trichloroethylene: long-standing changes in the brain after rehabilitation. *Neurotoxicology* 2(4): 659–73, 1981.

Ogata, M., Hasegawa, T. Effects of chlorinated aliphatic hydrocarbons on mitochondrial oxidative phosphorylation in rat with reference to effects of chlorinated aromatic hydrocarbons. *Ind. Health* 19(2): 71–75, 1981.

Thiele, D. L., Eisenbrodt, E. H., Ware, A. J. Cirrhosis after repeated trichlorethylene and 1,1,1-trichloroethane exposure. *Gastroenterology* 83(4): 926–29, 1982.

Winneke, G. Acute behavioral effects of exposure to some organic solvents—psychophysiological aspects. *Acta Neurol. Scand.* 92: 117–29, 1982.

Lindstrom, K. Behavioral effects of long-term exposure to organic solvents. *Acta Neurol. Scand.* 92: 131–41, 1982.

Lindstrom, K., Antti-Poika, M., Tola, S., Hyytiainen, A. Psychological prognosis of diagnosed chronic organic solvent intoxication. *Neurobehav. Toxicol. Teratol.* 4(5): 581–88, 1982.

Barret, L., Faure, J., Guilland, B., Chomat, D., Didier, B., Debru, J. L. Trichloroethylene occupational exposure: elements for better prevention. *Int. Arch. Occup. Environ. Health* 53(4): 283–89, 1984.

Phoon, W. H., Chan, M. O., Rajan, V. S., Tan, K. J., Thirumoorthy, T, Goh, C. L. Stevens-Johnson syndrome associated with occupational exposure to trichloroethylene. *Contact Derm.* 10(5): 270–76, 1984.

Haustein, U. F., Ziegler, V. Environmentally induced systemic sclerosis-like disorders. *Int. J. Dermatol.* 24(3): 147–51, 1985.

Stracciari, A., Gallassi, R., Ciardulli, C., Coccagna, G. Neurophysiological and EEG evaluation in exposure to trichloroethylene. *J. Neurol.* 232(2): 120–22, 1985.

Juntunen, J. Occupational toxicology of trichloroethylene with special reference to neurotoxicity. *Dev. Toxicol. Environ. Sci.* 12: 189–200, 1986.

Flindt-Hansen, H., Isager, H. Scleroderma after occupational exposure to trichloroethylene and trichloroethane. *Acta Derm. Venereol.* 67(3): 263–64, 1987.

Kulig, B. M. The effects of chronic trichloroethylene exposure on neurobehavioral functioning in the rat. *Neurotoxicol. Teratol.* 9(2): 171–78, 1987.

INDEX

A

acetylcholine, 158 ff.
acrylamide, 100
acrylics, 100, 221
  chemical structure, 77
acrylonitrile, skin, 59
Agent Orange, 22, 98–99, 228
agriculture, 224–225
  and illness, 224
AIDS, xenobiotics, 60
alcohol, structure, 78
aluminum, disease, 206
amitrole, 175
Anabuse, 10 (*see also* disulfiram)
  alcohol reaction, 101
anemia, lead, 124
anesthetics
  chemicals, 65
  disease, 206
antihistamines, 22-24
aplastic anemia, 15
arsenic, 3
  in batteries, 57
  and cancer, 28, 134
  and skin, 59
asbestos, 3, 7, 10, 19, 20, 226
  in brakes, 84, 117
  and cancer, 191, 210
  contamination, 74–75, 204
  and disease, 203
  exposure, 84–85, 87, 93, 191
  history of, 74, 210
  home exposure, 51
  insulation, 116
  mesothelioma, 51, 75
  non-cancer lesions, 214
  particles inhaled, 57
  smoking, 190
  state of the art, 210
  use of, 191
atrazine, 175
auto body works and cancer, 124
autopsy, importance of 121

B

beautician, exposure of, 118
behavioral testing, 31–32
Bendectin, 211
benomyl, 175
benzene
  bone marrow, 64
  cancer, 77
  chemical structures, 77
benzidine dyes, 15
  bladder cancer, 75
BHC (benzene hexachloride), 175, 177
Bhopal, India, 120
birth control pills, 68
  chemical components, 122
  legal, 198
  pituitary tumor, 45
  strokes, 210
birth defects, 15, 20, 30–32, 66, 130–
    131, 132, 175, 177
  medical costs, 131
bladder cancer, 75
blast furnace exposure, 116

chlordane
  absorption of, 148
  in animals, 146
  biology of, 146-147
  and blood dyscrasias, 64, 143
  and cancer, 147, 175
  disposal of, 141
  effects of, 146–147, 175
  extent of use, 150
  labeling, 149
  leukemia and, 143
  neurotoxicity, 147
  persistence of, 148
  recommendations, 139
  reproductive effects, 147
  structure of, 145
  suspension of, 150
  symptoms, 140–141, 142–143, 148, 193
chlorofluorocarbons, 221
chlorotrianisene, 129–130 (see also TACE)
chlorpyrifos (see Dursban)
cholinesterase, 158, ff.
chromium, 86
  and cancer, 109, 134
  effect of skin, 59
  exposure to, 109
  kidney and, 125
  and nasal septal perforation, 43, 112–113
chromium plating, 57
  heavier than air, 57
chromosomal damage, 134
chromosomes, 18
  mutations of, 33–34
chronic obstructive lung disease (see COPD)
circadian rhythms, 68
clindamycin, 65
coke industry, 85–86
  and cancer, 205
consumer chemicals, 220–223
contamination
  background of, 204
  or water, 26
COPD, 117
costs, 2, 9–10

D

2,4-D, (2,4-dichlorophenoxy acetic acid), 68, 175, 228
  structure for, 23
data collection, 5–6
  linkage, 76
  sources, 210, 233–237
DBCP, 15
  sterility, 75
DDD, 175, 177–178
DDE, 175, 177–178
DDT, 33, 175, 177–178
DES (Diethyl stilbesterol), 68, 175, 176–177
  agricultural use, 122–123
  in animals, 33
  cancer and, 129
  genital abnormalities in females, 46
  genital abnormalities in males, 46
  meat residues, 123
  and pregnancy, 33
  and reproduction, 33
diagnosis, 16–17
dibromochloropropane (see DBCP)
Dicofol, 175, 176
dioxins, 22–23, 31, 125, 175
  dose, 60
disease
  causation, 6, 203, 215–217
  clues to, 14
  distribution of, 189
  occupational incidence of, 94
  pattern of, 52
  prevention, 11
  sentinel, 75
disulfiram, 67, 101 (see also Anabuse)
dizziness, 41
DNA, 18–19
  damage, 214
  effects of, 34
  neoplasia, 214
Dow Chemical Co., 161, 163
drugs, information on, 235
dry cleaning
  chemicals used, 62, 91
Dursban, 144–145, 158 ff., 175
  absorption, 165 ff.
  animal tests, 162–163, 165–167
  dioxin question, 168–169

ADVANCES IN MODERN ENVIRONMENTAL TOXICOLOGY, VOLUME XXI
M.A. Mehlman, Series Editor

CHEMICALLY-INDUCED ALTERATIONS IN SEXUAL AND FUNCTIONAL DEVELOPMENT: THE WILDLIFE/HUMAN CONNECTION

edited by:
Theo Colborn and Coralie Clement

This book presents the findings of a multidisciplinary group of endocrine experts, who gathered at the Wingspread Conference Center in Racine, Wisconsin in July 1991, to study endocrine disruptors in the environment and their affects on fish, wildlife, and humans. The thesis of the book is that the fetus is extremely vulnerable to exogenous chemicals. The volume is divided into various sections—general background in comparative endocrinology and physiology; wildlife, laboratory, and human evidence; and evidence of exposure—all supporting the book's thesis. The long-term changes in human and mouse reproductive tracts that result from exposure to the endocrine disruptor diethylstilbestrol (DES) are presented; other chapters describe endocrine disruption in fish and the offspring of fish-eating animals. The discussions included in this book point out that as more chemicals are introduced into the environment, linking endocrine disruptors through maternal exposure to their effects on offspring becomes more difficult.

Following World War II, the Great Lakes region became home for one of the largest industrial and agro-chemical complexes in North America. The Lakes and surrounding lands also became major disposal sites for the wastes produced by these industries and other human activities. This book integrates the research on endocrine disruptors that grew out of environmental degradation of the Great Lakes with research from other regions of the Northern Hemisphere and other disciplines to provide an eclectic overview of a global problem.

TABLE OF CONTENTS

ISBN: 0-911131-35-3

460 pages (approximately)

Princeton Scientific Publishing Co., Inc.
P.O. Box 2155 · Princeton, New Jersey 08543 · (609) 683-4750 · fax: (609) 683-0838
Price $68.00 + $3.00 shipping and handling.   Prepayment required on all orders.

ADVANCES IN MODERN ENVIRONMENTAL TOXICOLOGY, VOLUME XX
M.A. Mehlman, Series Editor

PREDICTING ECOSYSTEM RISK

edited by:
John Cairns, Jr., B.R. Niederlehner, and David R. Orvos

Most ecosystems are experiencing anthropogenic changes. Global atmosphere, oceans, forests, agricultural lands, and surface and ground water are all at risk from diverse human activities. The recognition of these risks and their quantification are important first steps in sustaining a healthy environment.

Two major factors must be considered in predicting ecosystem risk: the probability of deleterious effects and the resilience or ability to recover following anthropogenic and/or natural disturbance. Predicting the probability of deleterious effects should depend upon a comparison of an estimate of a safe concentration that does not damage biological integrity to an independent estimate of the expected environmental concentration. In addition to predicting the probability of adverse effects, ecosystem resiliency must also be estimated.

Much attention has been given in recent years to the problem of restoring damaged ecosystems. Evidence is mounting that some ecosystems recover quite rapidly from pollutional effects (although recovery is not to their predisturbance condition), but others do not recover for tens, hundreds, or perhaps even thousands of years. Risk and resilience should be considered together because an ecosystem unlikely to recover or to be restored in time spans of interest to humans should be subjected to much less risk than those with considerably higher resiliency. Ideally, of course, the purpose of predicting ecosystem risk is to prevent damage, i.e., effective prediction of risk is far superior to reacting to oil spills, environmental pollution, and the like *after* damage has occurred.

Of course, all ecosystem risk estimation must consider various cost and benefit factors, and this volume does not intend to be an exhaustive methodological book; rather, its purpose is to furnish illustrative examples to introduce more detailed explanations of this fascinating topic of predicting ecosystem risk.

TABLE OF CONTENTS

ISBN: 0-911131-27-2   400 pages (approximately)

Princeton Scientific Publishing Co., Inc.
P.O. Box 2155 · Princeton, New Jersey 08543 · (609) 683-4750 · fax: (609) 683-0838
Price $65.00 + $3.00 shipping and handling.   Prepayment required on all orders.

PROCEEDINGS OF
The IXth UOEH International Symposium and
The First Pan Pacific Cooperative Symposium

Industrialization and Emerging Environmental Health Issues: Risk Assessment and Risk Management

edited by:

Takesumi Yoshimura • Kenzaburo Tsuchiya • S.D. Lee • L.D. Grant • M. A. Mehlman

In Kitakyushu, Japan, October 2-6, 1989, an international symposium on "Industrialization and Emerging Environmental Health Issues: Risk Assessment and Risk Management" was held. This five-day symposium addressed problems related to the impact of industrialization on the environment and human health. The symposium brought together internationally-recognized experts and scholars from around the world. The topics covered a wide range of interests and concerns for both industrialized and developing nations.

This volume documents the meeting by publishing the papers presented in Kitakyushu. These papers are valuable reading material for anyone involved in, or interested in, the scientific and medical fields addressing environmental health issues, and in particular the risk assessment and risk management of those issues in developing nations. The volume contains over 60 articles by internationally-known scientists and professionals, including:

ENVIRONMENTAL PROBLEMS: PAST, PRESENT, AND FUTURE K. Fuwa

CURRENT ISSUES IN THE EPIDEMIOLOGY AND TOXICOLOGY OF OCCUPATIONAL EXPOSURE TO LEAD P.J. Landrigan

GLOBAL POLLUTION E. Bingham

THE DISTRIBUTION, CYCLING, AND POTENTIAL HAZARDS OF INDUSTRIAL CHEMICALS IN MARINE ENVIRONMENTS M. Morita

COMPREHENSIVE ASSESSMENT OF OCCUPATIONAL AND ENVIRONMENTAL HEALTH PROBLEMS: AN OVERVIEW - A PROPOSAL OF MACRO APPROACH: INDUSTRIAL ECOLOGICAL SCIENCES K. Tsuchiya

SEMICONDUCTOR INDUSTRIES J. LaDou

LONG-TERM CARCINOGENICITY BIOASSAYS ON INDUSTRIAL CHEMICALS AND MAN-MADE MINERAL FIBERS : PREMISES, PROGRAMS, AND RESULTS C. Maltoni, F. Minardi, M. Soffritti, and G. Lefemine

INFORMATION AND MANAGEMENT SYSTEM TO REDUCE CHEMICAL RISKS N. Htun

ROAD TRANSPORT IMPACT ON THE ENVIRONMENTAL HEALTH M. Murakami, M. Ono, and K. Tamura

ASBESTOS DISEASE-1990-2020: THE RISKS OF ASBESTOS RISK ASSESSMENT I.J. Selikoff

HEALTH RISK ASSESSMENT OF RADIATION T. Sugahara

TOXIC EVALUAtiON OF CHLORINATED AROMATIC HYDROCARBONS IN HUMAN ENVIRONMENTS Y. Masuda

CARCINOGENICITY OF MOTOR FUELS: GASOLINE M.A. Mehlman

INDUSTRIALIZATION AND EMERGING ENVIRONMENTAL HEALTH ISSUES: LESSONS FROM THE BHOPAL DISASTER C.R. Krishna Murti

MUNICIPAL AND INDUSTRIAL HAZARDOUS WASTE MANAGEMENT: AN OVERVIEW L. Fishbein

THE EXPORT OF HAZARDOUS WASTE TO THE CARIBBEAN BASIN REGION W.H.E. Suite

WASTE MANAGEMENT IN *ASEAN* COUNTRIES F.A. Uriarte, Jr.

THE CURRENT STATUS OF SOLID WASTE MANAGMENT IN P.R. CHINA Shi Qing

MANAGMENT OF INDUSTRIAL WASTE - A EUROPEAN PERSPECTIVE T. Schneider

RISK REDUCTION MANAGEMENT FOR HAZARDOUS WASTE IN JAPAN M. Tanaka and K. Ueda

MANAGING THE RISK TRANSITION K.R. Smith

MULTIMEDIA RISK ASSESSMENT FOR ENVIRONMENTAL RISK MANAGEMENT S.D. Lee

RISK ASSESSMENT AND RISK MANAGEMENT IN JAPAN J. Kagawa

RISK ASSESSMENT/RISK MANAGEMENT OF MOTOR VEHICLE EMISSIONS M.P. Walsh

RISK ASSESSMENT AND RISK MANAGEMENT IN JAPAN E. Yokoyama

THE ROLE OF EPIDEMIOLOGY IN RISK ASSESSMENT T. Yoshimura

ENVIRONMENTAL AND OCCUPATIONAL HAZARDS IN EXPORT PROCESSING ZONES IN SOUTH AND EAST ASIA M. Thorborg

plus the many more papers by international authorities such as: J.J. Convery, E. L. Anderson , N.A. Ashford, A. Koizumi , D.J. Ehreth , B.D. Goldstein , D.G. Hoel , J. Higginson

Princeton Scientific Publishing Co., Inc.
P.O. Box 2155 • Princeton, New Jersey 08543 • (609) 683-4750 • fax: (609) 683-0838
Price: $65 + $5 shipping and handling.   Prepayment required on all orders.

VOLUME VI
Applied Toxicology
of Petroleum Hydrocarbons

Edited by H.N. MACFARLAND, Gulf Oil Corporation
C.E. HOLDSWORTH, American Petroleum Institute
J.A. MACGREGOR, Standard Oil of California
and M.L. KANE, American Petroleum Institute

Princeton Scientific Publishing Co., Inc.